AN INTRODUCTION TO
QUANTITATIVE
ECOLOGY

McGRAW-HILL
BOOK COMPANY

New York
St. Louis
San Francisco
Düsseldorf
Johannesburg
Kuala Lumpur
London
Mexico
Montreal
New Delhi
Panama
Paris
São Paulo
Singapore
Sydney
Tokyo
Toronto

ROBERT W. POOLE

Assistant Professor
Illinois Natural History Survey

Research Associate
Biometrics Unit
Cornell University

An Introduction to Quantitative Ecology

This book was set in Times New Roman.
The editors were William J. Willey and Barry Benjamin;
the production supervisor was Sam Ratkewitch.
The drawings were done by Reproduction Drawings Ltd.
Kingsport Press, Inc., was printer and binder.

Library of Congress Cataloging in Publication Data

Poole, Robert W.
    An introduction to quantitative ecology.

    (McGraw-Hill series in population biology)
    Bibliography: p.
    1. Ecology. 2. Biotic communities—Mathematical
models. 3. Animal populations—Mathematical models.
I. Title. [DNLM: 1. Ecology. 2. Mathematics.
QH541 P822i 1974]
QH541.P66   574.5   73–13694
ISBN 0–07–050415–6

**AN INTRODUCTION TO
QUANTITATIVE
ECOLOGY**

234567890 K P K P 7987654

To my mother

CONTENTS

Part Three

Part Four

PREFACE

Ecology is basically just an elegant word meaning natural history. Although ecology is academically one of the youngest sciences, the first ecologist was undoubtedly some paleolithic ape who first took note of the natural world surrounding him. Most early ecology consisted of purely descriptive studies, such as a listing of the species of plants along the cliffs of Dover. As soon, however, as someone began to count the number of individuals of each species and wondered why one species was more common than another, ecology became quantitative. Quantitative ecology, therefore, is the study of ecology with the use of measurements or counts. Sometimes the mathematical analysis consists of no more than counting the number of individuals in an area. At other times, however, sophisticated and complex mathematics may be needed to determine the causes of some ecological phenomena we have observed.

This book is intended to be a complete survey of ecology with a strong mathematical bias, provided the reader is well founded in the basics of natural history. Ecology is a broad subject, however, and it is impossible to discuss, even superficially, all its aspects. Therefore, I have been forced to ignore some important areas of study. I do hope, however, that a broad enough spectrum of ecology is presented to allow the reader to handle the majority of the ecological problems he may encounter. The book can be used either in an intensive first course in ecology, or in advanced courses stressing the more mathematical side of ecology. The text is written primarily for the nonmathematically inclined biologist. Most biologists are not particularly interested in or knowledgeable about mathematics, although this is not as true as it once was. Therefore, I have set as a minimum mathematical level high school algebra. All mathematical topics are presented in cookbook fashion, and I have avoided the use of calculus except in its simplest and most easily understandable forms. It has been necessary, unfortunately, to introduce a healthy segment of statistics, because it is impossible to analyze quantitative data without it. The sections on statistics are only complete enough to enable the reader to apply the method to the topic at hand, and are not intended to be introductions to statistics in general. Hopefully, these sections will be sufficient to allow the reader without statistics to understand the material. These sections may also serve as useful reviews for readers who have had a course in statistics at one time or another.

The emphasis throughout the book is on the presentation of a problem, a method of solving the problem, the analysis of an example, and a discussion of the ecological generalities and implications suggested by the analysis. The book progresses from relatively simple situations to the more complex problems found in nature. In almost all cases I have avoided deriving the models or statistical methods, and have concentrated on explaining the uses of the models, their implications, and most importantly their assumptions. It has sometimes been necessary to sacrifice the niceties of mathematics for the purposes of clarity, but in these cases I have tried to give references to more formal definitions and derivations in other articles and books.

I am deeply grateful to a number of people who have read all or part of the manuscript. Their comments and criticisms are responsible, in large part, for any merit this book may have. In particular I wish to thank Beverly R. Poole, Peter W. Price, Kenneth E. F. Watt, and Robert H. Whittaker.

ROBERT W. POOLE

INTRODUCTION

Ecology is a difficult word to define and has always meant different things to different people. Definitions range from " the relationship of an organism to its environment," a definition including such diverse fields as physiology, population genetics, evolutionary biology, and taxonomy, to more specific studies. I prefer a more limited definition of ecology as "the relationships between the density or biomass of a population or populations and the environment, interactions between populations and within populations, and the effect of the population on the environment." The environment includes all physical and biological variables affecting a population, including inter-actions between the individuals of a population and between individuals of different species. The group of individuals of a species in a limited, defined area is a *population*. Except in unusual cases a population is an arbitrary unit defined by the physical boundaries of the environment and the convenience of the ecologist. In genetics a population is defined as a group of individuals of a species having approximately equal probabilities of mating with any other individual in the group. Unfortunately, this definition is next to useless in ecology for practical reasons which should become clear later in the text.

The populations of all of the species in an area form a *community*, and a community and the physical environment together form an *ecosystem*.

Ecology is basically a quantitative science. A quantitative description of a set of observations such as 100 crows per acre is preferable to a verbal statement such as "many" or "few" crows per acre. A population changes in abundance because of the interactions of several factors. The behavior of a population or of several populations can be understood if the underlying factors and their importance are discovered by analysis of observational or experimental data, and the factors' interactions and actions rigidly formulated as a descriptive or mathematical model.

The steps in the study, analysis, and solution of an ecological problem are not much different than in any other science, except for the added complexity. These steps are:

1 Preliminary study and description of the problem. This study includes a listing of the species involved and a rough summarization of the probable species and physical environmental variable interactions.

2 The setting up of hypotheses based on the descriptive study and on a knowledge of the general ecological interactions most often observed.

3 The creation of a sampling program or laboratory experiments, and the gathering of data.

4 Analysis of the data, testing of the hypotheses, and quantification of the multitude of population parameters.

5 Modeling of the problem as a series of mathematical equations.

6 Testing of the model using new data, generation of new hypotheses, if needed, and the reformulation of the equations to include necessary changes. The model is successively improved by feedback between more data and changes in the equations of the model.

7 Use of the model predictively.

In many situations step *1*, the descriptive stage, is perfectly adequate, depending on what the ecologist is trying to discover. However, a mathematical model is a firm statement of how a process is thought to work and is amenable to testing. If the model representing our hypotheses proves to be wrong, as it often does, the equation or equations can be dissected, each of the parts reexamined, and the model reformulated. This does not mean that complicated formulations are to be preferred if simple models are adequate. The emphasis on quantitative is not meant to belittle the observational or natural history approach, because observations always form the basic first step of the quantitative analysis. Observation and analysis are integral parts of the same procedure. A completely theoretical approach is just as much to be avoided as is the gathering of data or descriptions for no specific purpose. Hypothesis and theory should always be related to actual observation and experiment. Only if the development of the model or theory is the result of complex feedback between hypothesis and observation can one be sure that the theory or equations really represent something approximately like the real world. A model is like a picture of a forest. The picture is not the forest but only illustrates it. A picture may be completely realistic or contain various grades of abstraction. Mathematical models are the same. Modeling is not necessarily a required part of the study of an ecological problem, although a model is often useful. Sometimes the gathering of data and the testing of hypotheses provide the understanding we are looking for. To answer a question such as, " Does rainfall significantly affect larval mortality?" a hypothesis is created, data gathered, and the hypothesis tested. In contrast, if it is necessary to determine exactly how rainfall affects larval mortality, a model is needed. Statistical methods for answering the latter question are based on a particular mathematical model and, whether one is creating a model *de novo* or using a statistical technique such as regression, we are still modeling in a strict sense.

Models serve two purposes. First, the model is a description of the process, and second, it can be used to predict the effects of changes in the variables and parameters of the process. With a model it is often possible to study the consequences of changing factors such as rainfall, food supply, or predation. If a model of an economically important plant or animal population has been formulated, the best means of increasing or decreasing the population by manipulating some of the factors affecting it can be evaluated and a plan of action arrived at prior to actual field work. Thus, the effect of doubling the numbers of a parasite in an area can be predicted before the fact. The great advantage of the quantitative approach is that it can be predictive and it is precise.

In some experiments simple models can often be used. In the majority of natural populations, however, there are usually too many important factors involved to use these models in a realistic way. The simple models, however, can be used to form preliminary hypotheses about the basic interactions within the population or between populations. These hypotheses are tested by observation or experiment, and with this background of information more suitable models can sometimes be derived to account for the factors that are unique to any particular situation. Simple models are useful in understanding the basic principles underlying a set of interactions, but each ecosystem is also affected by a series of factors unique to it. If there is one generality in ecology, it is that there are no generalities.

I have personally found statistical techniques to be the most important and necessary subject in the study of populations and communities. Therefore, statistical methods are greatly emphasized throughout this book. In addition, the events determining the changes in density or biomass of any group of organisms are determined in large part by chance. Models including chance events are more realistic than those that do not include the effects of chance. Presumably, if every minute factor, such as sunspots on Betelgeuse, were known, *stochastic* models would not be necessary. This is hardly possible. Stochastic models are, therefore, discussed whenever possible. The use of stochastic models is severely limited, however, by the great mathematical complexity of their derivation.

In walking through the woods we can observe that the orchids are disappearing or that the checkerspot butterfly is unusually common. The question is, "Why have these two events happened, what factors caused them to happen, and in what manner and degree did the environmental factors interact to cause these things to happen?" More importantly, if these factors continue at their present levels, or increase or decrease, will the orchids disappear and how soon, or will the butterfly population decrease next year, and if so, how much?

The purpose of this book is to try to present a survey of quantitative methods which might be used to answer these or similar questions, and to discuss in general the subject of the ebb and flow of populations and communities.

1

INCREASE OF A POPULATION IN AN UNLIMITED ENVIRONMENT

If there is a basic principle of population ecology, it probably is, "In an unlimited, constant, and favorable environment the number of individuals of a species will increase exponentially." Under these theoretical but unrealistic conditions mice, grasshoppers, flies, elephants, birds, ragweed, dandelions, and man would each eventually cover the earth completely. Of course no animal or plant species has ever covered the earth, although man seems to be trying. A point is reached, and the population stops increasing, either because a greater density of individuals leads to greater mortality or a decreased birthrate, or because conditions change and are no longer favorable. Chapters 1 and 2 consider a population either increasing or decreasing in an unlimited environment. In the third chapter survival and, conversely, mortality are considered to be dependent in part on the density of the population.

The most commonly used equations in population ecology to simulate population growth stem in large part from a book by Lotka (1925). However, another approach using matrix algebra has been developed by Leslie (1945, 1948) and Lewis (1943). The matrix approach has not been widely used in ecology, probably because of the unfamiliarity of many ecologists with matrices. The matrix approach is more versatile than the more commonly used models of unlimited population growth.   I will, therefore, approach the subject by first briefly introducing the simpler equations, then use them as a base to move on to the matrix solutions.   A brief section on matrix algebra is included in this first chapter and may be omitted by those already familiar with the subject.   Chapter 2 will be a discussion of some stochastic population models and of probability theory in general.

1-1   EXPONENTIAL POPULATION GROWTH

Figure 1-1 illustrates a common fate of a population of a plant or animal species living in an essentially unlimited, favorable environment.   As time passes the population density increases at a faster and faster rate until (preserving the myth of the unlimited environment) with time the number of individuals approaches infinity.   The equation of this curve is

$$N_t = N_0 e^{rt} \qquad t \geq 0 \tag{1-1}$$

where N_t is the number of individuals in the population at some time t, and N_0 is the population density at some arbitrarily set time $t = 0$.   The letter t represents units of time, in humans usually years and in fruit flies most often days.   The units of t are set by the experimenter and convenience.   In a

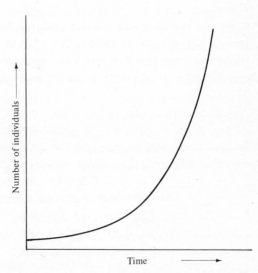

FIGURE 1-1
Diagrammatic representation of the exponential growth of population density in a constant and unlimited environment.

human population $t = 0$ might be set at 1920.   The year 1940 would then be equal to $t = 20$, if each unit of t equals 1 year.   Equation (1-1) states that the size of the population at time t is equal to the population at the beginning (at $t = 0$) times e, the base of natural logarithms ($e = 2.71828...$), raised to the rtth power where t is time and r is a constant discussed shortly. There are two numbers in Eq. (1-1) free to vary: N_t and t.   All other numbers always remain constant: e, r, and N_0.   Equation (1-1) could also be written as

$$N_t = f(t)$$

or the population density at time t is a *function* of t.   More succinctly, the population density depends in some way on how much time has elapsed. The notation $f(t)$ stands for "function of the variable t."   In Eq. (1-1) $f(t)$ stands for $N_0 e^{rt}$, but in other situations with a different curve could stand for other functions, perhaps $N_0^3 e^{r^2 t}$.   Because the density of the population depends on the amount of time elapsed, it is referred to as the *dependent variable*. Time is the *independent variable*.   Although in this case there is only one independent variable, there may be more than one.   If humidity were also being considered, the population density would be a function of both time and humidity, or

$$N_t = f(t, h)$$

letting h stand for humidity.

The meaning of the constant r can be illustrated by referring to Fig. 1-2, the same curve as Fig. 1-1, except at an arbitrarily chosen point X a line is drawn contiguous to the point.   The slope of the line is equal to the rate of change in the number of individuals in the population.   The point X does not occupy any length of the curve or any period of time, so the rate is instantaneous at exactly X.   Similar points Y and Z have also been placed on the curve, and the contiguous lines drawn.   At point Y the increase in the number of individuals is greater than at point X, and in the same way the population at point X in time is growing faster than at point Z.   The number of individuals produced between $t = 3$ and $t = 4$ would be much fewer than those produced between $t = 7$ and $t = 8$.   However, the ratio $t = 3/t = 4$ is equal to the ratio $t = 7/t = 8$, and indeed this ratio will always remain the same for a single interval of time.   The number of individuals added in one time interval rapidly increases with time, but the multiplication of the population in one time interval remains constant.   The rate at which N, the number of individuals, changes as time changes can be represented as dN/dt, and for this curve is

$$\frac{dN}{dt} = rN$$

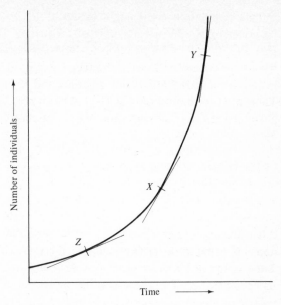

FIGURE 1-2
Diagrammatic representation of exponential population growth in a constant and unlimited environment, with lines drawn contiguous to the curve at three points to illustrate the rate of growth.

This is referred to as a *differential equation*. At time t the rate of increase in density is postulated to be equal to the constant r times the number of individuals present. When this equation is solved for explicit values of N_t, given an initial starting density N_0, Eq. (1-1) results. The solution of differential equations is beyond the scope of this book. The constant r is a measure of the rate of multiplication of the population at an instant of time and is a constant, as is the multiplication of the population during an interval of time equal to 1. On the other hand, dN/dt represents the rate of change in number of individuals, and this rate constantly changes from instant to instant. The constant r is usually termed the *intrinsic rate of increase*. The methods of estimating the value of r will be discussed in Sec. 1-3. The value of r is a constant for a constant set of environmental conditions. If these conditions change, usually so will r. Some conditions are more conducive to reproduction or cause less mortality for a specific species than others.

If the probabilities associated with the fecundity and mortality of each age group remain the same for every generation, the relative proportions of the age groups of the population approach with time a fixed distribution known as the *stable age distribution* (Lotka, 1925). For example, if the population were divided into three age groups, and the numbers of each of the three groups were in the proportions 6 : 3 : 1, these proportions would be an age distribution of the population. If in an unlimited environment with constant fertility and mortality statistics the proportions of the age groups remained 6 : 3 : 1, the age distribution of the population is said to be the stable age distribution.

A population has a birthrate and a death rate. The difference between the birthrate b and the death rate d is the instantaneous rate of increase of the population

$$r = b - d \tag{1-2}$$

The intrinsic rate of increase is defined only for a population with a stable age distribution. Equation (1-1) is valid only if there is a stable age distribution and only if the rates of birth and death determining r do not change. Rearranging Eq. (1-1) slightly and considering the change in number of individuals for only one time interval $t = 1$ gives

$$\frac{N_{t+1}}{N_t} = e^r = \lambda$$

the finite rate of increase. While r is the rate of increase of a population with a stable age distribution, λ is the multiplication of the population in one interval of time. It is also a constant, as already mentioned. If the population at time t consisted of 50 individuals and at time $t + 1$ of 75 individuals, λ equals 1.5. In other words, the population has multiplied by a factor of 1.5 during the time t to $t + 1$. Because

$$\log_e \lambda = r$$

the intrinsic rate of increase in the example is .405. The two statistics r and λ are appropriate, remember, only if the population has a stable age distribution.

1-2 THE LIFE TABLE

The basic information needed to study density changes and rates of increase or decrease is contained in a life table. A life table contains such vital statistics as the probability of an individual of a certain age dying or, conversely, the average number of offspring produced by a female of a given age.

The most reliable method of determining these statistics is to begin with a group of individuals all born at the same time, a cohort, and follow the

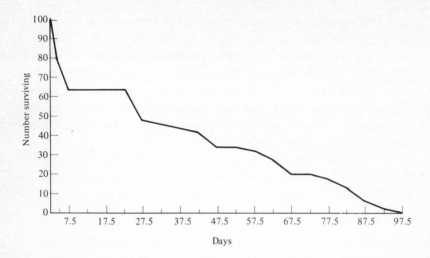

FIGURE 1-3
Survivorship curve for a cohort of 100 individuals of the muscid fly *S. nudiseta* reared at 20°C from egg to the death of the last individual. (*After Rabinovich, 1970.*)

life of the cohort, noting the deaths of individuals and the birth of offspring until the demise of the last individual. It is convenient in the calculations to use cohorts of 100 or 1,000 individuals. Rabinovich (1970) formulated a life table for a muscid fly, *Synthesiomyia nudiseta*, reared at 20°C. He began with 100 eggs and counted the number of individuals remaining after intervals of 5 days, with the exception of a first interval of only 2.5 days. The number of the original 100 individuals surviving at 5-day intervals until the last individual died approximately 95 days later is shown in Fig. 1-3. This curve is called a *survivorship curve* and demonstrates a high initial mortality followed by a steady decline in numbers to the death of the last individual. The number of individuals of the initial cohort alive at age x is termed l_x. The value of $l_{12.5}$, the number of individuals alive at day 12.5, is 64. The statistic l_x will be given a slightly different definition in Sec. 1.3 to conform to the standardly used notation, and the student should always be sure which definition is being used. The l_x values for the fly cohort are tabulated in Table 1-1. The number of individuals dying during each interval of time is usually noted as d_x and is simply the number of individuals alive at age x minus the number alive at age $x + 1$; that is, $d_x = l_x - l_{x+1}$. If $l_{2.5} = 75$ and $l_{7.5} = 64$, d_x for the 5-day interval between ages 2.5 and 7.5 days is 11. The d_x values for the cohort of flies are also listed in Table 1-1. The percentage of the cohort dying in the interval x to $x + 1$ of those alive at age x is q_x and is the value of d_x for the same interval of time divided by the number of flies alive at age x, l_x. The percentage of flies dying during the interval between 2.5 and 7.5 days, which were alive at 2.5 days, is 11/75, or 14.7 percent.

A slightly more complicated set of statistics to arrive at are the number of individuals alive during the interval between x and $x + 1$, L_x. Because individuals are dying continuously there is no exact answer except at an instant of time. Therefore, L_x in reality represents the average number of individuals alive during the interval. Strictly speaking, L_x is the total number of units of time during the interval x to $x + 1$ lived by all the individuals of the cohort and is equal to the area below the survivorship curve and

Table 1-1   THE LIFE TABLE FOR A COHORT OF 100 INDIVIDUALS OF THE FLY *S. NUDISETA* REARED AT 20°C FROM EGG TO THE DEATH OF THE LAST INDIVIDUAL

x days	q_x	l_x	d_x	L_x	T_x	e_x
0	.250	100	25	87	768	19.20
2.5	.147	75	11	69	681	45.40
7.5	.000	64	0	64	612	47.81
12.5	.000	64	0	64	548	42.81
17.5	.000	64	0	64	484	37.81
22.5	.250	64	16	56	420	32.81
27.5	.041	48	2	47	364	37.92
32.5	.043	46	2	45	317	34.46
37.5	.045	44	2	43	272	30.91
42.5	.190	42	8	38	229	27.26
47.5	.000	34	0	34	191	28.09
52.5	.058	34	2	33	157	23.08
57.5	.125	32	4	30	124	19.38
62.5	.285	28	8	24	94	16.79
67.5	.000	20	0	20	70	17.50
72.5	.100	20	2	19	50	12.50
77.5	.222	18	4	16	31	8.61
82.5	.000	14	8	10	15	5.35
87.5	.571	6	4	4	5	4.15
92.5	1.000	2	2	1	1	2.00
97.5	…	0				

SOURCE: After Rabinovich, J. E., Vital Statistics of *Synthesiomyia nudiseta* (Diptera: Muscidae), *Ann. Entomol. Soc. Am.*, vol. 73, 1970.

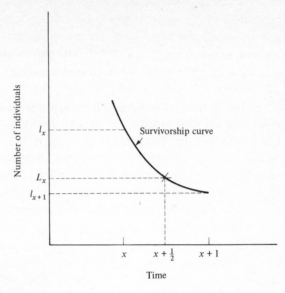

FIGURE 1-4
Graphic method of finding L_x when the decrease in number of survivors from time period x to $x + 1$ is not linear.

between the two points x and $x + 1$. Again strictly speaking, this area is found by integration of the l_x curve between x and $x + 1$, or

$$L_x = \int_x^{x+1} l_x \, dx \qquad (1\text{-}3)$$

where dx is not the d_x of the life table and is only part of the integral notation. There are two practical methods of determining the value of L_x for some interval x to $x + 1$. If it can be assumed that individuals are dying at a constant rate during the interval of time, that is, l_x and l_{x+1} can be connected by a straight line, L_x is the average of l_x and l_{x+1}

$$L_x = \frac{l_x + l_{x+1}}{2}$$

If this. cannot be assumed and the line connecting l_x and l_{x+1} in the survivorship curve is known to be curvilinear, not straight, an approximate value of L_x may be arrived at by finding the value of l_x on the l_x curve corresponding to $x + \frac{1}{2}$, the midpoint of the age interval as shown in Fig. 1-4. Usually nothing is known about the changes in rates of mortality during the age interval, and Eq. (1-3) will be the appropriate way of determining the series of L_x statistics.   The values of L_x are usually called the *stationary age distribution*.

One further statistic is T_x, the number of units of time, 5 days for the fly cohort, lived by the cohort from age x until all individuals are deceased. The values of T_x are found by adding up the L_x column from x to the end. From the T_x statistics the life expectancy of an individual of age x from age x, e_x, can be determined as

$$e_x = \frac{T_x}{l_x}$$

The life expectancy of a muscid fly 12.5 days old is $e_{12.5} = \frac{548}{64}$, or 8.56 age intervals of 5 days, which is equivalent to 42.8 days.

One final statistic, m_x, is the average number of offspring produced per female of age x from x to $x + 1$.   Like l_x, m_x is a continuously changing variable, and in its calculation the standard procedure is to take the number of offspring produced per female in the interval x to $x + 1$ as m_x.   The exact value of m_x cannot, in general, be determined, and the above method of estimation assumes that all the births occur instantaneously at the midpoint of the time interval.   Therefore, between $x = 4$ and $x = 5$, $m_{4.5}$ is taken to be the average number of offspring produced by a single female between $x = 4$ and $x = 5$.   An example of m_x statistics is given in Sec. 1-3.   When a life table contains m_x statistics, only the females of the cohort are considered, and l_x is defined as the probability of a female surviving to age x, where x is the midpoint of the interval x to $x + 1$ in order to conform with the m_x statistics. The reasons for this should become apparent later.   The complete life table of the cohort of 100 individuals of *S. nudiseta* reared at 20°C is shown in Table 1-1.

1-2.1   Natural Populations

Rarely can the cohort life table method be applied to natural populations. In human populations, and perhaps in many animal or plant populations, the two most easily observed sets of data are the number of individuals alive during a series of equal-length age group intervals and the number of deaths and births in each age group during an interval of time.   Provided the population is not increasing too rapidly relative to the time interval used, and if the age distribution is stable, the observed number of deaths in each age group during a length of time equal to the age interval is an estimate of d_x of the cohort life table.   Also, the number of individuals in the age group alive at the midpoint of the age interval is approximately equal to l_x.   If the population does not have a stable age distribution or does not remain at a constant size, these approximations do not necessarily hold. The reader should refer to Keyfitz (1968) for an excellent discussion of life tables as applied to human populations.

If the age of an individual at its death can be determined, an approximate life table can be formulated by finding the series of d_x values. Murie (1944) collected the skulls of 608 Dall mountain sheep, *Ovis dalli*, and from the number of annual rings in the horns was able to determine the age of the sheep at its death. The number of sheep dying in each of a series of age intervals in terms of 1,000 sheep provides the d_x column from which the remainder of the table may be derived, provided the skulls found represent a random sample of all the individuals which have died.

The mountain sheep case is perhaps fortuitous. In most natural populations the estimation of population size, size of age groups, fertility, and mortality rates presents serious problems in sampling. The sampling of natural populations for the estimation of these statistics will be discussed in Chap. 11.

1-3 ESTIMATION OF THE INTRINSIC RATE OF INCREASE

The calculation of the parameter r, the intrinsic rate of increase, is based only on the females of a population, and it is assumed that there are enough males to go around. The first step in the calculation of the intrinsic rate of increase is the formulation of a life table for the population. Unfortunately, a confusing nomenclatorial tangle centers around the definition of l_x. In Sec. 1-2, l_x was defined as the number of individuals of a cohort alive at age x. However, l_x is always changing, although usually it is known only at discrete intervals of time. Therefore, in the computations that follow it is necessary to ignore the continuous nature of l_x and m_x and treat them as numbers changing by jumps. In other words, mortality and fertility are assumed to occur instantaneously at the midpoint of the time interval. In contrast to the definition of l_x given in Sec. 1-2, it is standard notation in the literature when estimating r to define l_x as the percent of the cohort alive at the midpoint of the time interval. In terms of the previous section l_x as redefined is L_x divided by the initial size of the cohort. The student with some calculus may refer to Pielou (1969) for an understanding of the problem. Otherwise this confusing change in terminology should be accepted on faith to conform with the standard notation used in most papers dealing with the estimation of the intrinsic rate of increase. Birch (1948) has estimated the l_x and m_x statistics as defined above for a cohort of the weevil *Calandra oryzae* in an unlimited environment of wheat at 20°C and 14 percent relative humidity (Table 1-2).

In any one generation the total number of individuals produced per female per generation is equal to the fecundity of the female, m_x, as modified by the probability of her living to be that old, l_x. A female in the age group 8 to 9 weeks could produce 12.5 offspring (Table 1-2) but has only a 79 percent chance of reaching that age. Therefore, the total number of offspring produced per female during the interval of time is

$$l_x m_{x(8-9)} = (.79)(12.5)$$
$$= 9.88 \text{ offspring}$$

The total number of female offspring produced per female during a single generation is equal to the sum of the $l_x m_x$ products, or

$$R_0 = \sum_{x=0}^{n} l_x m_x$$

Table 1-2 A PARTIAL LIFE TABLE OF THE WEEVIL *C. ORYZAE*

Age in weeks, x	l_x	m_x	$l_x m_x$
4.5	.87	20.0	17.400
5.5	.83	23.0	19.090
6.5	.81	15.0	12.150
7.5	.80	12.5	10.000
8.5	.79	12.5	9.875
9.5	.77	14.0	10.780
10.5	.74	12.5	9.250
11.5	.66	14.5	9.570
12.5	.59	11.0	6.490
13.5	.52	9.5	4.940
14.5	.45	2.5	1.125
15.5	.36	2.5	.900
16.5	.29	2.5	.800
17.5	.25	4.0	1.000
18.5	.19	1.0	.190

$R_0 = 113.560$

SOURCE: From Birch, L. C., The Intrinsic Rate of Natural Increase of an Insect Population, *J. Ecol.*, vol. 17, 1948.

The sign Σ (sigma) stands for the sum of a series of numbers. The numbers summed are indicated by the limits of summation found above and below the sign. The first number to be summed, the first age group, is indicated by $x = 0$, 17.400 in Table 1-2. The x refers to the subscript of the numbers being added and can be thought of as a counter taking values from $x = 0$ to n, the last age group. The upper limit is n, sometimes written more completely as $x = n$, where n refers to the last number summed, .190. In the weevil example $n = 14$ because there are 15 numbers to be added. The limits of summation are not always from 0 to the last number. The limits of summation might be

$$\sum_{x=r}^{n-r}$$

meaning to sum up a series of numbers beginning at some rth number and ending at the $(n - r)$th number. If in some particular example n equals 20 and r is 5, this is equivalent to

$$\sum_{x=5}^{15}$$

In some situations $x = 1$ is the first number of a series rather than $x = 0$, depending on convention and the situation. In some cases if the limits of summation are obvious, they are left off the summation sign to conserve space.

In the weevil example in Table 1-2, R_0 (termed the *net reproduction rate*) equals 113.56, or the population is multiplying 113.56 times each generation under the given set of environmental conditions and in an unlimited environment. The net reproduction rate R_0 is analogous to λ, the finite rate of increase, except that λ is defined for an interval of time equal to 1, and R_0 for a length of time equal to the mean length of a generation, T. If the time interval t equals the generation time T, R_0 equals λ, and in general

$$R_0 = e^{rT}$$

The mean generation time can be roughly estimated by dividing the log to the base e of R_0 by the intrinsic rate of increase r.

$$T = \frac{\log_e R_0}{r} \tag{1-4}$$

If λ is close to 1.0, a simple approximation to Eq. (1-4) (Dublin and Lotka, 1925) is

$$T \approx \frac{\sum x l_x m_x}{\sum l_x m_x}$$

or the sum of the $xl_x m_x$ column (x is the midpoint age in weeks for the weevils) divided by R_0. In Birch's example $T \approx 943.09/113.56 \approx 8.3$ weeks. Given T, a rough estimate of r is

$$r = \frac{\log_e R_0}{T} \qquad (1\text{-}5)$$

or

$$r = \frac{\log_e 113.56}{8.30}$$

$$= .57$$

This is only a rough, crude approximation but may be be fairly accurate if λ is close to 1.0.

If mortality and fertility remain the same and the population has a stable age distribution, the value of r may be estimated from the equation

$$\sum_{x=0}^{n} e^{-rx} l_x m_x = 1 \qquad (1\text{-}6)$$

To find r, possible values of r are substituted into the equation, possibly beginning with the rough estimate calculated from Eq. (1-5), until the left side of the equation equals 1. This method of computation is called iteration or an iterative process. The computations in Eq. (1-6) are quick if a calculator with an e^x key is available. A general trick in iteration is to choose two numbers, one of which is sure to be larger than the true value of r, and the second sure to be smaller. Instead of 1, a number larger than 1 and a number smaller than 1 will result as the solution of the left-hand side of the equation. From an examination of the deviations of these two numbers from 1, an approximate value of r setting the left-hand side relatively close to 1 can be chosen. This rough estimate of r can be further improved by successive steps, iterations, until the deviation of the solution from the left-hand side of the equation is as small as is wanted.

In the weevil example the rough approximate value for the intrinsic rate of increase of *C. oryzae* was .57. Beginning with two estimates of r as .5 and .8

$$r = .5 \qquad \sum_{x=0}^{n} e^{-.5x} l_x m_x = 4.0920 \qquad \text{deviation from 1, } +3.0920$$

$$r = .8 \qquad \sum_{x=0}^{n} e^{-.8x} l_x m_x = .8215 \qquad \text{deviation from 1, } -.1785$$

In examining the two deviations r is seen to be much closer to .8 than to .5. By taking $r = .76$ the deviation is $+.0104$. The deviation is small and positive, indicating a number slightly larger than .76. A final estimate of r is about .762, the intrinsic rate of increase of C. *oryzae* reared at 20°C and 14 percent relative humidity.

If the life table is known and the intrinsic rate of increase calculated, the stable age distribution of the population can be computed. In a stable age population of N individuals the birthrate β during one interval of time t can be defined as the number of female births during the interval t to $t + 1$, B_t, divided by N_t. For computing the finite birthrate, which has the same relation to b as λ does to r, the following equation should be used:

$$\frac{1}{\beta} = \sum_{x=0}^{n} l_x e^{-r(x+1)}$$

In the weevil example $1/\beta = .81167$ and, dividing 1 by .81167, $\beta = 1.23067$. The proportion p_x of individuals in the age group x to $x + 1$ in the stable age distribution is given by

$$p_x = \beta l_x e^{-r(x+1)}$$

Table 1-3, from Birch (1948), shows the calculation of the stable age distribution of C. *oryzae* at 29°C and 14 percent relative humidity. The older age groups are not included because they make almost no contribution to the final age

Table 1-3 CALCULATION OF THE STABLE AGE DISTRIBUTION OF THE WEEVIL C. *ORYZAE*

Age group (x)	l_x	$e^{-r(x+1)}$	$l_x e^{-r(x+1)}$	Stable age distribution	
0–	.95	.4677	.4443150	54.740	
1–	.90	.2187	.1968300	24.249	
2–	.90	.10228	.0920520	11.341	immatures
3–	.90	.04783	.0430470	5.304	
4–	.87	.02237	.0194619	2.398	
5–	.83	.01046	.0086818	1.070	
6–	.81	.00489	.0039609	.488	
7–	.80	.002243	.0017944	.221	
8–	.79	.001070	.0008453	.104	
9–	.77	.000500	.0003850	.047	adults
10–	.74	.000239	.0001769	.022	
11–	.66	.000110	.0000726	.009	
12–	.59	.000051	.0000301	.004	
13–	.52	.000024	.0000125	.002	
14–	.45	.000011	.0000050	.001	

SOURCE: From Birch, L. C., The Intrinsic Rate of Natural Increase of an Insect Population, *J. Ecol.*, vol. 17, 1948.

distribution.   The stable age distribution is heavily loaded with the immature stages of the weevils.   Only 4.5 percent of the total number of individuals in the stable age distribution are adults.

The intrinsic rate of increase r is simply the difference between the instantaneous birthrate and the instantaneous death rate: $r = b - d$.   If r has been determined, these two rates can be found by first finding b as

$$b = \frac{\beta r}{\lambda - 1}$$

and then determining d from $r = b - d$.   In the weevil experiment $.76 = .82 - d$, or $d = .06$.   The finite death rate δ is found from the solution of

$$d = \frac{\delta r}{\lambda - 1}$$

for δ.

1-4   MATRIX ALGEBRA

Matrix algebra provides a convenient way of compiling, storing, and manipulating large amounts of data.   A knowledge of matrix algebra is essential, not only in the use of the matrix models of population growth introduced in the next section of this chapter, but also in many of the statistical methods used later in the book.   It is vital that the user of this text understand the basics of matrix algebra presented in this section.   Many aspects of matrix algebra, such as addition, subtraction, and multiplication, are relatively straightforward.   Others, including inversion, latent roots, and determinants, are more complicated.   In discussing these last-mentioned three subjects the operations are carried out on small matrices.   The computations on larger matrices become increasingly complicated and tedious and are almost always problems to be carried out using a computer and one of the many available computer programs.   It is hoped that by limiting the discussion to smaller matrices, the general principles involved can be illustrated without the necessity of going into the complications of the analysis of larger examples.   For those interested in more than just the simple treatment presented here, Searle (1966) has written an excellent book on matrix algebra for biologists.

1-4.1   The Matrix

A matrix is an array of numbers.   Suppose an area of lawn of 9 square feet is divided into square-foot quadrats and the number of earthworms in each quadrat censused.   A possible result might be

$$
\begin{matrix}
26 & 24 & 9 \\
52 & 61 & 19 \\
12 & 33 & 6
\end{matrix}
$$

These observations can be set up as a matrix

$$
\begin{bmatrix}
26 & 24 & 9 \\
52 & 61 & 19 \\
12 & 33 & 6
\end{bmatrix}
$$

Each element (number) in the matrix may be indicated as a_{ij}, where i is the number of the row and j the number of the column of the observation. Therefore, the number 19 is indicated as $a_{2,3}$ and the entire matrix is

$$
\begin{bmatrix}
a_{1,1} & a_{1,2} & a_{1,3} \\
a_{2,1} & a_{2,2} & a_{2,3} \\
a_{3,1} & a_{3,2} & a_{3,3}
\end{bmatrix}
$$

The entire matrix may be designated by a boldface letter, say **A**. Sometimes the matrix is indicated as $\mathbf{A} = \{a_{ij}\}$. This matrix has three rows and three columns, and so its *order* is 3×3. Because there are equal numbers of rows and columns the matrix is *square*. The elements of a square matrix, $a_{1,1}$, $a_{2,2}$, and $a_{3,3}$ in this example, are the *diagonal*.

Another example of a matrix of observations is the number of larvae of some insect found alive after spraying on each of four subsequent days. The matrix might be

	Field 1	Field 2	Field 3
Day 1	53	12	14
Day 2	22	17	23
Day 3	19	19	22
Day 4	21	12	19

The order of this matrix is 4×3, and it is not square.

A *diagonal matrix* is a square matrix with the only nonzero elements in the diagonal. An example is

$$
\begin{bmatrix}
6 & 0 & 0 & 0 \\
0 & 9 & 0 & 0 \\
0 & 0 & 11 & 0 \\
0 & 0 & 0 & 7
\end{bmatrix}
$$

A special form of the diagonal matrix is the *identity matrix* which has only 1s in the diagonal, e.g.,

$$\begin{bmatrix} 1 & 0 & 0 & 0 \\ 0 & 1 & 0 & 0 \\ 0 & 0 & 1 & 0 \\ 0 & 0 & 0 & 1 \end{bmatrix} = \mathbf{I}$$

A matrix may consist of only a single column, e.g.,

$$\mathbf{x} = \begin{bmatrix} 5 \\ 7 \\ 9 \\ 3 \end{bmatrix}$$

and is called a *column vector*.   A matrix of a single row is a *row vector*, e.g.,

$$\mathbf{x}' = \begin{bmatrix} 9 & 16 & 3 & 5 & 12 \end{bmatrix}$$

A matrix of a single element is a scalar

$$a = \begin{bmatrix} 3 \end{bmatrix}$$

Usually small boldface letters are used for row and column vectors.

The transpose of a matrix is a matrix with its rows and columns interchanged.

$$\mathbf{A}' = \begin{bmatrix} a_{1,1} & a_{2,1} & a_{3,1} \\ a_{1,2} & a_{2,2} & a_{3,2} \\ a_{1,3} & a_{2,3} & a_{3,3} \end{bmatrix}$$

The earthworm matrix was

$$\mathbf{A} = \begin{bmatrix} 26 & 24 & 9 \\ 52 & 61 & 19 \\ 12 & 33 & 6 \end{bmatrix}$$

The transpose of **A** is

$$\mathbf{A}' = \begin{bmatrix} 26 & 52 & 12 \\ 24 & 61 & 33 \\ 9 & 19 & 6 \end{bmatrix}$$

Sometimes a large matrix may be broken into smaller matrices, each termed a *submatrix*.   This process is called *partitioning*.   The matrix **A** could be partitioned as

$$\mathbf{A} = \left[\begin{array}{c|cc} 26 & 24 & 9 \\ 52 & 61 & 19 \\ 12 & 33 & 6 \end{array} \right]$$

resulting in a scalar, a row vector, a column vector, and a square matrix of order 2×2. Sometimes each submatrix of the partition is given a letter such as

$$\mathbf{A} = \left[\begin{array}{c|c} a & \mathbf{x} \\ \hline \mathbf{y} & \mathbf{B} \end{array}\right]$$

1-4.2 Addition of Two Matrices

Suppose the same 9-square-foot field plot was again censused a year later, resulting in a matrix of observations indicated as **B**. The total number of earthworms caught in each square-foot quadrat for both years is found by adding matrix **A** and matrix **B**.

$$\mathbf{A} + \mathbf{B} = \begin{bmatrix} 26 & 24 & 9 \\ 52 & 61 & 19 \\ 12 & 33 & 6 \end{bmatrix} + \begin{bmatrix} 31 & 42 & 9 \\ 17 & 63 & 12 \\ 13 & 14 & 7 \end{bmatrix}$$

In the addition of two matrices the elements in the same position in the two matrices are added, i.e.,

$$\mathbf{A} + \mathbf{B} = \begin{bmatrix} 26+31 & 24+42 & 9+9 \\ 52+17 & 61+63 & 19+12 \\ 12+13 & 33+14 & 6+7 \end{bmatrix} = \begin{bmatrix} 57 & 66 & 18 \\ 69 & 124 & 31 \\ 25 & 47 & 13 \end{bmatrix}$$

Two matrices can be added only if they are of the same order, i.e., are *conformable*.

1-4.3 Subtraction of Two Matrices

Subtraction proceeds in the same way as addition, except that respective elements are subtracted rather than added. Utilizing the same example, the differences in number of earthworms in each quadrat from the first to the second year are

$$\mathbf{A} - \mathbf{B} = \begin{bmatrix} 26-31 & 24-42 & 9-9 \\ 52-17 & 61-63 & 19-12 \\ 12-13 & 33-14 & 6-7 \end{bmatrix} = \begin{bmatrix} -5 & -18 & 0 \\ 35 & -2 & 7 \\ -1 & 19 & -1 \end{bmatrix}$$

Again subtraction is possible only if the two matrices are of the same order.

1-4.4 Multiplication

The simplest form of multiplication in matrix algebra is the multiplication of a scalar by a matrix. Every element in the matrix is simply multiplied by the scalar. This can be illustrated by an example:

$$4\begin{bmatrix} 12 & 11 \\ 9 & 3 \end{bmatrix} = \begin{bmatrix} 48 & 44 \\ 36 & 12 \end{bmatrix}$$

The multiplication of a matrix by a column vector can be illustrated by an example:

$$\begin{bmatrix} 12 & 7 & 9 \\ 6 & 3 & 2 \\ 5 & 4 & 2 \end{bmatrix}\begin{bmatrix} 2 \\ 1 \\ 3 \end{bmatrix} = \begin{bmatrix} (12)(2) + (7)(1) + (9)(3) \\ (6)(2) + (3)(1) + (2)(3) \\ (5)(2) + (4)(1) + (2)(3) \end{bmatrix} = \begin{bmatrix} 58 \\ 21 \\ 20 \end{bmatrix}$$

The result of the multiplication is another column vector. In this example the matrix is *postmultiplied* by the column vector. A matrix cannot be *premultiplied* by a column vector. A matrix can be postmultiplied by a column vector only if the number of elements in the column vector is equal to the number of columns in the matrix.

The premultiplication of a matrix by a row vector is another row vector, e.g.,

$$[6 \quad 12 \quad 17]\begin{bmatrix} 12 & 7 & 3 & 7 \\ 1 & 6 & 5 & 0 \\ 12 & 2 & 1 & 4 \end{bmatrix} = \begin{bmatrix} (6)(12) & (6)(7) & (6)(3) & (6)(7) \\ + & + & + & + \\ (12)(1) & (12)(6) & (12)(5) & (12)(0) \\ + & + & + & + \\ (17)(12) & (17)(2) & (17)(1) & (17)(4) \end{bmatrix}$$

$$= [288 \quad 148 \quad 95 \quad 110]$$

The premultiplication of a matrix by a row vector is possible only if the number of elements in the row vector is equal to the number of rows in the matrix. The premultiplication of a column vector by a row vector is a scalar, e.g.,

$$[12 \quad 7 \quad 6]\begin{bmatrix} 9 \\ 12 \\ 1 \end{bmatrix} = (12)(9) + (7)(12) + (6)(1) = 198$$

The postmultiplication of a column vector by a row vector is a matrix, e.g.,

$$\begin{bmatrix} 9 \\ 12 \\ 1 \end{bmatrix}[12 \quad 7 \quad 6] = \begin{bmatrix} (9)(12) & (9)(7) & (9)(6) \\ (12)(12) & (12)(7) & (12)(6) \\ (1)(12) & (1)(7) & (1)(6) \end{bmatrix}$$

$$= \begin{bmatrix} 108 & 63 & 54 \\ 144 & 84 & 72 \\ 12 & 7 & 6 \end{bmatrix}$$

The multiplication of two matrices **A** and **B** is possible only if the number of columns in the first matrix is equal to the number of rows in the second matrix. If this is true, the matrices **A** and **B** are *conformable for* *multiplication* in the order $\mathbf{A} \times \mathbf{B}$. In the multiplication of two matrices it is convenient to think of the second matrix as a group of column vectors. In the multiplication

$$\begin{bmatrix} 1 & 0 & 2 \\ 3 & 1 & 1 \\ 1 & 2 & 1 \\ -1 & 3 & 2 \end{bmatrix} \begin{bmatrix} 1 & 2 \\ 0 & 1 \\ 0 & -1 \end{bmatrix} = \mathbf{A} \times \mathbf{B}$$

the second matrix $\mathbf{B}$ can be thought of as two column vectors

$$\begin{bmatrix} 1 \\ 0 \\ 0 \end{bmatrix} \begin{bmatrix} 2 \\ 1 \\ -1 \end{bmatrix}$$

and $\mathbf{A}$ multiplied by each column vector of $\mathbf{B}$ separately:

$$\begin{bmatrix} 1 & 0 & 2 \\ 3 & 1 & 1 \\ 1 & 2 & 1 \\ -1 & 3 & 2 \end{bmatrix} \begin{bmatrix} 1 \\ 0 \\ 0 \end{bmatrix} = \begin{bmatrix} 1 \\ 3 \\ 1 \\ -1 \end{bmatrix}$$

and

$$\begin{bmatrix} 1 & 0 & 2 \\ 3 & 1 & 1 \\ 1 & 2 & 1 \\ -1 & 3 & 2 \end{bmatrix} \begin{bmatrix} 2 \\ 1 \\ -1 \end{bmatrix} = \begin{bmatrix} 0 \\ 6 \\ 3 \\ -1 \end{bmatrix}$$

The result of the multiplication of $\mathbf{A}$ by $\mathbf{B}$ is

$$\mathbf{AB} = \begin{bmatrix} 1 & 0 & 2 \\ 3 & 1 & 1 \\ 1 & 2 & 1 \\ -1 & 3 & 2 \end{bmatrix} \begin{bmatrix} 1 & 2 \\ 0 & 1 \\ 0 & -1 \end{bmatrix} = \begin{bmatrix} 1 & 0 \\ 3 & 6 \\ 1 & 3 \\ -1 & -1 \end{bmatrix}$$

Unlike two numbers the product $\mathbf{AB}$ is not the same as $\mathbf{BA}$. In fact, the product $\mathbf{BA}$ is not possible in this example because in that order the two matrices are not conformable.

1-4.5 Determinants

A determinant is a polynomial of the elements of a square matrix, and in square matrices consisting only of numbers the determinant is a scalar. The calculation of determinants for larger matrices is a complicated and tedious process best done by computers. Only the determinants of smaller matrices will be considered here, and the calculations will be discussed as a process.

For the general methods applicable to large matrices the reader is referred to Searle (1966).

The determinant of a matrix $\mathbf{A}$ is usually indicated as $|\mathbf{A}|$. For a 2×2 matrix such as

$$\begin{vmatrix} 4 & 9 \\ 3 & 7 \end{vmatrix}$$

the determinant is calculated as

$$(4)(7) - (3)(9) = 1$$

The determinant of a 3×3 matrix such as

$$|\mathbf{A}| = \begin{vmatrix} 1 & 2 & 3 \\ 4 & 5 & 6 \\ 7 & 8 & 10 \end{vmatrix}$$

can be calculated as

$$|\mathbf{A}| = 1(+1)\begin{vmatrix} 5 & 6 \\ 8 & 10 \end{vmatrix} + 2(-1)\begin{vmatrix} 4 & 6 \\ 7 & 10 \end{vmatrix} + 3(+1)\begin{vmatrix} 4 & 5 \\ 7 & 8 \end{vmatrix}$$

$$= 1(50 - 48) - 2(40 - 42) + 3(32 - 35)$$

$$= -3$$

The process is called expansion by minors. The determinant

$$\begin{vmatrix} 5 & 6 \\ 8 & 10 \end{vmatrix}$$

is the *minor* of the first element of the first row, 1. Similarly

$$\begin{vmatrix} 4 & 6 \\ 7 & 10 \end{vmatrix}$$

is the minor of the second element of the first row, 2. The one in parentheses determines the sign of each term of the expansion and is chosen according to the rule $(-1)^{i+j}$. The sign of the first term is therefore $(-1)^{1+1} = (-1)^2 = +1$. The sign of the first term is positive. The sign of the second term is $(-1)^{1+2} = -1^3 = -1$, or negative. The minor plus the coefficient $(-1)^{i+j}$ is the *cofactor* of the element.

The expansion could have been carried out on the second row as well and would have led to the same result, i.e.,

$$|\mathbf{A}| = 4(-1)\begin{vmatrix} 2 & 3 \\ 8 & 10 \end{vmatrix} + 5(+1)\begin{vmatrix} 1 & 3 \\ 7 & 10 \end{vmatrix} + 6(-1)\begin{vmatrix} 1 & 2 \\ 7 & 8 \end{vmatrix}$$

$$= -4(-4) + 5(-11) - 6(-6) = -3$$

In fact, the expansion can be carried out on any of the three rows or three columns of this 3×3 matrix. The calculation of the determinant of larger matrices becomes almost impossible without the use of a computer. There are many computer programs available for this purpose.

1-4.6   The Inverse

Division, in a strict sense, does not exist in matrix algebra. Instead division-like operations are carried out by the use of the *inverse* of a matrix. The inverse of a matrix $\mathbf{A}$ is indicated as $\mathbf{A}^{-1}$, just as the reciprocal of a number can be indicated as $28^{-1} = \frac{1}{28}$.

The properties of an inverse are:

1   The inverse exists only for square matrices.
2   The inverse of a matrix with a determinant equal to zero does not exist. If the determinant is zero, the matrix is said to be *singular*.
3 $\mathbf{A}^{-1}\mathbf{A} = \mathbf{A}\mathbf{A}^{-1} = \mathbf{I}$.
4   An inverse of a matrix is unique.
5   The determinant of the inverse of $\mathbf{A}$ is equal to the reciprocal of the determinant of $\mathbf{A}$; that is, $|\mathbf{A}^{-1}| = 1/|\mathbf{A}|$.

The calculation of an inverse will be illustrated by an example and involves the cofactors of all the elements of the matrix. Taking the following matrix column by column

$$\begin{bmatrix} 2 & 4 & 3 \\ 6 & 2 & 7 \\ 9 & 3 & 1 \end{bmatrix}$$

the cofactors of the first column are

$$(-1)^{1+1}\begin{vmatrix} 2 & 7 \\ 3 & 1 \end{vmatrix} \quad (-1)^{2+1}\begin{vmatrix} 4 & 3 \\ 3 & 1 \end{vmatrix} \quad (-1)^{3+1}\begin{vmatrix} 4 & 3 \\ 2 & 7 \end{vmatrix}$$

$$= -19, 5, 22$$

of the second column

$$(-1)^{2+1}\begin{vmatrix} 6 & 7 \\ 9 & 1 \end{vmatrix} \quad (-1)^{2+2}\begin{vmatrix} 2 & 3 \\ 9 & 1 \end{vmatrix} \quad (-1)^{2+3}\begin{vmatrix} 2 & 3 \\ 6 & 7 \end{vmatrix}$$

$$= 57, -25, 4$$

and of the third column

$$(-1)^{3+1}\begin{vmatrix} 6 & 2 \\ 9 & 3 \end{vmatrix} \quad (-1)^{3+2}\begin{vmatrix} 2 & 4 \\ 9 & 3 \end{vmatrix} \quad (-1)^{3+3}\begin{vmatrix} 2 & 4 \\ 6 & 2 \end{vmatrix}$$

$$= 0, 30, -20$$

These cofactors are set up as a matrix, and the matrix is multiplied by the reciprocal of the determinant of **A**, that is, $1/|\mathbf{A}|$.

$$\frac{1}{190} \begin{bmatrix} -19 & 5 & 22 \\ 57 & -25 & 4 \\ 0 & 30 & -20 \end{bmatrix}$$

This is the inverse of **A**, that is, $\mathbf{A}^{-1}$. Multiplication by the scalar can be carried out if desired, giving

$$\mathbf{A}^{-1} = \begin{bmatrix} -.1000 & .0263 & .1158 \\ .3000 & -.1316 & .0211 \\ .0000 & .1579 & -.1053 \end{bmatrix}$$

The reader can verify that $\mathbf{A}\mathbf{A}^{-1} = \mathbf{A}^{-1}\mathbf{A} = \mathbf{I}$.

The large matrices likely to occur in most situations can be inverted only with the use of a computer.

1-4.7   Latent Roots and Latent Vectors

The latent roots and vectors of a matrix, sometimes also referred to as eigenvalues and eigenvectors, will be used several times in this book. For the purposes of this brief introduction to matrix algebra the latent roots of a matrix **A** are those numbers satisfying the equation

$$|\mathbf{A} - \lambda\mathbf{I}| = 0$$

called the *characteristic equation*. The matrix **A** must be square, and if **A** is 4×4 there will be four values of λ satisfying the equation, i.e., four latent roots. The calculation of the latent roots will be illustrated by an example.

$$\mathbf{A} = \begin{bmatrix} 1 & 4 & 1 \\ 2 & 1 & 0 \\ -1 & 3 & 1 \end{bmatrix}$$

$$\left| \begin{bmatrix} 1 & 4 & 1 \\ 2 & 1 & 0 \\ -1 & 3 & 1 \end{bmatrix} - \lambda \begin{bmatrix} 1 & 0 & 0 \\ 0 & 1 & 0 \\ 0 & 0 & 1 \end{bmatrix} \right| = 0$$

and from this

$$\begin{vmatrix} 1-\lambda & 4 & 1 \\ 2 & 1-\lambda & 0 \\ -1 & 3 & 1-\lambda \end{vmatrix} = 0$$

The determinant of this matrix is found by *diagonal expansion*.   In a diagonal expansion of a matrix of the form

$$\begin{vmatrix} a_{1,1} + x_1 & a_{1,2} & a_{1,3} \\ a_{2,1} & a_{2,2} + x_2 & a_{2,3} \\ a_{3,1} & a_{3,2} & a_{3,3} + x_3 \end{vmatrix}$$

the expansion is

$$x_1 x_2 x_3 + x_1 x_2 a_{3,3} + x_1 x_3 a_{2,2} + x_2 x_3 a_{1,1} + x_1 \begin{vmatrix} a_{2,2} & a_{2,3} \\ a_{3,2} & a_{3,3} \end{vmatrix}$$

$$+ x_2 \begin{vmatrix} a_{1,1} & a_{1,3} \\ a_{3,1} & a_{3,3} \end{vmatrix} + x_3 \begin{vmatrix} a_{1,1} & a_{1,2} \\ a_{2,1} & a_{2,2} \end{vmatrix} + \begin{vmatrix} a_{1,1} & a_{1,2} & a_{1,3} \\ a_{2,1} & a_{2,2} & a_{2,3} \\ a_{3,1} & a_{3,2} & a_{3,3} \end{vmatrix}$$

Carrying out this expansion on the example matrix with $x_1 = x_2 = x_3 = \lambda$

$$-\lambda^3 + \lambda^2(1 + 1 + 1) - \lambda \left\{ \begin{vmatrix} 1 & 4 \\ 2 & 1 \end{vmatrix} + \begin{vmatrix} 1 & 1 \\ -1 & 1 \end{vmatrix} + \begin{vmatrix} 1 & 0 \\ 3 & 1 \end{vmatrix} \right\}$$

$$+ \begin{vmatrix} 1 & 4 & 1 \\ 2 & 1 & 0 \\ -1 & 3 & 1 \end{vmatrix} = 0$$

and this reduces to

$$\lambda(\lambda^2 - 3\lambda - 4) = 0$$

The first root is $\lambda_1 = 0$, and

$$\lambda^2 - 3\lambda - 4 = 0$$
$$(\lambda - 4)(\lambda + 1) = 0$$
$$\lambda_2 = 4 \qquad \lambda_3 = -1$$

The largest positive root is $\lambda_2 = 4$ and is called the *dominant latent root*. Determining the roots of an equation larger than the one above becomes more difficult the higher the order of the matrix and, consequently, numerical methods and a computer program are necessary in almost all cases.   The computer program to be used will depend on whether or not the matrix is *symmetric* (the elements below the diagonal mirror the elements above) or not.

A latent vector is a vector with a number of elements equal to the order of the **A** matrix.   There is a latent vector for each latent root.   If there are four latent roots, there are four latent vectors.   A latent vector satisfies the relationship

$$\mathbf{A\mu} = \lambda\mathbf{\mu}$$

where $\mathbf{\mu}$ is the latent vector. There is an infinite number of vectors satisfying this equation, but the elements of the vector always remain proportional to each other. Therefore, if one element of the vector is arbitrarily set, the other elements are automatically determined.

In the example the first latent root was zero. In calculating the latent vector for root i, the equation

$$(\mathbf{A} - \lambda_i \mathbf{I})\mathbf{\mu}_i = 0$$

is used. For the example this is

$$\begin{bmatrix} 1 - \lambda_i & 4 & 1 \\ 2 & 1 - \lambda_i & 0 \\ -1 & 3 & 1 - \lambda_i \end{bmatrix} \begin{bmatrix} \alpha_i \\ \beta_i \\ \gamma_i \end{bmatrix} = 0$$

letting α_i, β_i, and γ_i represent the three elements of each latent vector. For the root $\lambda = 0$, this multiplication results in three equations:

$$\alpha_1 + 4\beta_1 + \gamma_1 = 0$$
$$2\alpha_1 + \beta_1 = 0$$
$$-\alpha_1 + 3\beta_1 + \gamma_1 = 0$$

Because one of the elements is arbitrarily set, let $\alpha_1 = 1$, resulting in $\beta_1 = -2$ and $\gamma_1 = 7$. The latent vector corresponding to $\lambda = 0$ is

$$\mathbf{\mu}_1 = \begin{bmatrix} 1 \\ -2 \\ 7 \end{bmatrix}$$

For the other two roots, $\lambda_2 = 4$ and $\lambda_3 = -1$, if α_2 is set at 3 and α_3 assigned a value of 2

$$\mathbf{\mu}_2 = \begin{bmatrix} 3 \\ 2 \\ 1 \end{bmatrix} \qquad \mathbf{\mu}_3 = \begin{bmatrix} 2 \\ -2 \\ 4 \end{bmatrix}$$

1-5 A MATRIX MODEL OF POPULATION GROWTH

The use of matrices in population demography was largely developed in two articles by Leslie (1945, 1948). Three basic statistics are required for the use of this technique:

$n_{x,t}$ = the number of females alive in the age group x to $x + 1$ at time t.
P_x = the probability that a female in the x to $x + 1$ age group at time t will be alive in the age group $x + 1$ to $x + 2$ at time $t + 1$.

F_x = the number of female offspring born in the interval t to $t + 1$ per female aged x to $x + 1$ at time t that will be alive in the age group 0 to 1 at time $t + 1$.

As in Sec. 1-3, it is assumed that the population consists of discrete age groupings, in contrast to the reality of a continuously changing set of fertility and mortality statistics. For computational purposes P_x is equal to L_{x+1}/L_x, and F_x is equal to m_x.

Using these three statistics a series of equations can be created accounting for the change in numbers of individuals from one time, t, to the next, $t + 1$. Given F_x and the number of females in each age group at time t, the number of females alive in the 0 to 1 age group after one interval of time $t + 1$ is equal to the sum of the numbers of females produced by each age group

$$\sum_{x=0}^{n} F_x n_{x,0} = n_{0,1}$$

As a simple example with four age groups

$$n_{0,1} = F_0 n_{0,0} + F_1 n_{1,0} + F_2 n_{2,0} + F_3 n_{3,0}$$

If the four F_x values are 2, 6, 8, and 2, and the numbers of individuals in each of the four age groups at time t are 10, 6, 8, and 4, the number of individuals in the 0 to 1 age group in the next generation will be

$$n_{0,1} = (2)(10) + (6)(6) + (8)(8) + (2)(4)$$
$$= 128 \text{ individuals}$$

The number of females in the age interval 1 to 2 at time $t + 1$ is equal to the number of individuals in the age group 0 to 1 at time t times the probability that the female will survive to the next age group, P_x, or

$$P_0 n_{0,0} = n_{1,1}$$

The argument is the same for all higher age groups. There are, therefore, a set of $n + 1$ linear equations in which n is the last age group listed in the life table:

$$\sum F_x n_{x,0} = n_{0,1}$$
$$P_0 n_{0,0} = n_{1,1}$$
$$P_1 n_{1,0} = n_{2,1}$$
$$P_2 n_{2,0} = n_{3,1}$$
$$\vdots$$
$$P_{n-1} n_{n-1,0} = n_{n,1}$$

These equations can be represented in matrix form as

$$\mathbf{Mn}_0 = \mathbf{n}_1$$

where $\mathbf{n}_0$ is a column vector giving the distribution of individuals in age groups at time 0, and $\mathbf{n}_1$ is the column vector for the population after one interval of time. Written as matrices this set of equations becomes

$$
\begin{bmatrix}
F_0 & F_1 & F_2 & F_3 & \cdots & F_{n-1} & F_n \\
P_0 & 0 & 0 & 0 & \cdots & 0 & 0 \\
0 & P_1 & 0 & 0 & \cdots & 0 & 0 \\
0 & 0 & P_2 & 0 & \cdots & 0 & 0 \\
0 & 0 & 0 & P_3 & \cdots & 0 & 0 \\
\vdots & \vdots & \vdots & \vdots & & & \vdots \\
0 & 0 & 0 & 0 & \cdots & P_{n-1} & 0
\end{bmatrix}
\begin{bmatrix}
n_{0,0} \\
n_{1,0} \\
n_{2,0} \\
n_{3,0} \\
n_{4,0} \\
\vdots \\
n_{n,0}
\end{bmatrix}
=
\begin{bmatrix}
n_{0,1} \\
n_{1,1} \\
n_{2,1} \\
n_{3,1} \\
n_{4,1} \\
\vdots \\
n_{n,1}
\end{bmatrix}
$$

If the matrix $\mathbf{M}$ is postmultiplied by the column vector $\mathbf{n}_0$, the equations listed above are restored. The matrix $\mathbf{M}$ is of order $n + 1$ and is square.

As a simple example with four age groups with a mortality and fertility schedule as in the matrix $\mathbf{M}$ below and a distribution of individuals into age groups as in the column vector $\mathbf{n}_0$, the number of individuals in each of the four age groups after one time interval is given by the column vector $\mathbf{n}_1$.

$$
\begin{bmatrix}
0 & 2 & 4 & 6 \\
.6 & 0 & 0 & 0 \\
0 & .4 & 0 & 0 \\
0 & 0 & .2 & 0
\end{bmatrix}
\begin{bmatrix}
10 \\
6 \\
2 \\
1
\end{bmatrix}
=
\begin{bmatrix}
26.0 \\
6.0 \\
2.4 \\
.4
\end{bmatrix}
$$

The matrix model does not assume a stable age distribution. Leslie (1945) gives an example of the use of the matrix model in simulating population growth in a laboratory population of the Norway rat, *Rattus norvegicus*.

Because of the relationship

$$\mathbf{Mn}_0 = \mathbf{n}_1$$

then

$$\mathbf{Mn}_1 = \mathbf{n}_2 \quad \text{or} \quad \mathbf{M}^2\mathbf{n}_0 = \mathbf{n}_2$$

and

$$\mathbf{Mn}_2 = \mathbf{n}_3 \quad \text{or} \quad \mathbf{M}^3\mathbf{n}_0 = \mathbf{n}_3$$

In general

$$\mathbf{M}^r\mathbf{n}_0 = \mathbf{n}_r \tag{1-7}$$

The number and distribution of individuals after r intervals of time is equal to the matrix $\mathbf{M}$ raised to the rth power times the original $\mathbf{n}_0$ column vector, so long as the mortality and fertility statistics defining the rate of population growth remain the same.

To illustrate the use of Eq. (1-7) suppose that the $\mathbf{M}$ matrix is

$$\begin{bmatrix} .00 & 6.43 & 14.00 & 18.00 \\ .78 & 0 & 0 & 0 \\ 0 & .71 & 0 & 0 \\ 0 & 0 & .60 & 0 \end{bmatrix}$$

and the initial population distribution is

$$\begin{bmatrix} 0 \\ 0 \\ 20 \\ 0 \end{bmatrix}$$

This example is actually possible because many experiments start with an initial population of adults which are all of the same age group.   After one interval of time the number and distribution of individuals is

$$\mathbf{n}_1 = \begin{bmatrix} 280 \\ 0 \\ 0 \\ 12 \end{bmatrix}$$

and after two, three, and four intervals

$$\mathbf{n}_2 = \begin{bmatrix} 216 \\ 218 \\ 0 \\ 0 \end{bmatrix} \qquad \mathbf{n}_3 = \begin{bmatrix} 1402 \\ 168 \\ 155 \\ 0 \end{bmatrix} \qquad \mathbf{n}_4 = \begin{bmatrix} 3250 \\ 1094 \\ 119 \\ 93 \end{bmatrix}$$

The age group distribution is changing, although with time the population will reach a stable age distribution.   In other words, the oscillations in the proportions of the different age groups will eventually damp themselves out. Of more interest is the change in numbers of individuals with time.   The total number of individuals after each interval of time is plotted in Fig. 1-5.   Instead of a smooth, exponential curve, this curve oscillates slightly.   These oscillations will damp themselves out as the age distribution becomes stable.   This example demonstrates the utility of the matrix equations.   Equation (1-1) treats all individuals of the population as identical without regard to their age or assumes that the age groups form a stable age distribution.   In most real cases the individuals of a population do not have a stable age distribution.

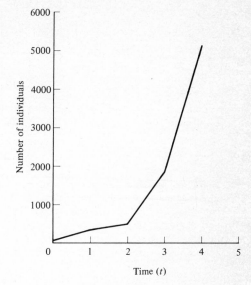

FIGURE 1-5
The population increase in a hypothetical population with relatively high survival probabilities for each age group.

These oscillations are even more pronounced if the mortality of the early age groups is high. If the P_x values in the **M** matrix are changed to $P_0 = .20$, $P_1 = .10$, and $P_3 = .05$, then

$$\mathbf{n}_1 = \begin{bmatrix} 280 \\ 0 \\ 0 \\ 1 \end{bmatrix} \quad \mathbf{n}_2 = \begin{bmatrix} 18 \\ 56 \\ 0 \\ 0 \end{bmatrix} \quad \mathbf{n}_3 = \begin{bmatrix} 360 \\ 4 \\ 6 \\ 0 \end{bmatrix} \quad \mathbf{n}_4 = \begin{bmatrix} 109 \\ 72 \\ 0 \\ 0 \end{bmatrix} \quad \mathbf{n}_5 = \begin{bmatrix} 463 \\ 22 \\ 7 \\ 0 \end{bmatrix}$$

The population is still increasing, even at these high mortalities, but the oscillations are pronounced (Fig. 1-6). Given enough time, the population will achieve a stable age distribution and stop oscillating.

1-5.1 Uses

In any real situation it is unlikely that the values of P_x and F_x will not change with time. But because $\mathbf{Mn}_0 = \mathbf{n}_1$ and $\mathbf{MMn}_0 = \mathbf{n}_2$, it is also possible to use

$$\mathbf{M}_2 \mathbf{M}_1 \mathbf{n}_0 = \mathbf{n}_2 \tag{1-8}$$

where $\mathbf{M}_1$ and $\mathbf{M}_2$ are two different matrices of the fertility and mortality statistics F_x and P_x.

To illustrate the use of Eq. (1-8) I have again used the matrix presented earlier as an example, but in the second time interval the mortalities have been

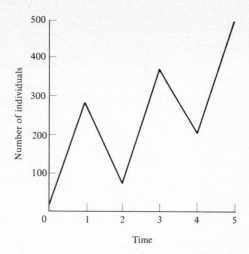

FIGURE 1-6
The population oscillations in a hypothetical population with an unstable age
distribution and low survival probabilities
of the age groups.

changed, perhaps because predation pressures increased. If the original
population of 100 individuals was distributed as

$$\begin{bmatrix} 50 \\ 30 \\ 15 \\ 5 \end{bmatrix}$$

and the fertility and mortality statistics for the two intervals of time were

$$\mathbf{M}_1 = \begin{bmatrix} 0 & 6.43 & 18.00 & 18.00 \\ .78 & 0 & 0 & 0 \\ 0 & .71 & 0 & 0 \\ 0 & 0 & .60 & 0 \end{bmatrix} \qquad \mathbf{M}_2 = \begin{bmatrix} 0 & 6.43 & 18.00 & 18.00 \\ .62 & 0 & 0 & 0 \\ 0 & .32 & 0 & 0 \\ 0 & 0 & .20 & 0 \end{bmatrix}$$

the population after two intervals of time will be $\mathbf{M}_2\mathbf{M}_1\mathbf{n}_0 = \mathbf{n}_2$, or

$$\begin{bmatrix} 791 \\ 343 \\ 12 \\ 2 \end{bmatrix} = \mathbf{n}_2$$

for a total of 1,148 individuals. This method can be used for any number of
different matrices $\mathbf{M}$ in the same way as $\mathbf{M}^r\mathbf{n}_0 = \mathbf{n}_r$.

This basic technique was used by Darwin and Williams (1964) to study
the effect of different hunting policies on a rabbit population in New Zealand.
Using 4-week intervals, they developed a series of 13 matrices, $\mathbf{M}_i$, representing
the probable fertility and survival statistics for each 4-week period of an

entire year.   Each of the 13 matrices was different because of seasonal differences in mortality and reproduction.   Starting with a population n_0 at the beginning of the year (the population was divided into 39 age groups), the predicted size and age distribution of the population at the end of the year was calculated as

$$\mathbf{M}_{13}\mathbf{M}_{12}\mathbf{M}_{11} \cdots \mathbf{M}_1 n_0 = n_{13}$$

As will be shown later, the dominant latent root of the matrix $\mathbf{M}$ for a population with a stable age distribution is the finite rate of increase. Therefore, the dominant latent root of the product of the 13 $\mathbf{M}$ matrices is a measure of the finite rate of increase for the entire year.   Darwin and Williams were interested in minimizing this finite rate of increase when it was possible to hunt rabbits only once during the year.   They knew the approximate percentage of each age group that would survive the hunting period.   These percentages were used to create a diagonal matrix, also of order 39×39, with the percentages of survival of each age group comprising the diagonal.   The diagonal matrix was inserted between the $\mathbf{M}$ matrices representing those months between which hunting could take place, and the finite rate of increase for the year was found by calculating the dominant latent root of the product of the 13 $\mathbf{M}$ matrices and the diagonal matrix, e.g.,

$$\mathbf{M}_{13}\mathbf{M}_{12}\mathbf{D}\mathbf{M}_{11} \cdots \mathbf{M}_1$$

The $\mathbf{D}$ matrix is moved around to all possible positions.   The best time of the year to hunt the rabbits to reduce their rate of increase is found by choosing the position of $\mathbf{D}$ resulting in the smallest dominant latent root.

1-5.2   The Finite Rate of Increase of a Population with a Stable Age Distribution

The dominant latent root of the matrix $\mathbf{M}$ is equal to the finite rate of increase of a population with a stable age distribution.   As shown earlier, the rate of increase changes and the numbers oscillate in a population without a stable age distribution until the population achieves a stable age distribution.

The dominant latent root of the matrix $\mathbf{M}$ may be found by the processes of diagonal expansion, as discussed in Sec. 1-4.   However, Leslie (1948) has proposed a simpler method of calculating the latent roots of $\mathbf{M}$. Starting with a hypothetical matrix

$$\begin{bmatrix} 0 & \frac{45}{7} & 18 & 18 \\ \frac{7}{9} & 0 & 0 & 0 \\ 0 & \frac{5}{7} & 0 & 0 \\ 0 & 0 & \frac{3}{5} & 0 \end{bmatrix}$$

form a diagonal matrix H with diagonal elements $h_{1,1} = P_0 P_1 P_2$, $h_{2,2} = P_1 P_2$, $h_{3,3} = P_2$, and $h_{4,4} = 1$

$$\begin{bmatrix} \frac{1}{3} & 0 & 0 & 0 \\ 0 & \frac{3}{7} & 0 & 0 \\ 0 & 0 & \frac{3}{5} & 0 \\ 0 & 0 & 0 & 1 \end{bmatrix}$$

and then form a new matrix B by taking

$$B = HMH^{-1}$$

where H^{-1} is the inverse of the matrix H. The inverse of a diagonal matrix such as H is another diagonal matrix with the elements in the diagonal the reciprocals of the elements of the original diagonal matrix, in this case

$$H^{-1} = \begin{bmatrix} 3 & 0 & 0 & 0 \\ 0 & \frac{7}{3} & 0 & 0 \\ 0 & 0 & \frac{5}{3} & 0 \\ 0 & 0 & 0 & 1 \end{bmatrix}$$

In the example

$$B = \begin{bmatrix} 0 & 5 & 10 & 6 \\ 1 & 0 & 0 & 0 \\ 0 & 1 & 0 & 0 \\ 0 & 0 & 1 & 0 \end{bmatrix}$$

Instead of decimals in the subdiagonal of the matrix as in M, in B there are all 1s. The top row of B gives the coefficients of the equation for the latent roots of M. The equation for the example is

$$\lambda^4 - 5\lambda^2 - 10\lambda - 6 = 0$$

Because there are four rows and columns, there are four latent roots. Although in this case the roots are relatively easy to find, in most actual situations: (1) some of the roots will be complex, i.e., contain imaginary components, and (2) the roots will be too many to find with elementary algebra. Therefore, a numerical computer program is ordinarily needed to find the roots of the equation. In the above case the largest positive root, the dominant latent root, is 3. Therefore, the finite rate of increase is 3. If the finite rate of increase exists, the population will eventually reach a stable age distribution, provided the largest positive root is larger than the absolute values of all the remaining roots.

1-5.3   The Stable Age Distribution

Returning to the definition of latent roots and vectors, there is not just one latent vector per root, but an infinite number.   However, the elements of every latent vector of a given root are always in the same proportion which, after all, is what a stable age distribution is.   The stable age distribution of a population is calculated as

$$\mathbf{M}\mathbf{n}_s = \lambda_1 \mathbf{n}_s \qquad (1\text{-}9)$$

where $\mathbf{n}_s$ is the stable age distribution of the population, and λ_1 is the dominant latent root of the matrix $\mathbf{M}$.   To help the computations the matrix $\mathbf{B}$ is used again.   Equation (1-9), using the $\mathbf{B}$ matrix, becomes

$$\mathbf{B}\mathbf{v}_s = \lambda_1 \mathbf{v}_s$$

where $\mathbf{v}_s$ is a latent vector of $\mathbf{B}$ but not of $\mathbf{M}$.   The vector $\mathbf{v}_s$ is calculated as

$$v_{0,s} = \lambda_1 v_{1,s} \qquad v_{1,s} = \lambda_1 v_{2,s} \qquad v_{2,s} = \lambda_1 v_{3,s} \qquad \text{etc.}$$

By arbitrarily taking any one of the age groups as 1, the rest can be calculated.   In the example the dominant latent root is 3, and if the last age group is taken as 1, $v_{2,s} = \lambda_1 v_{3,s}$, or $v_{2,s} = (3)(1) = 3$, and so forth, giving the vector $\mathbf{v}_s$ as

$$\mathbf{v}_s = \begin{bmatrix} 27 \\ 9 \\ 3 \\ 1 \end{bmatrix}$$

This vector is converted to the stable age distribution of the matrix $\mathbf{M}$ by the equation

$$\mathbf{n}_s = \mathbf{H}^{-1}\mathbf{v}_s$$

where $\mathbf{n}_s$ is the stable age distribution of the population.   In the example

$$\mathbf{n}_s = \begin{bmatrix} 3 & 0 & 0 & 0 \\ 0 & \frac{7}{3} & 0 & 0 \\ 0 & 0 & \frac{5}{3} & 0 \\ 0 & 0 & 0 & 1 \end{bmatrix} \begin{bmatrix} 27 \\ 9 \\ 3 \\ 1 \end{bmatrix} = \begin{bmatrix} 81 \\ 21 \\ 5 \\ 1 \end{bmatrix}$$

If the dominant latent root of the $\mathbf{M}$ matrix of fertility and mortality statistics is greater than 1, the population will increase; and if the root is less than 1, the population will become extinct.   For a game or pest animal for which hunting will cause a known mortality in each age group, the matrix $\mathbf{M}$ can be created and the dominant latent root calculated.   If the dominant latent root is less than 1, the regime of hunting will cause the animal to become extinct.

1-5.4 Some Consequences of an Unstable Age Distribution

A species introduced into an unfavorable constant environment where it cannot persist may still increase to great numbers and exist for some time at population levels above those of the population originally introduced. Consider a hypothetical **M** matrix similar to the others used as examples, but with much higher mortality rates for the early age groups.

$$\begin{bmatrix} 0 & 6 & 18 & 18 \\ .10 & 0 & 0 & 0 \\ 0 & .09 & 0 & 0 \\ 0 & 0 & .05 & 0 \end{bmatrix}$$

The dominant latent root of this matrix is equal to approximately .9, and therefore the population will decrease until it becomes extinct. Using this hypothetical matrix and an initial age distribution

$$\mathbf{n}_0 = \begin{bmatrix} 0 \\ 0 \\ 20 \\ 0 \end{bmatrix}$$

the number and distribution of individuals into the four age groups were calculated for successive intervals of time. The total numbers of individuals for each time interval are plotted in Fig. 1-7. Even though the environment will not support this species, the population increases in the beginning to high density levels and oscillates for several time intervals. The population does not drop below the original starting population of 20 individuals until 17 time intervals later.

It has been stated almost as a dictum that a population begun with an unstable age distribution in an unlimited environment will reach a stable age distribution in time, whether increasing, decreasing, or staying at the same density. Like most principles in ecology there is an exception to this rule. Bernadelli (1941) has postulated a hypothetical beetle population with the mortality and fertility matrix

$$\begin{bmatrix} 0 & 0 & 6 \\ .50 & 0 & 0 \\ 0 & .33 & 0 \end{bmatrix}$$

The dominant latent root of this matrix is 1, so the population will neither increase or decrease as a trend. Taking the powers of this matrix

$$\mathbf{M}^2 = \begin{bmatrix} 0 & 3 & 0 \\ 0 & 0 & 2 \\ \frac{1}{6} & 0 & 0 \end{bmatrix} \qquad \mathbf{M}^3 = \mathbf{I} \qquad \mathbf{M}^4 = \mathbf{M} \qquad \mathbf{M}^5 = \mathbf{M}^2 \qquad \mathbf{M}^6 = \mathbf{I}$$

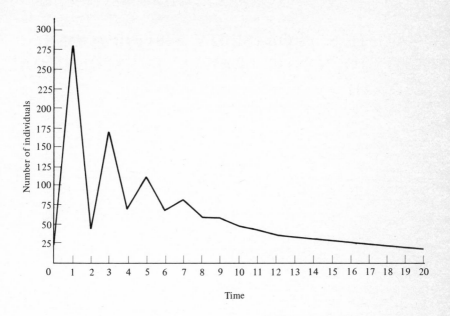

FIGURE 1-7
The change in numbers of a hypothetical population with an unstable age distribution becoming extinct because its finite rate of increase is less than 1.

and so forth, one should note that this population has a periodicity of 3. If each time interval were 1 month, the population would fluctuate with the same period every 3 months. Unless the initial population had a stable age distribution, the population will fluctuate once every 3 months and will never approach a stable age distribution. This behavior is particularly true of populations having only one of the F_x values greater than zero and is related to the characteristics of the roots of the matrix **M**, a subject too complex to introduce here (see Leslie, 1948). Mayflies exhibit this type of reproduction, mating en masse and dying shortly afterward.

2

STATISTICS, PROBABILITY DISTRIBUTIONS, AND STOCHASTIC MODELS OF EXPONENTIAL GROWTH

The purpose of this chapter is to introduce some fundamentals of statistics and probability distributions. From this base the concept of stochastic models will be explained in a general way without going into a discussion of how they are derived.

2-1 STATISTICS AND PROBABILITY DISTRIBUTIONS

Probably the two best-known statistics are the mean and the variance of a series of observations. Suppose that the following numbers of individuals had been observed in a series of 10 quadrats:

$$10, 20, 32, 24, 37, 62, 9, 12, 11, 42$$

The average number of individuals in a quadrat, the *mean*, is simply the sum of all the observations divided by the number of observations n

$$\bar{x} = \frac{\sum_{i=1}^{n} x_i}{n}$$

or for the example

$$\bar{x} = (10 + 20 + 32 + 24 + 37 + 62 + 9 + 12 + 11 + 42)/10 = 25.9 \text{ individuals}$$

It might also be desirable to find the average deviation of each observed value from the mean. The deviation of the third observation from the mean is $32 - 25.9 = 6.1$. The average deviation, if determined as the mean of all the deviations, is zero because the sum of the deviations is zero. To avoid this problem the squared deviations are usually used instead, i.e.,

$$\sum_{i=1}^{n} (x_i - \bar{x})^2$$

and when divided by $n - 1$ this is the variance s^2 of a series of observations. An equation for the variance useful in computation is

$$s^2 = \frac{\sum_{i=1}^{n} x_i^2 - [(\sum_{i=1}^{n} x_i)^2/n]}{n - 1}$$

The square root of s^2, that is, s, is the *standard deviation* and is a measure of the average deviation of the observations from the mean value. In the example $s^2 = (9423 - 6708)/9 = 301.67$. By taking the square root the standard deviation is ± 17.37.

2-1.1 The Normal Distribution

In Table 2-1 the weights of 2,119 individuals in a population are recorded. The observations are classified into 10-pound groupings. A total of 213 individuals was found to weigh between 120 and 130 pounds. In Fig. 2-1 the

Table 2-1   THE DISTRIBUTION OF THE OBSERVED WEIGHTS OF 2,119 INDIVIDUALS GROUPED INTO 10-POUND INTERVALS

Weight group	Number of observations	Percent of total
80–90	14	.0068
90–100	28	.0132
100–110	66	.0311
110–120	130	.0613
120–130	213	.1005
130–140	298	.1406
140–150	346	.1633
150–160	336	.1586
160–170	273	.1288
170–180	188	.0887
180–190	102	.0481
190–200	50	.0236
200–210	20	.0094
210–220	7	.0033
220–230	2	.0009

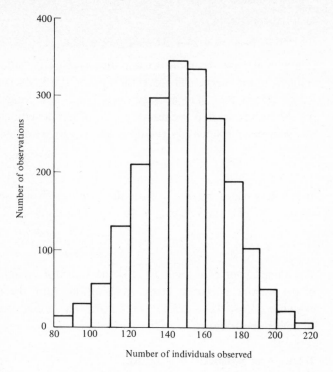

FIGURE 2-1
A histogram of the frequency distribution of the individuals in a hypothetical population as counted in 2,118 separate observations.

number of observations occurring in each weight group is plotted against groups. This is a *frequency distribution.* The histogram is roughly bell-shaped, containing many observations toward the middle groups and few at the two *tails* of the distribution. Groups of 5-pound intervals could have been used equally well, or 2-, or 1-pound intervals. As the size of the group interval becomes smaller, the discontinuities of the distribution become smaller and smaller. Finally, assuming infinitesimally small weight groups, the distribution would resemble the curve in Fig. 2-2.

Figure 2-2 is a *normal* curve given by a function known as the *normal probability density function.* The density function is continuous, the distribution in Fig. 2-1 approaching the normal curve in shape as a limit, i.e., as the groupings become smaller and smaller. Data plotted as in Fig. 2-1 that result in a bell-shaped frequency distribution approximated by a normal curve are said to be normally distributed. A normal distribution of data is

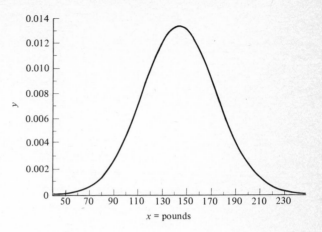

FIGURE 2-2
The normal curve with parameters $\mu = 145$ and $\sigma^2 = 900$.

commonly observed in biology, and many statistical tests are based on the assumption of normality in the data.

In Table 2-1 the probability of an individual observation falling into the 80- to 90-pound category is about .67 percent.   The probability of an observation falling into the 140- to 150-pound group is 16.33 percent.   The probability of an observation occurring in either the 80- to 90-pound or the 90- to 100 pound group is the sum of the two probabilities; .67 + 1.32 = 1.99 percent. The probability of an individual falling into at least one of the weight groups is 1.   It must belong to one of them.

The same is true of the normal distribution, but because the curve is continuous the probability of an observation being between 80 and 90 pounds is found by finding the area below the curve between the points 80 and 90. As in the frequency distribution the observations must belong to some weight no matter how large or small, and so the total area under the normal curve is equal to 1.   The two tails of the normal distribution taper off to infinity, approaching zero probability in both directions (ignoring the limit imposed by zero in this particular case).   The equation for the normal distribution is

$$y = f(x) = \frac{1}{(2\pi\sigma^2)^{1/2}} \exp \frac{-(x - \mu)^2}{2\sigma^2} \tag{2-1}$$

The independent variable is x, the observation, and x can range between plus and minus infinity.   Two parameters define the normal distribution: the mean μ and the variance σ^2.   The symbols μ and σ^2 are used to indicate the

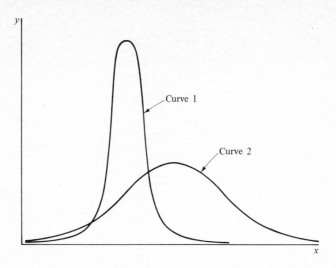

FIGURE 2-3
Two normal curves, curve 1 with a high mean relative to the variance, and curve 2 with a low mean relative to the variance.

population mean and variance, respectively, rather than the sample mean and sample variance $\bar{x}$ and s^2, for reasons discussed below. The mean determines the peak of the curve, and the variance is a measure of its breadth. In Fig. 2-3 curve 1 has a low mean and a low variance relative to curve 2. Conversely, curve 2 has a higher mean and a relatively high variance.

If in the artificial example the true population mean and variance were 145 and 900, respectively, the normal curve shown in Fig. 2-2 calculated from Eq. (2-1) would result. The probability of an individual weighing 200 pounds or more is equal to the area under the curve from 200 to plus infinity, or

$$\text{Probability of 200 pounds or more} = \int_{200}^{\infty} \frac{1}{(2\pi 900)^{1/2}} \exp \frac{-(x-145)^2}{(2)(900)} \, dx$$

To use this result the independent variable x is standardized as

$$z = \frac{(x - \mu)}{\sigma}$$

An observation of 150 pounds in standard form is $z = (150 - 145)/30 = .167$. This standardization reduces Eq. (2-1) to

$$y = \frac{1}{(2\pi)^{1/2}} e^{-z^2/2}$$

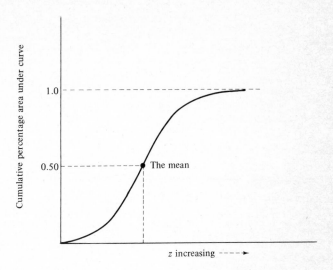

FIGURE 2-4
The cumulative normal distribution for the standardized normal distribution.

The *standardized normal distribution* has a mean of zero and a variance of 1. The probability of an observation of 200 pounds or more is the area under the standardized normal curve from $z = (200 - 145)/30 = 1.83$ to plus infinity. To find this probability the *cumulative distribution function* of the normal curve is used.   Moving from minus infinity toward plus infinity, the total percent of the area under the standardized normal curve between z and minus infinity can be plotted as in Fig. 2-4.   The cumulative normal distribution function $F(z)$ for the standardized normal distribution is tabulated in Beyer (1968) and in many other compendiums of statistical tables.   For $z = 1.83$ it is found in the table that 96.64 percent of the area under the normal curve is found between minus infinity and $z = 1.83$.   Subtracting 96.64 percent from 100 percent, that is, $1 - F(z)$, it is found that 3.36 percent of the area under the curve lies between $z = 1.83$ and plus infinity.   Translated in terms of the example the probability of an individual weighing 200 pounds or more is 3.36 percent.   For negative values of z, the probability of a number less than z is $1 - F(z)$, and the probability of an observation larger than z is $F(z)$.   The probability of an öbservation being less than -1.83 is $1 - .9664$, or 3.36 percent.   Conversely, the probability of an observation larger than $z = -1.83$ is 96.64 percent.   The probability of an individual weighing either more than 200 pounds or less than 90 pounds is 3.36 percent plus 3.36 percent, or 6.72 percent.   This means that the probability of an individual weighing between 90 and 200 pounds is $100 - 6.72$ percent, or 93.28 percent.

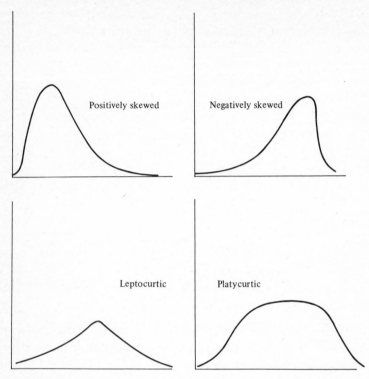

FIGURE 2-5
Some common deviations from normality in frequency distributions of data.

Sometimes a series of observations may deviate slightly from normality, although retaining a basically normal distribution. Some of these deviations from normality are shown in Fig. 2-5. Various measures of skewness and kurtosis may be found in most statistics books.

2-1.2 Sample and Population Means and Variances

It is sometimes possible to take measurements on all members of a population. Population is used here in the statistical, not ecological, sense. A statistical population is all the members or objects making up the group about which statistical inferences are being made. If the dry weight of bluebells in Illinois is being measured, the population is all the individual bluebell plants in Illinois. Often, as with the bluebells, it is impossible to take measurements on all the individuals in a population. Therefore, the mean and variance must be estimated from a sample taken from the population. Methods of sampling are discussed in Chap. 11. Because these two statistics are estimated from a sample, the true population mean and variance are not known. To distinguish the true population mean and variance from the estimated values of these two statistics the contrasting pairs of symbols μ and σ^2 versus $\bar{x}$ and s^2 are used.

If the mean is repeatedly estimated by successive samples from the population, it will vary. In sampling from a population not only must the mean be estimated but also the variance of the distribution of possible estimated means. If a mean is estimated from a sample but the variance of the distribution of means is large, the estimated mean is not as reliable as an estimated mean with a small variance. The variance of means $s_{\bar{x}}^2$ is

$$s_{\bar{x}}^2 = \frac{s^2}{n}$$

where s^2 is the sample variance, and n is the number of observations in the sample. The standard deviation of the estimated mean, usually called the *standard error* (SE), is the square root of the variance of the means and is usually indicated as $SE(\bar{x})$. The standard error is a measure of the average deviation of an estimated mean from the true population mean.

2-1.3   Confidence Intervals

Because of the variability of the estimated mean, it is not possible to make definite statements about the value of the population mean, only probabilistic ones. We can state that the population mean lies between two values, with a certain probability of being correct. This is known as placing a *confidence interval* around the population mean.

In presenting a sample mean it is often useful to be able to say that the true population mean lies within some given interval of values with a high probability of being correct. In the standard normal curve the probability of an observation exceeding $z = 1.96$ is 2.5 percent. Similarly the probability of an observation being less than $z = -1.96$ is also 2.5 percent. Placing 2.5 percent probability at both ends of the normal curve there is a 95 percent chance that an observation will fall between 1.96 and -1.96.

If the estimated mean is $\bar{x}$, and its standard error is $SE(\bar{x})$, using a procedure similar to that above, a confidence interval is constructed as

$$P\{\bar{x} - (1.96)[SE(\bar{x})] \le \mu \le \bar{x} + (1.96)[SE(\bar{x})]\} = 95 \text{ percent}$$

If $\bar{x} = 20$ and $SE(\bar{x}) = 5$

$$P(10.2 \le \mu \le 29.8) = .95$$

This expression states that the probability of the true population mean lying between 10.2 and 29.8 is 95 percent.   If you want to be 99 percent sure, then 1.96 should be replaced by 2.57.

$$P(7.15 \leq \mu \leq 32.85) = .99$$

Naturally, as you increase the probability of being correct, the confidence interval widens.

To utilize this form of confidence interval the true population standard deviation must be known.   In most cases only a sample estimate of the standard deviation is available.   In order to formulate a confidence interval around $\bar{x}$ when the true standard deviation is not known, the *Student's t statistic* is used.   The t distribution is determined by a single parameter, the number of degrees of freedom.   The statistic t is tabulated in many statistics texts.   For a confidence interval the degrees of freedom are $n - 1$, the number of observations minus one.   The reader should consult a statistics text for a discussion of degrees of freedom.   Let α represent the probability that the true mean will fall outside the confidence interval.   If the confidence interval contains the true mean 95 percent of the time, $\alpha = .05$.   The usual t table places all the probability in the upper tail of the distribution, but the confidence interval places half of the probability at each end.   Therefore, consult the table for $n - 1$ degrees of freedom and find the value of $t_{\alpha/2}$.   For an α of 5 percent find t under .975.   If there are 20 observations, there are 19 degrees of freedom.   Listed under .975 the correct value of $t_{\alpha/2}$ is 2.093. The confidence interval is

$$P\{\bar{x} - t_{\alpha/2}[\text{SE}(\bar{x})] \leq \mu \leq \bar{x} + t_{\alpha/2}[\text{SE}(\bar{x})]\} = .95$$

If $\bar{x} = 20$ and $\text{SE}(\bar{x}) = 5$, the confidence interval is

$$P(9.535 \leq \mu \leq 30.465) = .95$$

2-2 STOCHASTIC MODELS OF EXPONENTIAL GROWTH

One of the primary reasons for replication in experiments is to average out the effect of chance events.   The deterministic equation of exponential population growth

$$N_t = N_0 \, e^{rt}$$

predicts that a population with a stable age distribution in an unlimited. environment will increase as a smooth, exponential curve.   Although the growth of the population will be roughly exponential, the curve will rarely be perfectly smooth.   Sometimes the population will be slightly larger than predicted, and at other times slightly smaller.   If the experiment is run again, the result will again be roughly exponential, but other chance events will have occurred, causing a different set of deviations from the curve.   Therefore, in real situations after some time t has passed the number of individuals in the population may not necessarily be N_t individuals as predicted by the exponential equation.   Let us say that the exponential equation predicts that at time $t = 20$ there will be 100 individuals in the population.   However, because of chance events, such as random fluctuations in the temperature of the laboratory, there may be only 99 or perhaps 102 individuals.   If the experiment were run often enough, it might be possible to determine the probability that the population will consist of 95 individuals and likewise the probability of any given number of individuals.   A stochastic model, in contrast to the deterministic model, gives the probability associated with each possible outcome or population size at time t.   Population growth is inherently a probabilistic phenomenon. For instance, during some period of time there may be one birth, no births, or perhaps more than one birth.   The stochastic model of exponential growth is derived by considering the probabilities associated with births and deaths, although the situation is considerably more complex than this.   Therefore, unlike the deterministic model, a stochastic model does not predict that the population size will be exactly N_t, but rather that the probability of the population consisting of exactly N individuals at time t is p, where p is the probability.   These probabilities constitute a probability distribution, and like the normal distribution this probability distribution has a mean and a variance.   For the purposes of this book we can define the mean of this distribution as its *expectation*, or the expected number of individuals at time t. If the number of individuals in the population is symbolized by N, the number expected is written as $E(N_t)$.   Provided the assumptions used in the derivation are the same, $E(N_t) = N_0 e^{rt}$, the same as the deterministic result.   Because the probability distribution of the number of individuals at time t is known, the variance of the number of individuals observed can be calculated, in this case

$$\text{var}(N_t) = N_0 \frac{b + d}{b - d} [e^{(b-d)t}][e^{(b-d)t} - 1]$$

The square root of $\text{var}(N_t)$ is a measure of the average deviation of the actual population density at time t from the expected value $E(N_t)$.   Setting $b = .82$ and $d = .06$ with an initial population of 100 individuals, the expected population size after seven intervals of time is

$$E(N_7) = 100e^{(.76)(7)}$$

$$= 20{,}438 \text{ individuals}$$

The variance of the population size after seven time intervals is

$$\text{var}(N_7) = 100 \frac{.88}{.76} [e^{(.76)(7)}][e^{(.76)(7)} - 1]$$

$$= 4{,}813{,}181.69$$

Taking the square root of the variance, the standard deviation of the population size from its expected size at $t = 7$ is 2,193.89. On the whole the average deviation of the population from its expected value at $t = 7$ is about 2,200 individuals.

As a sweeping generality the stochastic models most commonly used to simulate population growth can be divided into two types: (1) continuous time models, and (2) discrete time models. Discrete time models estimate the mean and variance for intervals of time equal to 1. Continuous time models measure the mean and variance for any length of time t. Discrete time models can arise naturally, for example, in determining the probability of N seeds being set by *Phlox divaricata* 1 year hence. More commonly, however, the discrete time models are used as approximations to the continuous time situation in cases in which the continuous time models are difficult or impossible to derive. Setting t equal to 1 results in the discrete time model of exponential growth

$$E(N_{t+1}) = e^r N_t \tag{2-2}$$

and

$$\text{var}(N_{t+1}) = \frac{b + d}{b - d} (e^{b-d} - 1)(e^{b-d})N_t \tag{2-3}$$

In the *Calandra oryzae* experiment discussed in Chap. 1, b equaled .82 and d equaled .06, the same parameters used in the example of the continuous time model. In starting with a population of 100 individuals with a stable age distribution, the expected number of individuals after one interval of time is $E(N_{t+1}) = 213.8$, and $\text{var}(N_{t+1}) = 258.189$.

Although the superiority of continuous time models to their discrete approximations is clear, they are unfortunately in most cases very difficult if not impossible to derive. Discrete time approximations are slightly less difficult to derive, and it is sometimes possible to derive a discrete time model and, by using the methods discussed below, simulate the process for several time intervals. Note, however, that in the discrete time model the time variable t is assumed to advance by jumps of 1, although t is in reality a continuous variable. If, as later happens, the birthrate or death rate is postulated to be a function of time, it will also be a continuous variable.

The discrete time approximation in these cases must assume the birthrate or death rate to be constant during a single interval of time, changing by a single jump from one time interval to the next.   The situation is analogous to the discrete approximations of l_x and m_x used in Chap. 1.

The advantage of the stochastic model over the equivalent deterministic model is its greater reality.   The greater complexity of stochastic models, particularly in their derivation, tends in some cases to outweigh their advantages.   This is particularly true if the variance in number of individuals about the expected number is small.   There are cases, however, in which chance deviations can push the results of a process either one way or the other.   Two important examples can be found in the chapters on competition and population extinction.   Stochastic models have the advantage of being more realistic than deterministic models, predicting only what can happen and the probability of it happening, not what will happen.

2-2.1   Monte Carlo Methods

The methods employed to arrive at an answer using a discrete time stochastic approximation to a continuous time situation are roughly the same as playing a game involving chance, thus the name Monte Carlo techniques.   The expected answer is calculated and modified by a factor representing a random sample from the probability distribution to which the answer belongs.

Equations (2-2) and (2-3) give the expected value of the population size at $t + 1$, plus the variance of the expected value.   If the effect of chance variations is not taken into consideration, the increase in density of a population started with 20 individuals with the parameters $b = .52$ and $d = .48$ is shown in column 1 of Table 2-2.   The calculations in column 1 are strictly deterministic.   To make the results stochastic the following steps are taken:

1   Calculate the expected number at time $t + 1$ (Eq. 2-2).
2   Calculate the variance of the estimate (Eq. 2-3).
3   Calculate the standard deviation, i.e., the square root of the variance.
4   Pick a number at random from a table of random normal deviates and multiply it by the standard deviation.
5   Add the answer from step *4* to the answer from step *1*.   This is a possible size of the population at time $t + 1$.

This procedure assumes that the distribution of possible answers is normally distributed.   In this case (e.g., see Leslie, 1948) the distribution of possible answers is approximately normal, except at low population densities.   It should be emphasized, however, that the distribution is not necessarily always normal, e.g., it might have an exponential distribution.   An example with the parameters above is:

1 $E(N_{t+1}) = 20.816.$
2 $\text{var}(N_{t+1}) = 21.232.$
3 Standard deviation $= 4.608.$
4 A random normal deviation, -1.38, times the standard deviation, $4.608 = 5.359.$
5 $20.816 - 5.359 = 14.641.$

A table of random normal deviates in Beyer (1968) was used. A random normal deviate is a number drawn at random from the standardized normal distribution.   Unlike random numbers where each number is equally likely to occur, random normal deviates are normally distributed.   Even though the birthrate is greater than the death rate, although they are nearly equal, a chance negative deviation has caused the population to decrease rather than increase as expected.

The course of stochastic population growth can be simulated by repeatedly carrying out the above calculations.   For example, the stochastic estimate of the population after a second interval of time can be arrived at by taking 14.641 as the population density at time *t*, recomputing the standard deviation, selecting another random normal deviate, and so on.   A stochastic representation of population growth for 10 intervals of time is given in column 2 of Table 2-2 and plotted along with the deterministic result in Fig. 2-6.   Because deviations are random, if this procedure were repeated,

Table 2-2 A DETERMINISTIC AND TWO STOCHAS-
TIC SIMULATIONS OF POPULATION
GROWTH IN AN UNLIMITED ENVIRON-
MENT ($b = .52$, $d = .48$, $N_0 = 20$)

t	Deterministic	Stochastic no. 1	Stochastic no. 2
0	20.000	20.000	20.000
1	20.816	14.641	27.129
2	21.665	18.352	32.422
3	22.549	20.381	22.832
4	23.469	21.678	18.889
5	24.426	18.159	17.287
6	25.423	25.091	10.795
7	26.460	26.838	7.478
8	27.540	23.823	10.319
9	28.663	15.339	10.012
10	29.833	17.861	15.277

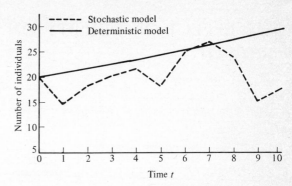

FIGURE 2-6
Population growth in an artificial population with parameters $b = .52$, $d = .48$, and $N_0 = 20$.   Solid line, the deterministic model; dashed line, a Monte Carlo simulation of the discrete time stochastic model of exponential population growth.

the exact path of the growth of the population would not be the same as it was the first time.   The calculation of a second set of 10 intervals of time is shown in column 3 of Table 2-2.

If long periods of time are involved in the simulation, as they often are, or if large numbers of replications of the simulation are needed to estimate the variance of the population at some time t, the calculations become tedious. The calculations are easily programmed for a computer, and a subroutine can be used to generate random normal deviates.   Other forms of the Monte Carlo technique will be discussed later in the text.

3

THE EFFECT OF DENSITY ON POPULATION GROWTH

Individuals are born and they die, and in order to live they need food, water, air, and other resources which may sometimes be in limited supply.   If there is not enough food, the birthrate must decrease or some individuals must die. There is probably no more fundamental fact of life than this.   Chapter 3 will consider this principle and try to treat it mathematically as a measurable component of the growth of a population.   I will start by giving a brief discussion of the effect of density on the growth of a population and then go on to consider modeling these effects in animal and plant populations.

3-1 THE EFFECT OF DENSITY ON POPULATION GROWTH

In Chap. 1 the calculation of the intrinsic rate of increase of the weevil *Calandra oryzae* was studied.   The calculation of r assumes food is unlimited. But if food were not unlimited, what would be the course of population growth?   Birch (1953) introduced a female and a male weevil into 12 grams of wheat at 29.1°C and 70 percent relative humidity and followed the change

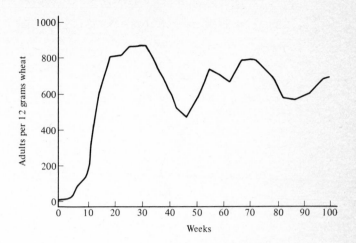

FIGURE 3-1
Trends in the densities of a population of *C. oryzae* reared at 29.1°C and 70 percent relative humidity in wheat.   The 12 grams of wheat were renewed at 2-week intervals.   Observed data points are means of 15 replicates, and only adults were counted.   (*After Birch, 1953.*)

in number of adults for 100 weeks.   The food supply was replenished every 2 weeks and assumed to be a constant 12 grams, although there were of course fluctuations in the food supply as it was eaten by the beetles. The changes in the number of adults over the course of 100 weeks are shown in Fig. 3-1.   There are three striking features of this graph:

1 At the beginning of the experiment the growth of the population is almost exponential.   The population is probably increasing at a rate close to the intrinsic rate of increase.

2 After the tenth week the rate of increase begins to slow, and at about the twentieth week the density of the population plateaus.

3 After the twentieth week and up to the one-hundredth week the population fluctuates around a mean of about 700 individuals, neither increasing nor decreasing as a trend.

It is possible, of course, that this is an exceptional case, so a series of replicates of an experiment performed by Pratt (1943) on the population growth of the cladoceran *Daphnia magna* in a limited environment at 18°C is graphed in Fig. 3-2.   Both the cladoceran and the weevil begin by increasing at a nearly exponential rate, both slow down and eventually level off.   Both populations go into a series of oscillations after the population

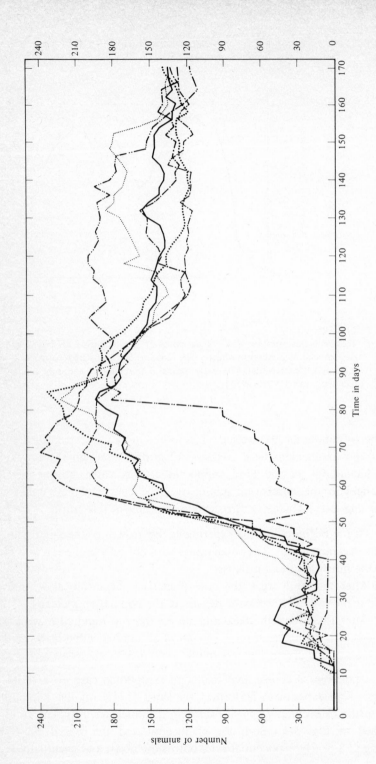

FIGURE 3-2
Population growth of six replicate populations of *D. magna* reared at 18°C in a limited environment. (*After Pratt, 1943.*)

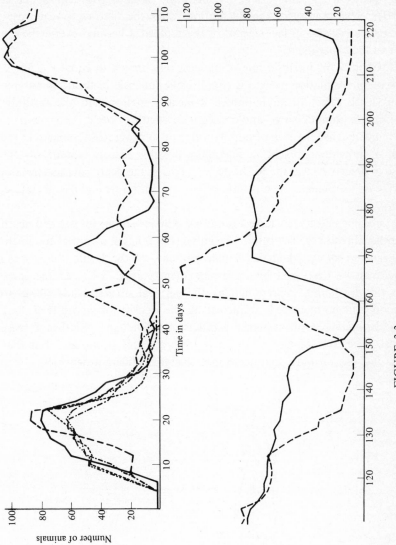

FIGURE 3-3
Population growth in five replicates of *D. magna* reared at 25°C in a limited environment. (*After Pratt, 1943.*)

size has plateaued. Pratt also conducted the same experiment at 25°C (Fig. 3-3). The severity of the fluctuations after the plateau is reached is greatly increased, and in fact several of the replicates became extinct before the end of the experiment.

In Chap. 1 the intrinsic rate of increase was shown to equal $r = b - d$, or the rate of population growth is equal to the intrinsic birthrate minus the intrinsic death rate. In an unlimited, constant environment the birthrate and death rate do not change, and consequently neither does r. But observation 2 indicates that the rate of population growth progressively slows as the number of individuals reaches a maximum possible density. Somehow the increase in density causes the birthrate to decrease, the death rate to increase, or both. At the plateau the birthrate equals the death rate and r is, therefore, equal to zero.

Increasing density can have a profound influence on birthrates and death rates, either directly or indirectly. Crombie (1942) has shown that the birthrate, measured as eggs laid per female per day, of the beetle *Rhizopertha dominica* can be decreased by increasing density (Fig. 3-4). In the same species the death rate, represented by the survival of first-instar larvae to adulthood, can be increased by increasing the density of larvae (Fig. 3-5), survival dropping sharply at higher densities of individuals. Neither of these examples measures changes in the true birthrate and death rate, but they indicate that the birthrate and death rate change at different densities.

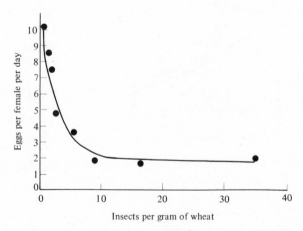

FIGURE 3-4
The relationship between the density of the population of adults of the beetle *R. dominica* and their fecundity. The population was reared at 30°C and 70 percent relative humidity. The curve is fit by eye. (*After Crombie, 1942.*)

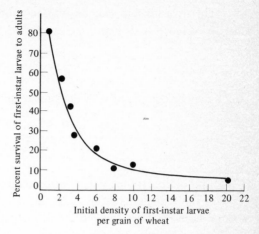

FIGURE 3-5
The relationship between the initial density of first-instar larvae of *R. dominica* and the percent survival to adults when reared in wheat. (*After Crombie, 1944.*)

3-2 THE LOGISTIC EQUATION

The initial growth of a population up to a plateau is roughly an s-shaped curve. The earliest attempt to simulate this curve was the logistic equation of Verhulst (1839). The logistic equation is much maligned because of its many unrealistic assumptions. However, it is a good starting point for a discussion of density effects. In spite of the failings of the logistic equation it is easily understood and often adequately mimics the early stages of the growth of a population up to a plateau.

The reasoning behind the logistic curve is simple. The population density increases, but at higher densities the rate of increase decreases. At some maximum density it does not increase at all. If the rate of increase depends on the density of individuals, the simple relationship

$$a = r - bN$$

can be postulated, where a represents the rate of population increase given a population density N. The letter b reflects the degree to which density decreases the rate of increase of the population. This parameter should not be confused with the intrinsic birthrate of Chap. 1. As the number of individuals increases, the value of bN increases until $a = 0$, the r and bN terms canceling out. The rate of increase a in the above equation is a linear function of the population density N; that is, the equation $a = r - bN$ is the equation of a straight line.

This relationship can be written more exactly as the change in the number of individuals with time as

$$\frac{dN}{dt} = N(r - bN) \tag{3-1}$$

which may be compared with the similar differential equation for exponential growth

$$\frac{dN}{dt} = rN$$

The term r of the exponential growth equation has been replaced by the $r - bN$ term of the logistic equation. As the number of individuals increases, the $r - bN$ term becomes smaller, the rate of increase in density decreasing until it is zero. The density of individuals at which the rate of increase is zero is often referred to as the *carrying capacity* of the environment and is represented by K. Equation (3-1), using K, becomes

$$\frac{dN}{dt} = rN \frac{K - N}{K} \tag{3-2}$$

As N approaches K, the right-hand side of the equation becomes zero and the population stops increasing. In fitting the logistic equation to empirical data Eq. (3-3) is used.

$$N_t = \frac{K}{1 + e^{c - rt}} \tag{3-3}$$

This is the explicit solution of the differential equation, and c is a constant acquired during integration. Equation (3-3) can be rearranged as

$$\ln \frac{K - N}{N} = c - rt$$

The notation ln stands for logarithm to the base e, the natural logarithm, and is equivalent to $\log_e$. This is the equation of a straight line. By using the regression methods discussed in Chap. 9, $\ln[(K - N)/N]$ can be regressed on t. The parameter c is the intercept of the regression equation and r the regression coefficient. Time t is the independent variable. The intrinsic rate of increase r can also be determined from life tables to check on the accuracy of the computations. The value of K is estimated from the data by trial and error until the calculated logistic curve gives a good fit to the observed data points. More accurate but complex means of fitting the logistic equation to data are discussed in Bliss (1970) and Watt (1968).

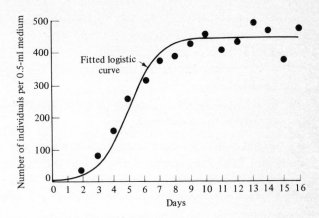

FIGURE 3-6

Population growth in a population of *P. aurelia* raised in 5 milliliters of Osterhaut's medium. The points are the observed data points, and the curve is the fitted logistic curve given by the logistic equation $N_t = 450/(1 + e^{5.041 - 1.022t})$. (*After Gause, 1934.*)

Andrewartha and Birch (1954) have fitted the logistic equation to a population of *Paramecium aurelia* growing in a limited environment; the data were taken from an experiment performed by Gause (1934). Gause introduced 20 *P. aurelia* into a tube of 5 milliliters of Osterhaut's medium. A constant amount of bacteria was added each day. The bacteria could not multiply in the medium. The numbers of individuals at daily intervals are plotted in Fig. 3-6 as is the fitted logistic curve. The equation of this curve is

$$N_t = \frac{450}{1 + e^{(5.041 - 1.022t)}}$$

The logistic curve fits the empirical data fairly well, although it does not account for the oscillations in population density after the carrying capacity of the environment has been reached.

3-2.1 Assumptions of the Logistic Equation

The chief criticisms of the logistic equation stem from the assumptions it makes. These assumptions are:

1 The population has a stable age distribution.
2 The rate of increase *a* is decreased by a constant amount for every individual added to the number already present.

3 The response of the rate of increase to an increase in density is instantaneous; i.e., there are no time lags.
4 The environment is constant.
5 Crowding affects all individuals and life stages of the population equally.
6 The probability of mating in sexual animals does not depend on the density of the population.

Populations can be started with a stable age distribution, but they rarely are. The fluctuations in Fig. 3-1 at the beginning of the experiment are due to a nonstable age distribution. The culture was started with all adults. However, populations started with an unstable age distribution tend to become stable with time, as shown in Chap. 1.

3-2.2 Effect of Density on the Birthrates and Death Rates

Assumptions *2* and *3* of the logistic equation are more serious difficulties than number *1*. Assumption *2* states that the relationship between the rate of increase and the density of the population is linear, the rate of increase being decreased by a constant amount for every individual added to the population. No distinction is made between the effect of density on the birthrates and death rates treated separately.

The effect of density can operate on the birthrate, the death rate, or both. Crombie (1942, 1944) studied the effect of increased density on the fecundity and mortality of the grain beetle *Rhizopertha dominica*. Some of these results were discussed in Sec. 3-1. The effect of an increased density of adults on the number of eggs laid per female per day is shown in Fig. 3-4. The relationship is not linear but curvilinear. Similarly the effect of increasing the number of first-instar larvae initially introduced into the grain has a curvilinear relationship on the percentage survival of the larvae to the adult stage (Fig. 3-5). Neither of these two experiments exactly measured the effect of density on the intrinsic rates of birth and death. In fact, the parameter *b* of the logistic equation may be influenced by many factors, such as density of adults, density of larvae, density of pupae, etc., and the environmental conditions of the experiment. In Fig. 3-7 the effect of the initial adult density on the rate of increase in one generation of the Azuki bean weevil, *Callosobruchus chinensis*, is plotted (Utida, 1941). The relationship is not linear but curvilinear. This implies that *b* in Eq. (3-1) is not a constant but a variable dependent on *N*. If the relationship between increased density and *b* is a simple curvilinear one, as in the *R. dominica* experiment, it can possibly be represented as some simple function of *N*. Sometimes, however, the curvilinear relationship may

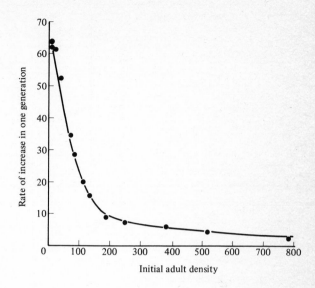

FIGURE 3-7
The effect of different adult densities on the rate of increase in the population of the beetle *C. chinensis*. The cultures were all reared at 24.8°C and 74 percent relative humidity. (*After Utida, 1941.*)

not be simple. The effect of the number of adults of the moth *Anagasta kühniella* on the number of eggs laid per female per 2 days is shown in Fig. 3-8 from data reported by Ullyett (1945).

Because there are cases in which the effect of density on fecundity or mortality may not be linear or a simple curvilinear function, it may sometimes be desirable to have a general equation for predicting the fecundity of a female or the mortality rate of individuals in a population at different densities. As a partial answer to this problem, Watt (1960) derived an equation predicting the mean fecundity of a female at different densities of individuals of a species laying eggs.

3-2.3 Time Lags and the Equality of Individuals

The logistic equation treats all members of the population as equal, and assumes that the response of the rate of increase of a population is instantaneous with changes in density. It is not difficult to conceive of situations, however, in which at a relatively high density females may lay more eggs or bear more offspring than can be supported later as larvae or immatures. Even at this point the population has a density greater than the carrying capacity of the

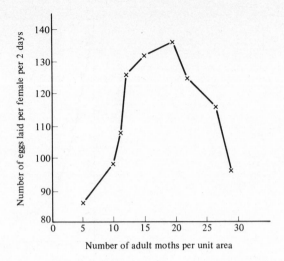

FIGURE 3-8
The effect of adult moth density on the number of eggs laid per female in a population of *A. kühniella.* (*After Ullyett, 1945.*)

environment once the individuals represented by the eggs grow to older larvae and adults. The effect of this increase will not be felt until later. In the interim the females will continue to lay more eggs, further pushing the population level over the limits of the available resources. Later, as the newborn grow older, the individuals of the population will begin to feel the effects of the increased density and the demand for food or some other limited resource, such as space or water. But because the density of the population is larger than the carrying capacity can support, the number of individuals must decline to or below the carrying capacity. The time lag or lags, therefore, can produce oscillations in population density after the population has reached the carrying capacity of the environment.

The logistic equation can be modified to include a simple time lag of the sort discussed above

$$\frac{dN}{dt} = rN_t \frac{K - N_{t-\tau}}{K}$$

where τ is the length of the time lag between an increase in density and the response of the rate of increase to its effects. For example, the increased density may affect the death rate of the larvae rather than decrease the number of eggs laid by the adults giving rise to these larvae. Therefore, the population feels the effects at time t of events that took place at $t - \tau$. This type of lag is well illustrated by the classic experiment by Nicholson (1957). Nicholson reared the fly *Lucilia cuprina* in a limited and constant amount of ground liver. The number of adult flies rose sharply to great numbers, but at this peak density of adults the number of eggs laid reached a point where the food available was insufficient to feed the developing larvae. Therefore, the developing larvae consumed all the available food, and because no more food was available almost all of them died. The population then crashed almost to extinction, the few surviving larvae producing a few adults to start the cycle over again. The result is a series of fluctuations similar to those observed in the cladoceran population in Fig. 3-3. This particular form of intraspecific competition has been called *scramble competition*. The magnitude of the effect of the time lag can be lessened if some of the individuals of the population can somehow protect a sufficient amount of food to insure their development. The excess individuals die, but a fairly constant proportion of the population will survive, provided the resources available remain constant. This is termed *contest competition* and is exemplified by territoriality in birds.

The effect of a simple time lag of the type discussed above on the behavior of the population depends on the product of the time lag and the intrinsic rate of increase. In general the larger this product the more violent the oscillations of the population. Wangersky and Cunningham (1957) point out that if $r\tau$ is between 0 and .7 the population will not fluctuate. A product of $r\tau$ between .7 and 1.8 results in damped oscillations around the carrying capacity, as in Fig. 3-12. Products of $r\tau$ larger than 1.8 result in a series of violent nondamping oscillations around the carrying capacity. The difference in amplitude of the oscillations in the two *Daphnia magna* populations discussed at the beginning of the chapter is probably due in part to an increase in the intrinsic rate of increase in the cladoceran at the higher temperature of 25°C. The general effect of time lags in the response of different age groups to increases in density is to cause the population to oscillate, sometimes quite severely, about the theoretical carrying capacity of the environment.

Although the individuals of some species such as *Paramecium bursaria* can be treated as equal, in almost all cases the different age groups of a population respond to increasing density in different ways. For example, greater density may increase the mortality in larvae but certainly cannot change their fertility which is zero in any case. Adults may have their fertility lessened by increased density, or their mortality, or perhaps both. In some cases larvae may react to greater density by lessening their rate of growth. In addition, the individuals of different age groups may interact with each other, the interactions leading to a series of different time lags. The complexity of these interactions and their effect on the rate of population increase can be illustrated by a population of the grain beetle *Tribolium castaneum*. In a population begun with all adults in a limited environment, egg laying is heaviest during the first 4 days, tapering off later as the first-instar larvae hatch. The larvae are cannibalistic toward eggs, their voracity increasing with age. As the larvae grow older and increase in density, a point is reached where almost no eggs survive. When this group of larvae reaches pupation, eggs can again survive to hatch. Adults are also cannibalistic toward eggs but less so than the larvae. This type of life history leads to the production of waves of pupae, because it tends to produce larvae in cohorts. These waves are about 35–40 days apart. However, adult density tends to be much less variable because the adults are cannibalistic toward the pupae; the commoner the adults and pupae, the more pupae are eaten. As the number of adults increases, the number of pupae surviving to adulthood decreases, and the process comes full cycle. There are several time lags inherent in this system of interactions. In general the process is so complex as to invalidate almost every assumption of the logistic equation. Every population is a case unto itself. In this case the limiting resource could almost be said to be space and not food. Therefore, if one wishes a truly realistic representation of population growth in a limited environment, the only recourse is to formulate a model taking into account the differences between age groups and the interactions between age groups. Taylor (1967) has created just such a model of population growth in *Tribolium confusum*. Unfortunately, the methods employed in formulating such models are beyond the scope of this book. Although the matrix model and the stochastic model discussed in Secs. 3-3 and 3-4 alleviate to some extent the restrictiveness of the logistic equation, they are at most only modifications of it. Therefore, the logistic equation and its matrix and stochastic counterparts should not be considered more than a convenient, but crude, approximation to a very complicated process.

3-3 A STOCHASTIC MODEL OF POPULATION GROWTH IN A LIMITED ENVIRONMENT

Bartlett (1960) has developed a stochastic model of limited population growth. This model is derived under almost the same assumptions as the logistic equation except: (*1*) the effect of chance events is included, and (*2*) the effect of density on the birthrate and on the death rate is considered separately.

If a population has N individuals in it, two possible events can happen next; there will be a birth or there will be a death. Because the rate of population growth r depends on both the birthrate and the death rate, the change in density with time

$$\frac{dN}{dt} = N(r - bN)$$

is written more exactly as

$$\frac{dN}{dt} = N(b - \alpha_1 N) - N(d + \alpha_2 N)$$

where b is the birthrate in the second equation, d is the death rate, α_1 is the effect of density on the birthrate, and α_2 is the effect of density on the death rate. Both α_1 and α_2 are assumed to be constant, and all individuals are treated as equal.

The probability of a population of size N increasing by one individual next is (Bartlett, 1960)

$$\Pr(N \to N + 1) = \frac{bN - \alpha_1 N^2}{(b + d)(N) - (\alpha_1 - \alpha_2)(N^2)}$$

and the probability of the population decreasing by 1 next is

$$\Pr(N \to N - 1) = \frac{dN + \alpha_2 N^2}{(b + d)(N) - (\alpha_1 - \alpha_2)(N^2)}$$

To simulate the growth of a population with this model a Monte Carlo approach is needed. The procedure is:

1 Calculate $\Pr(N \to N + 1)$ and $\Pr(N \to N - 1)$.
2 Pick a random number from a table of random numbers.
3 If the random number as a decimal is larger than $\Pr(N \to N + 1)$, the population decreases by 1; if it is smaller, the population increases by 1.
4 Recalculate the probabilities and continue.

If $N_0 = 20$, $b = .9$, $d = .5$, $\alpha_1 = .006$, and $\alpha_2 = .002$

$$\Pr(N \to N + 1) = \frac{(.9)(20) - (.006)(20^2)}{(1.4)(20) - (.004)(20^2)}$$

$$= .5909$$

and likewise $\Pr(N \to N - 1) = .4091$. These steps were carried out in Table 3-1 for five steps, and in Fig. 3-9 for 90, until the carrying capacity of 49 individuals was reached.

So far there is no time reference. The probability of an event, either a birth or a death, during some time interval Δt is

$$\rho_N = (b + d)(N) - (\alpha_1 - \alpha_2)(N^2)$$

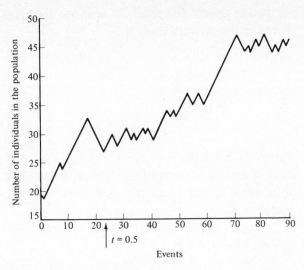

FIGURE 3-9

Simulated population growth in a limited environment using the event-by-event discrete time stochastic model. The parameters used are $N_0 = 20$, $b = .9$, $d = .5$, $\alpha_1 = .006$, and $\alpha_2 = .002$. The point t marks the accumulation of a .5 time interval.

For each event a random amount of time is calculated as

$$t = -\frac{1}{\rho_N} \ln(1 - R)$$

where R is a random number drawn from a table of random numbers and expressed as a decimal. If $N = 24$ and $R = .80697$, then $\rho_N = 31.296$ and $t = -(1/31.296) \ln (.19303) = .0526$. These calculations are repeated for each event, and the total time elapsed up to some event is the sum of the t values.

Table 3-1 A SIMULATION OF POPULATION GROWTH USING THE DISCRETE TIME STOCHASTIC MODEL $(N_0 = 20,\ b = .9,\ d = .5,\ \alpha_1 = .006,\ \alpha_2 = .002)$

Number of individuals	$P(N \rightarrow N + 1)$	$P(N \rightarrow N - 1)$	Event by taking random number
20	.5909	.4091	Death
19	.5936	.4064	Birth
20	.5909	.4091	Birth
21	.5881	.4119	Birth
22	.5854	.4146	Birth

The point where $t = .5$ in the example simulation is marked in Fig. 3-9, and the calculations for the first five events are listed in Table 3-2.

3-3.1   Fluctuations about the Carrying Capacity Due to Chance Events

After the population has reached the carrying capacity of the environment, it will remain at a constant density K if the assumptions of the logistic process are true.   However, the population will fluctuate purely because of chance events.   Other things being equal, the fluctuations will be largest relative to the mean density for small populations.

The probability of the population having a certain number of individuals in it at some point after the carrying capacity of the environment has been reached can be calculated from an equation due to Bartlett (1960).

$$AP(N + 1) = AP(N) \frac{N(b - \alpha_1 N)}{(N + 1)[(d + \alpha_2)(N + 1)]} \qquad (3\text{-}4)$$

The first step in using Eq. (3-4) is to take $AP(N)$ as any number, say 100. With parameters $b = .9$, $d = .5$, $\alpha_1 = .006$, $\alpha_2 = .002$, and $N_0 = 60$, $AP(N + 1)$ is calculated.   The choice of N_0 as a starting point is arbitrary but should be within the possible range of population densities.   For example, with the above parameters

$$AP(N + 1) = 100 \frac{60[.9 - (.006)(60)]}{61[.5 + (.002)(61)]}$$

$$= 85.4$$

Taking 85.4 as $AP(N)$ calculate $AP(N + 2)$, and so forth, until the calculated values become negligibly small.   To go back from $AP(N)$

$$AP(N - 1) = AP(N) \frac{N(d + \alpha_2 N)}{(N - 1)[b - \alpha_1(N - 1)]} \qquad (3\text{-}5)$$

Table 3-2   CALCULATION   OF   ACCUMULATED
TIME   FOR   THE   LOGISTIC   PROCESS
SIMULATION OF TABLE 3-1

Population size	ρ_N	t	Accumulated time
20	26.400	.0213	.0213
19	25.156	.0372	.0585
20	26.400	.0269	.0854
21	27.640	.0053	.0907
22	28.864	.0107	.1014

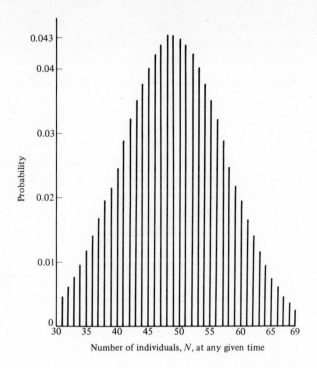

FIGURE 3-10
Distribution of probabilities of the population containing N individuals after the carrying capacity has been reached.   The population parameters are $b = .9$, $d = .5$, $\alpha_1 = .006$, $\alpha_2 = .002$, and $N_0 = 60$.

Again calculate backward until negligible values are reached.   After small values, the two tails of the distribution, have been reached at both ends the values of $AP(N \pm 1)$ are added

$$\sum AP(N \pm 1) \approx A$$

Each value of $AP(N \pm 1)$ is divided by the sum of all the values, and this is the probability of the population having N individuals at some point in time after the carrying capacity of the environment has been reached.   The sum of the $AP(N \pm 1)$ values approximated 5,104.   Therefore, $AP(51) = 224.24$, and the probability of 51 individuals in the population is $\frac{224.24}{5,104}$, or 4.39 percent. These calculations were carried out with the parameters listed above, and the frequency distribution of probabilities is shown in Fig. 3-10.

The distribution of probabilities is slightly skew, but if the carrying capacity is large enough, the skewness is slight and the distribution of possible densities has a normal form, although N, the density of the population, can take only discrete values.

The mean and variance of the density of a population after it reaches the carrying capacity of the environment are

$$\operatorname{var}(N) = \frac{r\gamma}{2\alpha^2}$$

$$E(N) = \frac{r}{\alpha} - \frac{\alpha \operatorname{var}(N)}{r}$$

where

$$\gamma = (b + d)\frac{\alpha}{r} - (\alpha_1 - \alpha_2)$$

The symbol α is the combined effect of density on the birthrates and death rates and is the sum of α_1 and α_2. When $b = .9$ and $d = .5$, $r = .4$, and if $\alpha_1 = .006$ and $\alpha_2 = .002$, $\gamma = (.008/.4)(1.4) - (.004) = .024$, so

$$\operatorname{var}(N) = \frac{(.4)(.024)}{(2)(.008)^2}$$

$$= 75$$

and

$$E(N) = \frac{.4}{.008} - \frac{(.008)(75)}{.4}$$

$$= 48.5 \text{ individuals}$$

3-4 AN AGE GROUP MODEL OF POPULATION GROWTH IN A LIMITED ENVIRONMENT

The matrix model of age-specific fertility and mortality rates can be modified to reflect the effect of density on the rate of population increase. The matrix model takes into consideration the age groups of the population and eliminates the necessity of assuming a stable age distribution for the population. Otherwise the assumptions of the matrix model are essentially the same as those of the logistic equation.

3-4.1 Density Affects the Probability of Survival

The matrix model discussed in Chap. 1 is

$$\mathbf{Mn_0 = n_1}$$

or the population vector after one unit of time is equal to the matrix of fertility and mortality statistics P_x and F_x times the population vector at time t.

Leslie (1948), using the same reasoning behind the logistic equation, postulated a quantity q_t representing the effect of density on the survival rates for each age group

$$q_t = 1 + \alpha N_t$$

where

$$\alpha = \frac{\lambda - 1}{K}$$

As in the logistic equation the effect of density on the probability of survival in each age group is a linear function of density. The parameter α is analogous to b in the logistic equation and is a constant. By assuming density acts only on the survival probabilities and affects all age groups equally, Leslie formed a diagonal matrix with the value of q_t in each position of the diagonal. This diagonal matrix is inverted, and the population after one interval of time is

$$\mathbf{MQ}^{-1}\mathbf{n}_0 = \mathbf{n}_1$$

The effect of this operation is to modify the survival probabilities P_x of the matrix $\mathbf{M}$. If the elements of $\mathbf{Q}$ are all 1s, the multiplication $\mathbf{MQ}^{-1}$ equals $\mathbf{M}$; that is, density does not affect the probabilities of survival, which would be true only if α equals zero. If, however, q_t is greater than 1, as it will be if density increases toward an upper limit, the probability of survival of each age group will be decreased by the multiplication $\mathbf{MQ}^{-1}$, and the rate of increase of the population will decline.

Suppose $\alpha = .000185185$, then $q_t = 1 + .000185185 N_t$. Starting with an initial population of 108 individuals distributed among four age groups as $\mathbf{n}_0' = (81\ 21\ 5\ 1)$, $q_t = 1.020$ and

$$\mathbf{Q}_0 = \begin{bmatrix} 1.020 & 0 & 0 & 0 \\ 0 & 1.020 & 0 & 0 \\ 0 & 0 & 1.020 & 0 \\ 0 & 0 & 0 & 1.020 \end{bmatrix}$$

and

$$\mathbf{Q}_0^{-1} = \begin{bmatrix} .98 & 0 & 0 & 0 \\ 0 & .98 & 0 & 0 \\ 0 & 0 & .98 & 0 \\ 0 & 0 & 0 & .98 \end{bmatrix}$$

If the matrix $\mathbf{M}$ is

$$\mathbf{M} = \begin{bmatrix} 0.00 & 6.43 & 18.00 & 18.00 \\ .78 & 0 & 0 & 0 \\ 0 & .71 & 0 & 0 \\ 0 & 0 & .60 & 0 \end{bmatrix}$$

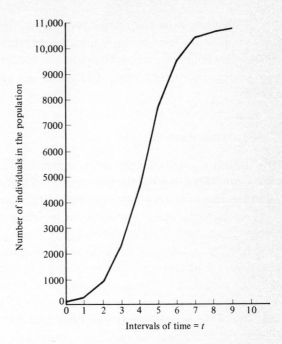

FIGURE 3-11
Simulated density-dependent growth using the matrix model $MQ^{-1}n_0 = n_1$ when the population has a stable age distribution.   The parameters are listed in the text.

the population vector after a single interval of time is

$$
\begin{bmatrix}
0.00 & 6.43 & 18.00 & 18.00 \\
.78 & 0 & 0 & 0 \\
0 & .71 & 0 & 0 \\
0 & 0 & .60 & 0
\end{bmatrix}
\begin{bmatrix}
.98 & 0 & 0 & 0 \\
0 & .98 & 0 & 0 \\
0 & 0 & .98 & 0 \\
0 & 0 & 0 & .98
\end{bmatrix}
\begin{bmatrix}
81 \\
21 \\
5 \\
1
\end{bmatrix}
=
\begin{bmatrix}
238 \\
62 \\
15 \\
3
\end{bmatrix}
$$

These calculations have been carried out for 10 time intervals in Fig. 3-11. If the initial population has a stable age distribution, the population densities predicted by the matrix method are approximately equal to those of the logistic curve.

If the initial age distribution is not a stable one, the population will oscillate, but the oscillations will damp with time. However, if the age distribution is very unstable, the oscillations may continue past the point where the carrying capacity of the environment is reached, although given time these oscillations will damp out.   Leslie (1948) has also derived a model in which density affects the fertility rate F_x.

As it has so far been developed, the matrix model of limited population growth is hardly less restrictive than the logistic equation.   Increasing density may affect either survival probabilities or fertility but not both.   In addition, increasing density is assumed to affect all age groups equally and in a linear fashion.

3-4.2   Time Lags

By using the matrix model the assumption of a stable age distribution has been eliminated, but not the assumptions of a linear response between density and the rate of increase and the implied absence of time lags.   Leslie (1959) has modified the age group model to include a time lag, provided increasing density modifies only the probability of survival of each age group.   Leslie modified the quantity q_t to be dependent not only on the population density at time t but also on the size of the population when each of the age groups were born.   Therefore, there is a different q_t value for each age group. The probability of survival, therefore, of an individual depends not only on the density of the population at time t but also on the density of the population when the individuals of each age group were born.   If the density of the population was high when the individuals of an age group were born, it may have stunted their development and made them more likely to die under later crowding than individuals who had developed during a period of low density. The q_t value for each age group is

$$q_{xt} = 1 + \beta N_{t-x-1} + \alpha N_t$$

where β is the effect of density at birth on the probability of survival under increasing density at some later time t, N_{t-x-1} the density of the population at the birth of the age group x, α the effect of density on survival at time t, and q_{xt} the overall effect of density on the probability of survival for each age group x.   This model assumes that density affects only the probabilities of survival.

As before, the q_{xt} values form a diagonal matrix.   With four age groups

$$\mathbf{Q} = \begin{bmatrix} q_{0t} & 0 & 0 & 0 \\ 0 & q_{1t} & 0 & 0 \\ 0 & 0 & q_{2t} & 0 \\ 0 & 0 & 0 & q_{3t} \end{bmatrix}$$

and, as before, $\mathbf{MQ}^{-1}\mathbf{n}_0 = \mathbf{n}_1$.   The quantity α is measured as before, but estimation of β will probably prove to be rather elusive.

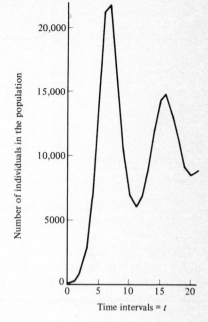

FIGURE 3-12
Simulated population growth in a limited environment, the matrix model containing a time-lag factor relating the population age groups to the density of the population at birth.

Using the constants $\beta = .000148148$ and $\alpha = .000037037$, the same matrix **M** of fertility and mortality statistics, and an initial population vector $\mathbf{n}_0 = (81\ 21\ 5\ 1)$, these calculations were carried out for 21 intervals of time and are plotted in Fig. 3-12. The time lag produces oscillations in the population density after the carrying capacity of the environment has been reached. However, with time the oscillations damp out. The period of the oscillations and their intensity depend on the relative sizes of α and β. In the example β is rather large compared to α, and the oscillations are great.

3-5 INTRASPECIFIC COMPETITION IN PLANTS

In the preceding sections the numerical response of a population of individuals to a limited resource, perhaps most often food, has been studied. The decrease in the rate of increase of a population with increasing density may or may not, depending on the circumstances, represent increasing intraspecific competition among individuals for the limited resource. The models, however, take into account only the observed numerical changes and do not provide

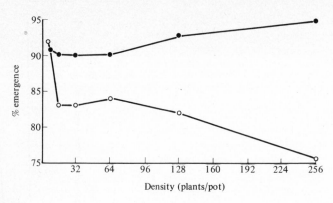

FIGURE 3-13
Relationship between density and percent emergence in pure stands of *A. fatua* (○) and *A. barbata* (●).   (*From Marshall and Jain, 1969.*)

much insight into the existence or manner of intraspecific competition should it exist.   Plant ecologists have traditionally not been much concerned with numerical changes in populations, although this is not as true as it used to be.   Instead their primary interest is usually in the effect of one individual on the growth or efficiency of another.   Because plants have definite static spatial dispersion patterns, the measurement of intraspecific competition in plants almost always depends on the relative distance of one plant from another rather than the overall density of the population.   Because of the importance of spatial patterns, this section should not be attempted until the material in Chap. 5, specifically Sec. 5-2, has been studied.

Marshall and Jain (1969) have provided experimental verification of the sensitivity of plant fertility and growth to changes in density.   In pure stands of the two grass species *Avena fatua* and *A. barbata*, the density of individuals in a pot was varied, and certain population parameters such as percent emergence of seedlings, proportion of fertile plants, dry weight per plant, and other variables were measured.   In Fig. 3-13 the effect of increasing density of plants per pot on the percentage of seeds emerging as seedlings is plotted. *Avena barbata* seems to be rather insensitive to increasing density in these terms, but in *A. fatua* there is a marked reduction in percent seedling emergence.   In Fig. 3-14 the effect of increasing density on the percentage of plants yielding at least one viable seed is shown.   Fertility is reduced by increasing density in both species of grass, slightly less so in *A. fatua* than in *A. barbata*.   Marshall and Jain also found significant effects of density on dry weight per plant, number of tillers produced, plant height, and number of spikelets per plant.

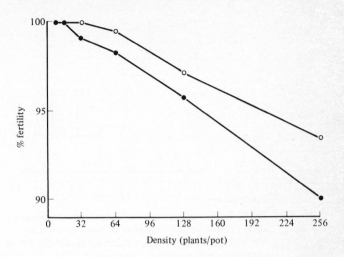

FIGURE 3-14
The relationship between density and percent fertility in pure stands of *A. fatua*
(o) and *A. barbata* (●).   (*From Marshall and Jain, 1969.*)

3-5.1   Pielou's Measures of Intraspecific Competition in Plants

Most studies on intraspecific competition in plants have been concerned with
field crops, postulating some specific spatial arrangement of the individual
plants.   Examples of such studies are those by Berry (1967) and Mead (1967).
In this section a more general method of detecting intraspecific competition
applicable to natural plant populations will be discussed.   In practice the
method requires that the individuals of the population occur in almost pure
stands.

Pielou (1962b) suggests that competition between two neighboring plants
may manifest itself in two ways: (*1*) the distance between any plant and its
nearest neighbor may be positively correlated with the sum of their sizes,
and (*2*) there may be a lower limit to the distance between any plant and
its nearest neighbor.   The first sign indicates that the closer two individuals
are the more they inhibit each other's growth.   In the latter case each
successful plant may have a "territory" within which no new colonizer can
establish itself.   Detection of the first indication of competition is accomplished
by calculating the correlation coefficient (see Chap. 9) between the distance
between a plant and its nearest neighbor, and the sum of the sizes of the two
individuals.   Detection of the second indication of competition is more
complicated.

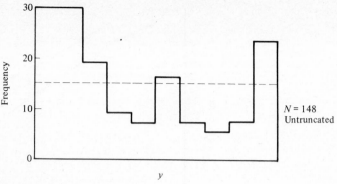

FIGURE 3-15
The frequency distribution of 148 observed squared distances between a randomly chosen individual and its nearest neighbor in a ponderosa pine forest. The dotted line is the expected frequency based on hypothesis that the pines have a random dispersion pattern. (*From Pielou, 1962.*)

If the distance between a randomly chosen individual and its nearest neighbor is r, the frequency distribution of these distances squared is

$$f(w) = \lambda e^{-\lambda w}$$

if the population has a random dispersion pattern. The letter w represents r^2, the squared distance between the individual and its nearest neighbor, and λ is the mean density of the population in number of individuals per circular area of unit radius. In an aggregated population the number of small and large values of w will be greater than expected, and the number of intermediate values too few. Figure 3-15 illustrates just such a case.

Competition will manifest itself in the plants closest together, i.e., the lower values of w. Therefore, some value of w is chosen, c, and only distances equal to c or smaller are considered. This eliminates the high values of w, resulting in a new, truncated frequency distribution for w between 0 and c as

$$f(w \,|\, 0 \le w \le c) = \frac{\lambda e^{-\lambda w}}{1 - e^{-\lambda c}}$$

It is now necessary to divide the range of possible values of w into i equal class intervals so that the expected proportion in each class is $1/i$. The boundary values of the classes are

$$w_r \begin{cases} = 0 & r = 0 \\ = \left(\dfrac{-1}{\lambda}\right)\log\left[1 - \dfrac{r}{i}\left(1 - e^{-\lambda c}\right)\right] & 0 < r < i \\ = c & r = i \end{cases}$$

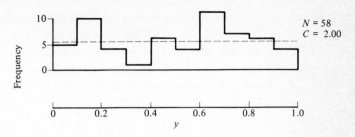

FIGURE 3-16
The truncated frequency distribution on a transformed y scale of squared distances between a randomly chosen ponderosa pine tree and its nearest neighbor. The distribution has been truncated at $w = 2.0$. The dotted line is the expected frequency in each class. (*From Pielou, 1962.*)

where r represents each class interval. Having divided the lower squared distances into each class interval, the observed frequency of observations in each class may be plotted as in Fig. 3-16. If competition is manifesting itself in manner 2, the number of observations falling into the class interval $r = 0$ to $r = 1$ should be less than expected because at the lower limit of distances between plants the presence of one individual should inhibit the establishment of another. The expected frequency distribution in the transformed y scale used in creating the class intervals is

$$f(y) = \frac{1 - e^{-\lambda w}}{1 - e^{-\lambda c}}$$

Dividing the truncated distribution into 10 classes, y ranging from 0 to 1.0 ($i = 10$), the significance of a deviation of the frequency of the first class f_1 from the expected frequency can be tested using the standardized normal variate

$$z = \frac{N - 10f_1}{3N^{1/2}}$$

where f_1 is the number of observations in the first class, and N is the total number of observations, i.e., squared distances, between 0 and c. A one-tailed test is used because the hypothesis is that the observed frequency f_1 is less than the expected frequency.

An example will illustrate the computations. Pielou (1962) studied a population of *Pinus ponderosa* in a 4,055-square-meter area. All trees larger than 2 meters tall were considered part of the population. A total of 148 distances between a plant chosen at random and its nearest neighbor was measured, and the untruncated frequency distribution is shown in Fig. 3-15.

A truncated frequency distribution was created by disregarding all values of w greater than 2.00 ($c = 2.00$). This left a sample size of $N = 58$. The range of w from 0 to 2.00 was divided into 10 equal classes, that is, $y = 0$ to $y = 1.0$, and the number of observations falling into each class plotted in Fig. 3-16. To test the first frequency class $f_1 = 5$, the z statistic is

$$z = \frac{58 - (10)(5)}{(3)(58)^{1/2}}$$
$$= .358$$

Comparing the z statistic to a table of the standardized normal distribution using a one-tailed test it is found that the probability of a deviation from zero, the expected value of z, as large as .358 is about 36 percent. It is concluded that the observed number of observations in the first class is not significantly smaller than expected. There is no reason to conclude that competition is being exhibited by an established individual, excluding invading individuals within some minimum distance. This is not to say that intraspecific competition is not taking place, because indication number 1 is still open. In fact Pielou (1960) found a significant positive correlation between distance from an individual to its nearest neighbor and the sum of their circumferences in these same trees. This finding indicates that close neighbors in a ponderosa pine forest inhibit each other's growth.

3-6 DENSITY DEPENDENCE AND DENSITY INDEPENDENCE

A density-dependent factor is an aspect of the environment whose effect is in some way related to the density of the population. An example is an increase in the death rate at high densities of the population due to conditioning of the food by metabolic wastes. The two-species predator-prey models of Chap. 8, in which predation increases and takes a greater percentage of the prey population as prey density increases, are also examples of density dependence. A density-independent factor does not vary in intensity as density changes. An example is the killing of a constant proportion of a population by a heavy rainfall or drought, independent of the density of the population.

A good deal of ink has been spilled in argument over the relative importance of density-dependent and density-independent factors to natural populations. Although density-dependent factors are of obvious importance in experimental laboratory cultures, their importance in natural populations is much more difficult to assess.   A comprehensive summary of both sides of the question may be found in the 1957 *Cold Spring Harbor Symposium on Quantitative Biology*.   My personal opinion is that on the whole populations are so strongly influenced by density-independent factors that rarely do they reach a size where density-dependent factors become important.   This is not necessarily true of all populations, as there are several known cases in which population density is apparently quite dependent on the amount of food available.   However, should the population reach a density where some resource becomes limiting, density-dependent factors will become the deciding answer to the impossible increase of the population past its food supply.   A population may be " regulated " by density-independent factors at one time and density-dependent factors at another.   In fact a density-dependent factor in some situations may act as a density-independent factor in others.   It does not appear profitable to try to generalize when every specific population is likely to be different from any other.

4

DISPERSAL

The preceding three chapters dealt with populations occurring in discrete areas closed to the outside. Theoretically the individuals of these populations participate in random mating, provided, of course, they are of opposite sexes. Glass jars in a laboratory fit this description, but unfortunately there are few other situations that do. In the field there are a few cases in which a population is truly closed and individuals neither leave nor enter the area. These cases are a decided minority in most ecological studies.

In a closed population as described above the changes in the number of individuals comprising the population are dependent on time, changes in the environmental factors determining the birthrate and death rate, and chance events. However, in most real populations in nature the distance between individuals and the mobility of the organisms become additional factors. If space is considered, the environmental factors determining the birthrate and death rate may, and usually do, change from place to place. In addition, individuals are lost to the population in one place because of migration, and individuals are gained because of immigration from other areas.

In the simplest of possible cases the population may be divided into a number of discrete units such as fish in ponds or pitcher plants in bogs.

A model of the entire population consists of considering population change in each unit. The simplest possible approach is to use for each subpopulation i the exponential growth equation

$$N_{t,i} = N_{0,i} e^{rt}$$

and redefine r as $b - d - v + u$, where v is the rate of loss of individuals to the subpopulation because of emigration, and u is the rate of gain because of immigration. This is not a satisfactory approach, because u depends on the relative densities of the surrounding units. A more satisfactory approach is to simulate the effects of dispersal by making it occur at an instant of time. A discrete time model of population growth is applied to each subpopulation. At the end of each time interval a given proportion of the population is assumed to be lost to the unit by emigration, and a number of individuals is gained from the surrounding units by immigration. The allocation of individuals leaving a unit for surrounding areas depends upon its proximity to each unit, and other considerations such as the average distance traveled by an individual and any directional movement the individual may make. After dispersal is simulated the process is repeated for as many time units as desired. The exact form of the simulation procedure developed for any specific population depends on the characteristics of the population, the number of units, the absence or presence of density effects, and so forth. This subject is discussed in detail by Watt (1968), and the reader may refer to his book for specific suggestions on how such simulations can be carried out.

Although it is convenient to consider the population to be comprised of discrete subpopulations, it is more common for a population to be continuously distributed over an area, e.g., dandelions in a park. The environmental factors affecting both the birthrate and death rate are continuously changing in a continuously dispersed population from place to place. The continuous nature of these changes in density creates incredible mathematical complications which cannot be discussed in this book. In fact many of these problems appear to be insolvable at the moment. The derivation of even the simplest models usually involves the use of grossly unrealistic simplifying assumptions which are rarely likely even remotely to be found in any real population.

Dispersal, quite apart from other considerations, is of fundamental importance in terms of the persistence of a population. If a population occurs in a number of relatively discrete units, the probability of the extinction of the individuals in each unit may be relatively great, either because of chance events, overexploitation of available resources, local catastrophic events, or elimination by a predator or parasite. The wide distribution of the population, however, almost assures that at least one unit will persist, and the dispersal of individuals from this area or areas to refound the populations lost will insure the continuation of the species.

In light of the complexities besetting the study of dispersal, the treatment of dispersal in this chapter is, at best, superficial. The chapter begins with a discussion of some of the biological aspects of dispersal as illustrated by an experiment with the fruit fly *Drosophila pseudoobscura*. Next, dispersal independent of births and deaths is considered, and some relatively simple, if unrealistic, models of the situation are proposed. In the terminal section a model taking both dispersal and births and deaths into consideration is presented.

4-1 SOME GENERAL COMMENTS

All organisms move in some way during at least part of their lifetime. Even the sessile barnacle has a motile stage. Some species move more than others, but all species do move. The distance an individual may move is not always dependent on size. Many small insects or seeds may be carried hundreds of miles by winds, while many large mammals may spend their entire life in a small, restricted area. I have tried to give below a short list of some of the different manners of dispersal found among species of plants and animals. The list is not meant to be complete, and includes only a few of the commoner forms of dispersal:

1 During their entire lifetimes the individuals may wander at random.
2 One stage of the individual moves at random; the others remain in one place.
3 One stage may be sessile and another carried passively.
4 An animal may move randomly in a given spot of ground, but not go beyond the boundaries of the area.
5 The individuals may make a directional movement once during a lifetime.

Other forms of dispersal could be added to this list, only serving to underscore the inherent complexity of the situation.

4-1.1 An Experimental Study of Dispersal in a Fruit Fly

I would like to start by reviewing one of the excellent experiments on dispersal performed by Dobzhansky and Wright (1943) on the fruit fly *D. pseudoobscura*. In four experiments a total of 14,026 orange-eyed mutant

North

Trap	Orange	Wild
32	0	11
33	2	27
34	2	26
35	3	43
36	2	17
37	1	26
38	1	8
39	4	19
40	9	16
41	6	12
42	8	46
43	3	8
44	5	9
45	7	4
46	54	28

West — central horizontal row — **East**

West arm (Trap | Orange | Wild), from outer to inner:

Trap	Orange	Wild
31	1	34
30	0	27
29	1	34
28	3	18
27	5	21
26	3	19
25	4	33
24	2	21
23	2	28
22	10	38
21	8	21
20	23	47
19	29	23
18	10	14
17	129	26

Center (release point): **215**, 25 | 16 ... 25

East arm (Trap | Orange | Wild), from inner to outer:

Trap	Orange	Wild
15	125	27
14	7	4
13	9	7
12	9	25
11	9	14
10	5	32
9	4	30
8	3	15
7	5	14
6	7	40
5	2	6
4	1	12
3	0	27
2	2	29
1	2	29

South

Trap	Orange	Wild
47	27	13
48	22	19
49	10	11
50	8	13
51	4	25
52	3	17
53	11	57
54	0	28
55	1	57
56	1	33
57	0	27
58	1	44
59	3	44
60	0	0
61	0	37

FIGURE 4-1

The results of the first day of collecting in experiment 4. The figure at the left in each cell is the number of the trap; the figure in the middle is the number of orange-eyed flies caught; the figure on the right is the number of wild-type flies caught. (*After Dobzhansky and Wright, 1943.*)

flies was released at points near Idyllwild, California. In each experiment a number of flies was released at a central point and recaptured by a series of traps in a cross pattern; the traps were spaced 20 meters apart.

In the fourth experiment a total of 4,810 orange-eyed flies was released in a field south of Idyllwild. The number of flies captured in each of the 61 traps in the cross pattern 1 day later is shown in Fig. 4-1. On the

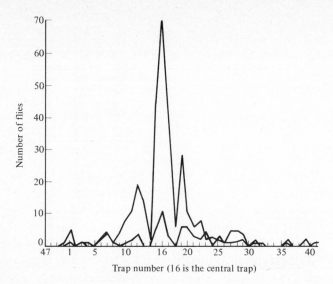

FIGURE 4-2
The number of orange-eyed flies caught in each of the traps of the linear array of traps on days 2 and 4 of experiment 4. The upper line is day 2 and the lower day 4. (*Data from Dobzhansky and Wright, 1943.*)

second day the north-south set of traps was taken down, and the east-west set of traps increased to 55. The total number of orange-eyed flies in the traps on days 2 and 4 is graphed in Fig. 4-2. Figure 4-2 reveals three characteristics of the dispersal of the flies:

1 The distribution of the flies in traps is roughly normal.
2 There is variability in the number of flies in each trap; i.e., the curve is not a smooth one.
3 The variance of the distribution increases with time.

Observation *3* means biologically that the distance of the average fly from the center increases with time.

The variation in each of the curves in Fig. 4-2 is due to the differential selection of habitat by the flies. To the flies some of the sites where traps had been placed were more attractive than others. Part of the variation is also due to changing temperature and humidity from day to day, and even from hour to hour. If the variations due to habitat selection and changing climate are smoothed out, the hypothesis is that the dispersion of flies from a central point along the transect can be fit by a normal curve. The reasons for hoping that this is true will become clear in the next section.

Unfortunately the distribution of flies on the first and sometimes the second and later days was leptokurtic, not normal. There were too many flies still in the center after a day, too many far away, and not enough in the traps in the middle. The kurtosis k of the distribution can be measured by the statistic

$$k = \frac{n \sum r^4 f}{(\sum r^2 f)^2} \qquad (4\text{-}1)$$

where r is the distance of the trap from the central release point, f is the number of flies caught in the trap, and n is the total number of flies caught in all the traps. A value of k greater than 3 indicates that the distribution of flies is leptokurtic. If k equals 3, the curve is normal. Table 4-1 lists the kurtosis values for each day of the experiment. The distribution of flies is leptokurtic at the beginning of the experiment, but with time approaches normality.

Dobzhansky and Wright conclude that the flies used in the experiment were of two types: those with a tendency to disperse fast, and a second group tending to disperse more slowly. If the distribution of each group around the center trap were normal, the two normal curves superimposed on each other would give a leptokurtic normal distribution. The curve tended to normality with time, because the faster-dispersing flies had moved beyond the range of the last trap during the later days of each experiment.

If the rate of dispersal is constant, the variance will increase by a constant amount each day. However, the flies dispersed at different rates as the temperature changed from day to day. Dobzhansky and Wright calculated

Table 4-1   THE KURTOSIS OF THE
DISTRIBUTION OF FLIES
IN EXPERIMENT 4

Day	k
1	8.3
2	5.9
3	4.2
4	4.0
5	3.0

SOURCE: After Dobzhansky, T., and Wright, S., Genetics of Natural Populations. X. Dispersion rates of *Drosophila pseudoobscura, Genetics*, vol. 28, 1943.

the variance of the distribution of flies around the center trap. This variance is termed the mean-square radial distance from the center and is

$$\text{msd} = \frac{\sum r^3 \bar{f}}{\sum r\bar{f} + \left(\dfrac{c}{2\pi}\right)} \tag{4-2}$$

where $\bar{f}$ is the average number of flies in the traps equidistant from the center, and c is the number of flies in the center trap. If it is possible to assume that the distribution is always normal, which it was not in these experiments, Eq. (4-2) becomes

$$\text{msd} = \frac{\sum r^2 \bar{f}}{n}$$

The calculated mean-square radial distances for each day of experiment 4 are listed in Table 4-2. The mean-square radial distance increases, and the increase is roughly linear, i.e., a constant amount per day, although there is a great deal of variation. Dobzhansky and Wright measured r in intervals of 20 meters, e.g., $r = 2 = 40$ meters.

Dispersal in some cases tended to be greater in one direction than in the others. If more flies dispersed north than elsewhere, it was said that the *drift* was to the north. In some of the experiments the drift was significant and was attributed by Dobzhansky and Wright to active habitat selection by the flies rather than to wind or some other climatic factor.

Table 4-2 THE OBSERVED MEAN-SQUARE RADIAL DISTANCE FROM THE POINT OF RELEASE OF THE ORANGE-EYED FLIES IN EXPERIMENT 4

Day	Mean-square radial distance
1	51.3
2	75.8
3	119.3
4	197.2
5	142.8

SOURCE: After Dobzhansky, T., and Wright, S., Genetics of Natural Populations. X. Dispersion rates of *Drosophila pseudoobscura, Genetics*, vol. 28, 1943.

In summary the following conclusions can be made about the *D. pseudoobscura* flies studied by Dobzhansky and Wright:

1 The population of flies was heterogeneous, some flies dispersing faster and further than others.   Apparently there were two types of flies in the population in relation to their tendency to disperse.

2 The distribution of flies from the release point was leptokurtic because of these two types, but tended to normality with time.

3 In some cases there was significant drift in the direction of dispersal.

The measurement of dispersal in the flies was complicated by changing climatic conditions, nonrandom habitat selection, and the death of some of the flies during the experiment.   Despite the difficulties involved, the *Drosophila* experiment is relatively simple as natural situations go.   Any mathematical model trying to simulate dispersal should consider the observations just made.

4-2   DENSITY-INDEPENDENT DISPERSAL

4-2.1   Empirical Models of Dispersal

Most of the models of dispersal treated later in the chapter are derived from specific sets of hypotheses.   It is often useful, however, also to have purely empirical models to fit an observed frequency distribution of individuals around a point of release or origination.   Regression equations are often used for this purpose.   The use of regression and its limitations are discussed in Chap. 9.

Paris (1965) studied the dispersal rates of the isopod *Armadillidium vulgare* in a Californian grassland.   In his experiments a large number of radioactively marked isopods were released at a center board.   Because of its radioactivity the presence of an isopod could be detected with a radiation detector.   To measure the dispersal of the isopods boards were arranged in concentric circles in 16 radii at 1-meter intervals around the central release area.   The number of isopods at a distance r from the center was determined after different intervals of time.   The results of one experiment 12 hours after release are illustrated in Fig. 4-3.   The curve drawn through the data points is given by the regression model.   Paris (1965) tried several regression models and found the most satisfactory model to be

$$Y = \alpha + \beta_1 \log r + \beta_2 \frac{1}{r}$$

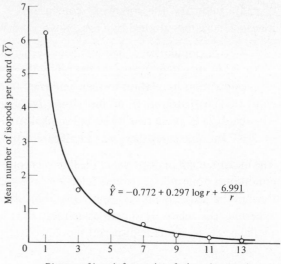

FIGURE 4-3
Regression of the number of isopods under boards serving as traps on distance of boards from point of release for day 1 in experiment 2.   Points indicate the mean number of tagged individuals per board.   The curve is a least-squares fit of the regression model to the data.   (*From Paris, 1965.*)

where Y, the dependent variable, is the number of isopods, and α, β_1, and β_2 are the regression coefficients.   Using the standard techniques the least-squares fit of the model to the data is

$$Y = -.772 + .297 \log r + \frac{6.991}{r}$$

The model is appropriate only 12 hours after the time of release.   The parameters of the model change if the time interval is changed.

In other situations some other regression model may be more appropriate. As pointed out in Chap. 9, the "best fit" is determined by which regression model results in the smallest residual variance.   Other regression models which might be tried are

$$Y = \alpha + \beta \log r$$

$$Y = \alpha + \beta \frac{1}{r}$$

$$\log Y = \log \alpha + \beta \log r$$

4-2.2 Probabilistic Dispersal Models

If an individual wanders at random, then for every point in the habitat there will be an associated probability of the individual being at that point. The probability depends on how much time has passed and the distance of the point from the starting point, as well as how fast the individual is moving. To make dispersal a *random walk*, dispersal is assumed to be independent of the density of individuals in an area; there are no births, and there are no deaths. To illustrate the principle of the random walk an analogy will be made to an animal living along the bank of a stream. Every day he moves 10 meters, either upstream or downstream, and moves in either direction with equal probability. If his position on day 0 is taken as 0, he has a 50 percent chance of being 10 meters upstream and a 50 percent chance of being 10 meters downstream on day 1, or

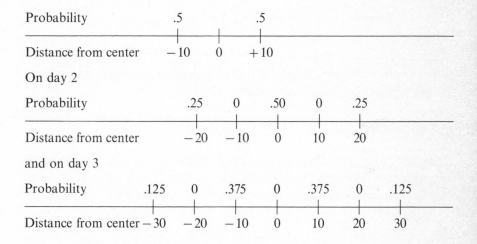

Probability   .5   .5

Distance from center   − 10   0   + 10

On day 2

Probability   .25   0   .50   0   .25

Distance from center   − 20   − 10   0   10   20

and on day 3

Probability   .125   0   .375   0   .375   0   .125

Distance from center − 30 − 20 − 10   0   10   20   30

A simulation of the random walk with equal probability of movement in either direction, that is, $p = q = .5$, as shown in Fig. 4-4. The movement of the animal up- or downstream was decided by flipping a coin. The probability that the animal will be any place r on the stream bank after n moves, $\Pr(r|n)$, is a *binomial probability*. The expression $\Pr(r|n)$ is read, "The probability of event r given n," or in this case the expression means the probability that the animal will be at position r after n moves have occurred. This probability is

$$\Pr(r|n) = \binom{n}{x} p^n q^{n-x}$$

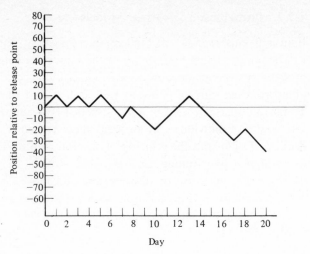

FIGURE 4-4
Simulation of a random walk with $p = q = .5$.

where $x = (n + r)/2$, p is the probability of moving in one direction, and q is the probability of moving in the other. Also

$$\binom{n}{x} = \frac{n!}{x!\,(n - x)!}$$

If p equals q, these probabilities approach with time a normal distribution with a mean of zero.   Given a number of animals moving in the same way, the mean of their distribution along the stream band after some length of time t is

$$m_t = \frac{t}{\Delta t}\,(p - q)\Delta x$$

In our example $\Delta t = 1$ day, $\Delta x = 10$ meters, and $p = q = .5$.   After 22 days

$$m_t = \tfrac{22}{1}(.5 - .5)10$$
$$= 0$$

If $p = q$, the mean will always be 0.   The variance at time t is

$$\sigma_t^{2} = \frac{t}{\Delta t}\,4pq(\Delta x)^2$$

or for the example $\sigma_t^{2} = (\tfrac{22}{1})(1)(10)^2 = 2{,}200$. If p is larger than q, the trend with time will be for more animals to move upstream than downstream, and the position of the mean of the distribution, the mean distance from the center, will be farther and farther upstream.   This is drift.   Because drift occurs as a constant amount each day, $m_t = 2ct$, where $2c$ is a constant.   The parameter c is termed the coefficient of drift.   The value of $2c$ equals $(p - q)(\Delta x/\Delta t)$.   With time the population will diffuse outward from the center, and the amount of diffusion for each time interval will be the mean-square dispersion per time interval, or a^2, and will be equal to $(\Delta x)^2/\Delta t$.

Returning to our hypothetical animals along the stream bank, note that on day 2 no animals can be at positions 10 units up or downstream because $x = \frac{3}{2}$ is not defined as a factorial and there is, therefore, zero probability of an individual being at either of these two positions on day 2.   In the field this particular type of random-walk dispersal pattern is never found, simply because an individual does not move a constant Δx in a constant Δt.   In other words Δx is variable from individual to individual and day to day.   The fruitfly experiment, however, leads us to believe that the dispersion of individuals about the release point may be approximately normal.   In fact if we allow Δx and Δt to become infinitesimally small, the binomial probability distribution of the random walk approaches as a limit the normal distribution.

$$f(y|t) = \frac{1}{(2\pi a^2 t)^{1/2}} \exp\left[-\frac{1}{2a^2 t} (x - 2ct)^2 \right] \tag{4-3}$$

The normal distribution is a continuous distribution and, therefore, the probability of an individual being at point x is zero because there are an infinite number of possible points x.   Therefore, to find the probability that an individual is between points a and b, we find the area under the curve given by Eq. (4-3) from a to b, i.e.,

$$\text{Probability of being between } a \text{ and } b = \int_a^b f(y|t)\, dy$$

Consider the following example.   In experiment 4 of the fruitfly experiment, the estimated mean-square dispersion per time interval was 30.12 in 20-meter units.   Therefore the variance of the normal distribution is $a^2 t$ at time t is $30.12\ t$.   On day 2 the variance is $(30.12)(2) = 60.24$.   Assuming that the drift of the population was zero, the distance $x = 1$ can be standardized as

$$z = \frac{1 - 0}{(60.24)^{1/2}} = .1289$$

Similarly for $x = 2 = 40$ meters, $z = .2579$.   Utilizing a table of the standard normal distribution we find that 55.12 percent of the standard normal distribution occurs between minus infinity and .1258.   For $z = .2579$ the cumulative area is 60.17 percent.   The difference in the two areas,

60.17 − 55.12 = 5.05 percent, is the percent area under the standard normal distribution between $z = .1289$ and $z = .2579$. Therefore, in a linear habitat 5.05 percent of the fruitflies should be between 20 and 40 meters from the release point in one direction on day 2 and, because of the symmetry of the normal distribution, 5.05 percent of the flies will be between 20 and 40 meters from the release point in the opposite direction.

In a two-dimensional situation with no drift, the position of an individual is marked on a pair of axes x and y, and Eq. (4-3) becomes

$$f(x, y \mid t) = \frac{1}{(2\pi a^2 t)^{1/2}} \exp\left[\frac{-(x^2 + y^2)}{2a^2 t}\right] \tag{4-4}$$

In two-dimensional problems, however, it is easier to work with radial distances from the release point. If r is the distance of an individual from the release point, Skellam (1951) shows that the proportion of the individuals released lying outside a circle of radius R is

$$p = \exp\left[\frac{-R^2}{a^2 t}\right] \tag{4-5}$$

For example, on day 2 the proportion of the released fruitflies lying outside a circle of $r = 4 = 80$ meters is

$$p = \exp\left[\frac{-4^2}{(30.12)(2)}\right]$$

$$= .7667$$

This proportion appears to be far too high. The discrepancy is probably due to the great leptokurtosis of the distribution of flies on the second day and the variability of the mean-square dispersal rate.

4-3 DENSITY-DEPENDENT DISPERSAL

In Sec. 4-2 the rate of dispersal and the distance traveled by an individual were independent of population density. The probabilistic model of Sec. 4-2.2, in fact, was first developed in physics to mimic the diffusion of gas molecules. The existence of density-dependent rates of dispersal in vertebrates is generally recognized. In many bird species exhibiting territorial behavior territory size can be compressed by increasing density only to a certain limit. At this limit excess birds are forced to leave the area. Therefore, if the population density is below this limit, dispersal will be small, sharply increasing once the density passes this critical limit.

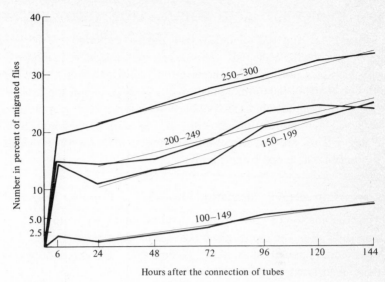

FIGURE 4-5

Percent of flies migrated from the central vial plotted against time for four densities of flies.   The fine straight lines running from 24 to 144 hours are regression lines fitted to the data.   (*From Sakai et al., 1958.*)

The existence of density-dependent dispersal rates in invertebrates is not so well documented, although Sakai et al. (1958) provide evidence of its existence in a laboratory strain of the fruitfly *Drosophila melanogaster*.   In their experiments populations of different densities were confined in a central vial connected by plastic tubing to three peripheral vials.   For each density the percentage of the population moving from the central vial to the peripheral vials as a function of time was measured.   The results of these experiments are shown in Fig. 4-5.   In the lowest density experiment, 100 to 149 flies, the percentage of migrated flies rises linearly with time.   However, with increasing density, i.e., densities over 150 flies, there appears to be an initial rapid dispersal of flies from the central vial followed by a linear rise after about $\frac{1}{2}$ day.   The linear rise of migrated flies at the lowest density corresponds roughly with random movement of the flies.   The higher densities show a distinct increase in dispersal with increasing densities.   Sakai et al. (1958) found significant differences among strains of *D. melanogaster* in the "critical density" above which dispersal became density dependent rather than simply random.   The critical density in the laboratory strain used was about 150 flies, but in several wild stocks the critical density ranged from 40 to 80 flies.

4-4   DISPERSAL AND POPULATION GROWTH

Sections 4-2 and 4-3 assumed that during a period of dispersal neither births nor deaths occurred.   This section will introduce a model taking into consideration not only the dispersal of individuals but also the growth or decline of the population.   This topic is perhaps one of the least understood but most essential subjects in population ecology.   Two important articles on dispersal and population growth, viz., Skellam (1951) and Bradford and Philip (1970), are not discussed, because their level of mathematics is far beyond that assumed by this book.

4-4.1   Exponential Population Growth and Random Dispersal

The combination of both dispersal and population growth has proved to be very difficult.   The model given below allows for both random dispersal and exponential population growth.   To illustrate the computations involved, the one-dimensional form of the model is discussed before the more general two-dimensional case is presented.   The model, as developed by Bailey (1968), was intended to be stochastic, but for reasons of simplicity only the deterministic components are given.

In the random walk Bailey (1968) postulates a linear habitat centered on zero and stretching in both directions to plus and minus infinity.   The habitat, if wide enough, need not be infinite, but at the boundaries the calculations become garbled because of edge effects.   If individuals are lost to the population at a given point in the habitat at a rate v because of dispersal, and gains from adjacent habitats are neglected, $r = b - d - v$. The population at each point will also receive individuals from adjacent points by dispersal.   The habitat is postulated to consist of a series of discrete and equally placed points.   Assuming a population at the center, $i = 0$, at $t = 0$ and with a rate of dispersal v, Bailey shows that the number of individuals at time t at any point i, if dispersal is only to the two positions on either side of i during Δt, is

$$m_i(t) = ae^{(b - d - v)t}I_i(vt) \tag{4-6}$$

The quantity $I_i(vt)$ is a function termed a modified Bessel function and is tabulated in Abramowitz and Segun (1965).   The letter a is the initial size of the population at the center.   If $b = .9$, $d = .5$, $v = .2$, and $t = 10$, the number of individuals four steps away from the center in either a positive or negative direction, $i = \pm 4$, is

$$m_4(10) = (10)(e^2)(.050729)$$
$$= 3.7484 \text{ individuals}$$

or 3.7484 individuals are expected at a point 4 intervals away from the center point after 10 intervals of time.   The quantity .050729 is the tabulated value of the Bessel function $I_4(5)$.

The general form of Eq. (4-6) for any number of initial colonies anywhere along the transect is

$$m_i(t) = e^{(b-d-v)t} \sum_{j=-\infty}^{\infty} a_j I_{i-j}(vt)$$

For example, if there are initial colonies of 10 individuals at $j = 3$ and 20 individuals at $j = 2$, the mean number of individuals at $i = 4$ at $t = 10$ is

$$m_4(10) = e^2[(10)(1.5906) + (20)(.6889)]$$
$$= 219.3367 \text{ individuals}$$

If there is only a population of size a, the origin initially, and if $b = d$

$$m_i(t) = ae^{-vt}I_i(vt)$$

In Fig. 4-6 the densities at each point along the linear habitat from $+5$ to -5 with $a = 100$ are plotted for $t = 2$ and 20.   If the birthrate equals the death rate, the population at the origin immediately begins to decline.   For other points the initial population is zero, and as time increases the number of individuals rises rapidly to a maximum and then dies slowly away.   The maxima occur later and are smaller for points progressively further from the origin.   If b is less than d, the same damped wave moving out from the origin occurs, but the damping occurs more rapidly.

If the birthrate is greater than the death rate, after a long period of time the density will approach an asymptotic value at all places in the linear habitat

$$m_i(t) \sim a(2\pi vt)^{-1/2} e^{(b-d)t} \quad t \to \infty$$

In other words, with time the population will become evenly spread out through the habitat.

Bailey has extended Eq. (4-6) into two dimensions.   The general equation for the expected number of individuals after some period of time t at point ij is

$$m_{ij}(t) = e^{(b-d-v)t} \sum_k \sum_l a_{kl} I_{i-k}(\tfrac{1}{2}vt) I_{j-1}(\tfrac{1}{2}vt)$$

where on a coordinate system ij is the position of the population relative to the origin, kl the position of the initial population relative to the origin, and a_{kl} the size of each of the initial populations.   If initially there are two populations, 20 individuals at 3,2 and 10 individuals at 5,1, the population at 1,1 after 10 intervals of time ($b = .9$, $d = .5$, $v = .2$) will be

$$m_{1,1}(10) = e^2[(20)(.1357)(.5652) + (10)(.0027)(1.2661)]$$
$$= 11.5871 \text{ individuals}$$

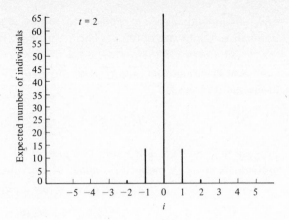

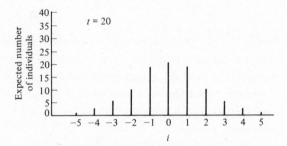

FIGURE 4-6
The expected number of individuals at position i at $t = 2$, 4, 10, and 20. $a = 100$, $b = d$, and $v = .2$.

5

DISPERSION PATTERNS AND SPATIAL DISTRIBUTIONS

The number of individuals in a population continually changes with time and distance. The dispersion pattern of the population, i.e., the positions of the individuals in the environment, at any instant represents the culmination of a history of births, deaths, and movements. In field populations it is difficult to define sharply the population interactions, but sometimes by observing the dispersion pattern of the individuals some insight into the biological characteristics of the species and the reasons behind the changes in density of the population can be gained. The purpose of this chapter is to look at some of the ways in which dispersion patterns can be studied.

If the individuals of a population occur in a series of discrete habitats such as ponds, the habitats can be treated as a series of units in a frequency distribution. The classes of the frequency distribution are the number of units with a given number of individuals in them. More often a population may not be dispersed among discrete units, but continuously dispersed throughout the area, although perhaps more densely in some places than in others. Because of the added difficulties of sampling in a continuous habitat, measurement of the dispersion pattern of a population consisting of discrete units will be treated first. Later some methods of studying dispersion patterns in a continuous environment are presented. Concluding the chapter will be a short section trying to relate dispersion and population growth.

5-1 DISPERSION PATTERNS IN DISCRETE UNITS

5-1.1 The Poisson Distribution

If you went out and counted the number of mosquito larvae of a particular species in 25 small pools of water, you might obtain a series of numbers as shown in Fig. 5-1. The total number of individuals in all 25 pools is 33. The mean number of larvae per pool is 1.3. The number of units, pools, with a given number of individuals has the following frequency distribution:

Number of individuals	0	1	2	3	4	5	6
Frequency	8	8	4	2	1	1	1

If each individual is assigned to a pool randomly and independently of the other individuals, and if the total number of individuals in a pool is much smaller than the total number the pool can hold, the number of pools with a given number of individuals is a *Poisson variate*. The frequency distribution is a *Poisson distribution* given by the function

$$P_x = \frac{a^x e^{-a}}{x!}$$

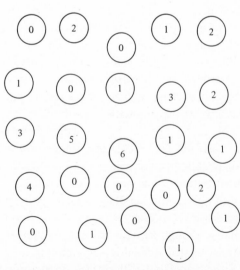

FIGURE 5-1
The distribution of mosquito larvae among 25 pools of water; an artificial example.

where x is the number of individuals per unit, and a is the mean number of individuals per unit. The symbol $x!$ stands for x factorial, and $5!$ is $5 \times 4 \times 3 \times 2 \times 1 = 120$. In the hypothetical example $a = 1.3$, so the expected percent number of units with two individuals is

$$P_2 = \frac{1.3^2 e^{-1.3}}{2!}$$

$$= .230$$

The expected number of units with two individuals is, therefore, $(.230)(25) = 5.75$ units, or in the example ponds.

McGuire, Brindley, and Bancroft (1957) fit a Poisson distribution to the frequency distribution of the number of corn-borer larvae, *Pyrausta nubilalis*, in corn stems in a field in Iowa. The observed and expected frequencies of larvae in stems are given in Table 5-1. The expected frequencies are those given by the assumption that the larvae are independently and randomly distributed among the stems, i.e., the Poisson distribution.

Table 5-1   THE OBSERVED AND EXPECTED FREQUENCIES OF NUMBER OF CORN-BORER LARVAE PER STEM IN A CORNFIELD IN IOWA. THE EXPECTED FREQUENCIES ARE GIVEN BY THE POISSON DISTRIBUTION

Number of larvae	Frequency observed	Frequency expected	$\frac{(O - E)^2}{E}$
0	355	240.20	54.87
1	600	622.35	.80
2	781	806.25	.79
3	567	696.35	24.03
4	441	451.04	.22
5	245	233.74	.54
6	135	100.93	11.50
7	42	37.37	.57
8	17	12.08	2.00
9	11	3.49	16.16
10+	11	.89	114.84
Total	3,205 stems		$\chi^2 = 226.32$

SOURCE: McGuire, J. U., Brindley, T. A., and Bancroft, T. A., The Distribution of European Corn Borer Larvae *Pyrausta nubilalis* (Hbn) in Field Corn, *Biometrics*, vol. 13, 1957.

The fit of the observed and expected frequency distributions may be tested by a *chi-square goodness-of-fit* test calculated as

$$\chi^2 = \sum_{x=0}^{N} \frac{(O - E)^2}{E}$$

where O is the observed frequency of x, E is the expected frequency, and N is the number of frequency groups.   In this example $\chi^2 = 226.32$.   The chi-square distribution is determined by its degrees of freedom.   The degrees of freedom are equal to the number of frequency groups N minus 2.   In this case df $= 11 - 2 = 9$.   The observed chi-square value is compared to a chi-square table with $N - 2$ degrees of freedom.   For the corn-borer larvae $\chi^2 = 226.32$ with 9 degrees of freedom.   The probability of a chi-square this large arising by chance is far smaller than .001.   It is concluded that the expected frequencies of the Poisson distribution do not fit the observed frequencies.   Consequently, the larvae are not randomly and independently dispersed among the plants.   The frequencies most responsible for the lack of fit can be detected by looking for the largest values of $(O - E)^2/E$.   In this case frequency groups 0 and 10 or more contribute the most to the total chi-square.   There are too many stems with no larvae and too many stems with many larvae.

As a general rule none of the observed frequency classes should be much less than 5.   If there are a number of very small frequency classes, the chi-square statistic will be biased and the test misleading.   Although the rule is not quite this strict and may be stretched without being greatly misleading, the user should avoid calculating chi-squares for tables with many classes containing a single individual.   Single-individual classes can be avoided by *pooling*, i.e., by adding adjacent frequency classes until the total frequency is 5 or more.   For every two groups added a degree of freedom is lost because the number of classes compared is decreased by 1.   In Table 5-1 all the frequency classes containing 10 or more individuals have been pooled into a single frequency class.

One of the interesting characteristics of the Poisson distribution is the equality of the mean and the variance.   Therefore, the ratio $s^2/\bar{x}$ is equal to 1, if the individuals of the population are randomly dispersed among the units. If the ratio is larger than 1, there are too many units with few or many individuals.   If the ratio is smaller than 1, the individuals are too equally divided among the units.   To test if the ratio is truly 1 calculate $(n - 1)s^2/\bar{x}$, where n is the number of units.   In the hypothetical mosquito population the variance equals 3.14, and the mean 1.32.   The ratio equals 2.38, and with $n = 25$ the test statistic is equal to 57.12.   This quantity has a chi-square distribution with $n - 1$ degrees of freedom. In the example the probability of a chi-square this large arising by chance if the true ratio is 1 is less than .001. The ratio is significantly different from 1. Because the ratio is larger from 1, we conclude that there are too many units with few and many individuals in them for the population to have a random dispersion pattern.

The individuals in the mosquito population are not randomly distributed in the pools. There are too many pools with too few individuals, and too many pools with several individuals. If individuals are not randomly allocated among the sample units, they may occur as a constant number per unit, or there may be too many units with few or no individuals and too many units with several individuals. In sampling from a continuously dispersed population these dispersion patterns are referred to as *regular* and *aggregated*, respectively, contrasting with the *random* dispersion pattern postulated by the Poisson distribution. If a population has an aggregated dispersion pattern, the mathematical distribution function or functions fitting the observed number of units with a given number of individuals is termed a *contagious* distribution. It is important for semantic reasons to make the distinction between the dispersion of individuals and the mathematical distribution, distribution referring to a mathematical function.

5-1.2 The Negative Binomial Distribution

In nature populations are rarely found to be randomly dispersed, and the probability of finding x individuals in a unit can rarely be predicted by using the Poisson distribution. In fact true randomness represents an infinitely small point on the continuum between regular and aggregated dispersion patterns. Inevitably if the sample is large enough, it will always be possible to prove the population is not randomly dispersed except under the most improbable circumstances. A number of mathematical distributions have been proposed to fit an observed frequency distribution if the individuals tend to be aggregated in a few of the discrete units. In addition to environmental heterogeneity, aggregated dispersion patterns can arise quite naturally. Suppose a female insect lays clusters of eggs randomly among discrete units and the number of egg clusters laid in each unit has a Poisson distribution. Also suppose that the number of larvae hatching from each egg cluster is also random with a Poisson distribution. Even though both phenomena are random, the dispersion pattern of the larvae will be aggregated and will have a distribution known as Neyman's type A. Another contagious distribution, the negative binomial, was based on two similar assumptions: (1) the dispersion of egg clusters has a Poisson distribution, and (2) the number of larvae hatching has a logarithmic distribution (see Chap. 14). The negative binomial has been shown to give a good fit to many actual frequency distributions and is widely used.

The negative binomial distribution function is

$$P_x = \left(\frac{1}{q^k}\right)\left(\frac{(k)(k+1)(k+2)\cdots(k+x-1)}{x!}\right)\left(\frac{p}{q}\right)^x$$

where $p = \bar{x}/k$, $q = p + 1$, and k is a parameter of the distribution.

Bliss and Fisher (1953) give an example of the application of the negative binomial to counts of adult, female red mites on 150 apple leaves. The distribution of female mites among leaves is given in Table 5-2. The average number of mites per leaf is $\bar{x} = 1.14667$, and k, as shown later, was estimated to be 1.02459. Therefore $p = 1.14667/1.02459 = 1.11915$, and $q = 2.11915$. The probability of finding four mites on a leaf picked at random is

$$P_4 = \left(\frac{1}{2.11915^{1.02459}}\right)\left(\frac{(1.02459)(2.02459)(3.02459)(4.02459)}{(4)(3)(2)(1)}\right)\left(\frac{1.11915}{2.11915}\right)^4$$

$$= (.46325)(1.05212)(.007779)$$

$$= .03791$$

Of 150 leaves, 3.791 percent, or 5.69, will be expected to have four mites on them if the dispersion pattern of the mites fits the negative binomial distribution. The expected probabilities from 0 to 8 and above are also

Table 5-2 OBSERVED AND EXPECTED FREQUENCIES OF ADULT FEMALE RED MITES ON 150 APPLE LEAVES. THE EXPECTED FREQUENCIES ARE BASED ON THE NEGATIVE BINOMIAL DISTRIBUTION

Mites per leaf	Number of leaves observed	Number of leaves expected	P_x	A_x
0	70	69.49	.4633	80
1	38	37.60	.2507	42
2	17	20.10	.1340	25
3	10	10.70	.0713	15
4	9	5.69	.0379	6
5	3	3.02	.0201	3
6	2	1.60	.0107	1
7	1	.85	.0057	
8 or more	0	.95	.0063	

SOURCE: Bliss, C. I., and Fisher, R. A., Fitting the Negative Binomial Distribution to Biological Data and a Note on the Efficient Fitting of the Negative Binomial, *Biometrics*, vol. 9, 1953.

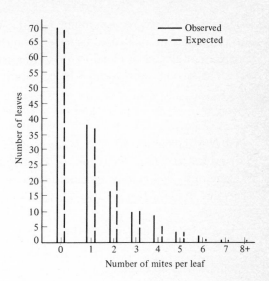

FIGURE 5-2
A comparison of the observed frequency distribution of adult female red mites among 150 apple leaves with the expected frequencies based on the negative binomial distribution. (*After Bliss and Fisher, 1953.*)

listed in Table 5-2. The fit of the expected and observed frequencies is quite good (Fig. 5-2), although the statistical tests for goodness of fit will be deferred until after the estimation of k has been discussed.

Bliss and Fisher (1953) point out three possible ways to estimate k. The first and simplest is

$$\hat{k}_1 = \frac{\bar{x}^2}{s^2 - \bar{x}} \qquad (5\text{-}1)$$

where $\bar{x}$ and s^2 are the sample mean and variance, respectively. The caret over the letter k indicates that k is being estimated and is not the population value of k which remains unknown. In the same way $\bar{x}$ might be indicated as $\hat{\mu}$, although it rarely is. The estimated value of k_1 from Eq. (5-1) is

$$\hat{k}_1 = \frac{1.14667^2}{2.27365 - 1.14667}$$

$$= 1.16670$$

This estimator of k is adequate if: (*1*) $\bar{x}$ is small and $\hat{k}/\bar{x}$ is greater than 6, or (*2*) $\bar{x}$ is large and $\hat{k}$ is greater than 13, or (*3*) $\bar{x}$ is moderate and $(\hat{k} + \bar{x})(\hat{k} + 2)/\bar{x}$ is equal to or greater than 15. For a definition of small, moderate, and large, see Anscombe (1950).   When in doubt Eq. (5-1) should not be used.   In this case, using criterion 3, $(2.313)(3.167)/(1.147) = 6.39$, which is not greater or equal to 15, so this estimator of k is not suitable.

The second estimator of k is the iterative equation

$$\hat{k}_2 \log\left(1 + \frac{\bar{x}}{\hat{k}_2}\right) = \log\left(\frac{N}{f_0}\right)$$

where N is the number of samples, e.g., leaves, and f_0 is the number of samples with no individuals.   The value of $\hat{k}_2$ is found by iteration, putting in an estimate of k_2 and modifying it until the two sides of the equation are equal. The value of $\log_{10}(N/f_0)$ is .330993, so

$$\hat{k}_2 \log\left(1 + \frac{\bar{x}}{\hat{k}_2}\right) = .330993$$

Beginning with an estimate of k_2 as 1.000

$\hat{k}_2$	$\hat{k}_2 \log(1 + \bar{x}/\hat{k}_2)$
1.000	.331767
.980	.329744
.992	.33092

The value of $\hat{k}_2$ is between .992 and .993, and by further iteration is found to be equal to .99231.   This estimate of k is good if $\bar{x}$ is small; if $\bar{x}$ is not small, k is small.   For the exact limits see Anscombe (1950).

The third and best estimator of k is the maximum likelihood estimator, also iterative

$$N \log_e\left(1 + \frac{\bar{x}}{\hat{k}_3}\right) - \sum\left(\frac{A_x}{\hat{k}_3 + x}\right) = 0$$

where A_x is the sum of the observed frequencies of units containing more than x individuals.   In our example in which the last x is 7, $A_6 = f_7$, $A_5 = f_7 + f_6$, $A_4 = f_7 + f_6 + f_5$, and so forth.   The values of A_x are listed in Table 5-2. Maximum likelihood is a statistical method of deriving optimal estimators of parameters.   Starting with $\hat{k} = 1.000$

$\hat{k} = 1.000$	$114.5875 - 114.9262 = .3387$
$\hat{k} = 1.050$	$110.4045 - 110.7227 = -.3182$
$\hat{k} = 1.026$	$112.5247 - 112.5432 = -.0185$
$\hat{k} = 1.023$	$112.7961 - 112.7752 = .0209$

The value of $\hat{k}_3$ is between 1.023 and 1.026.   By interpolation Bliss and Fisher (1953) give a final value of 1.02459 for $\hat{k}_3$.   To calculate the standard error of the estimated value of $\hat{k}_3$, designate the last two values of k used in the iteration h_1 and h_2, and the two discrepancies z_1 and z_2. In the example $h_1 = 1.026$ and $h_2 = 1.023$, where h_1 is always the higher of the two estimates, and $z_1 = -.0185$ and $z_2 = .0209$. The variance of the estimate of k_3 is

$$\text{var}(\hat{k}_3) = \frac{h_1 - h_2}{z_2 - z_1}$$

In the example

$$\text{var}(\hat{k}_3) = \frac{(1.026 - 1.023)}{(.0209 + .0185)}$$
$$= .07614$$

The square root of the variance is the standard error of the estimate; that is, $\text{SE}(\hat{k}_3) = .2759$. Bliss and Fisher (1953) also give the standard errors for the estimates of k_1 and k_2.

The goodness of fit of the negative binomial distribution to the observed frequency distribution may be tested by the use of two statistics U and T. The statistic to use depends on the values of $\bar{x}$ and $\bar{x}/k$, and the line dividing the two tests is shown in Fig. 5-3. If the negative binomial is a good fit to the data, the statistics U and T will have expected values of zero. The test consists of calculating the statistic U or T and its standard error. The estimated statistic plus or minus its standard error should contain the value zero if the negative binomial does fit the observed frequency distribution. If the interval does not contain the value zero, the negative binomial does not provide an adequate fit to the data.

The estimate of U is

$$U = s^2 - \left(\bar{x} + \frac{\bar{x}^2}{\hat{k}}\right)$$

In the example

$$U = 2.27365 - \left(1.14667 + \frac{(1.14667)^2}{1.023459}\right)$$
$$= -.1563$$

The variance of U is calculated as

$$\text{var}(U) = \frac{2\bar{x}(k+1)pq^2\left(1 - \dfrac{R^2}{-\ln(1-R) - R}\right)}{N + p^4\,\text{var}(\hat{k}_3)}$$

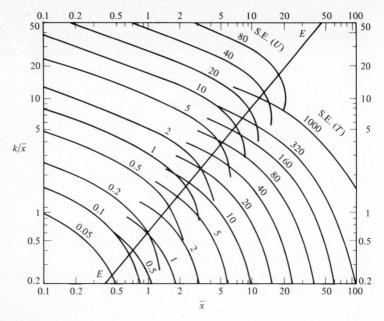

FIGURE 5-3
A diagram for the rapid estimation of the standard errors of the statistics U and T. The standard errors shown are for $N = 100$. For other values of N, the standard error is multiplied by $1 \cdot /N^{1/2}$. (*From Evans, 1953.*)

Because of the complexity of the standard error equations for both U and T approximate values may be read from Fig. 5-3 given values of $\bar{x}$ and $\bar{x}/k$. In the above equation $R = \bar{x}/(k + \bar{x})$, $p = \bar{x}/k$, and $q = p + 1$. In the mite example $U = -.1563$, and its standard error is $\pm.3747$. The estimate of U plus its standard error contains the value zero, and it is concluded that the negative binomial does provide a good fit to the observed frequency distribution of mites among leaves. In this case U is the correct statistic to use. For illustration T will also be calculated.

The statistic T is calculated as

$$T = \left(\frac{\sum f_x^3 - 3\bar{x} \sum f_x^2 + 2\bar{x}^2 \sum f_x}{N} \right) - s^2 \left(\frac{2s^2}{\bar{x}} - 1 \right)$$

and the variance of the estimated T is

$$\text{var}(T) = \frac{2\bar{x}(k + 1)p^2 q^2 [2(3 + 5p) + 3kq]}{N}$$

Approximate values of the standard error of T are also given in Fig. 5-3. The statistic T for the example (remember U is the appropriate statistic) is 1.553, and its standard error is ± 2.032. Again there is no reason to reject the negative binomial fit of the data.

These two statistics, although exact, can sometimes be bypassed by testing the fit between observed and expected frequencies by using a chi-square test, as was done for the Poisson distribution. Bliss and Fisher (1953) do not recommend the use of the chi-square test, because chance irregularities can distort the chi-square statistic.

The parameter k of the negative binomial can range between 0 and plus infinity. As k approaches infinity the negative binomial distribution simplifies to the Poisson. Therefore, if k is small, up to about 8, the population is aggregated, and the smaller the value of k the greater the aggregation. It is sometimes more convenient to have a value that increases with increasing aggregation. The reciprocal of k is sometimes used for this purpose.

If deaths occur at random in a population, i.e., if the population size decreases, the value of k will not change. This assumes, however, that the possibility of death for an individual is the same when density is high as when density is low. If this assumption can be made, k can be considered an invariable measure of aggregation as the population decreases because of the death of individuals. The reason for wanting a measure of aggregation independent of density is to measure a tendency to aggregate as an intrinsic part, and perhaps as a characteristic, of the species. Even if it is possible to assume that deaths occur randomly, k is determined not only by aggregation tendencies intrinsic to the species, but also by environmental heterogeneity. Separating one from the other remains a problem.

5-2   CONTINUOUS DISPERSION PATTERNS

In Fig. 5-4 are illustrations of random, regular, and aggregated dispersion patterns of individuals in a continuous habitat. The reasons for the three names are now apparent. The dots might represent grasshoppers in a field or dock in a lawn. The measurement of dispersion in a continuous habitat usually, but not always, revolves around creating artificial units and carrying out the procedures discussed in Sec. 5-1. There are also additions to the repertoire. The sampling procedure may involve repeatedly placing a sample quadrat at random in the habitat, or dividing the entire area, if it is small enough, into a gridlike series of squares. However, because of the creation of artificial sample units, a serious problem arises of how to take the samples and what size quadrat to use.

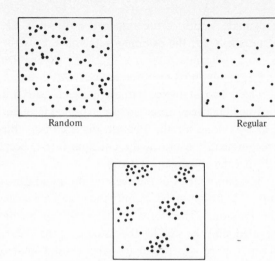

FIGURE 5-4
Three artificial populations to illustrate random, regular, and aggregated dispersion patterns.

5-2.1　Effect of Quadrat Size on the Detection of Aggregation

Figure 5-5 illustrates the problem of sample unit size. The detection of randomness or aggregation depends on the size of the sample unit used. If the dispersion of the population in Fig. 5-5 were measured by repeated sampling with quadrat B, the marked aggregation of the population would show. If quadrat A were used, the comparison of the number of individuals per quadrat with the Poisson distribution would indicate that the population has a regular dispersion pattern, the real dispersion pattern being highly aggregated. This has been illustrated rather graphically by Greig-Smith (1952) by using a series of progressively larger quadrats to measure dispersion in an artificial situation in which individuals were represented by colored disks. The results of the experiment are shown in Table 5-3. At the smaller quadrat sizes the variance : mean ratio indicates a random dispersion pattern, but by using larger quadrats the ratio begins to show a significant amount of aggregation.

5-2.2　Pattern Analysis

The scales of pattern in a population may be detected by dividing the area into a large number of small square units by a grid with a scale of powers of 2, for example, 16×16 or 8×8. The number of individuals or some

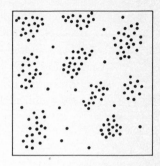

FIGURE 5-5
An illustration of the effect of quadrat size
on the detection of aggregation. A quadrat of size *A* indicates a regular dispersion
pattern, even though the true pattern is
highly aggregated.

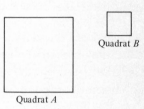

Quadrat *B*

Quadrat *A*

Table 5-3   **MEASUREMENT OF DIS-
PERSION IN AN ARTI-
FICIAL POPULATION TO
I L L U S T R A T E   T H E
EFFECT OF DIFFERENT
SIZE QUADRATS ON THE
DEGREE OF DISPERSION
INDICATED**

Quadrat size, cm	Variance/mean ratio
10	.9394
15	1.0527
20	.9958
25	1.1342
30	1.3928
35	1.5674
40	1.6340

SOURCE: From Greig-Smith, P., The Use of
Random and Contiguous Quadrats in the Study
of the Structure of Plant Communities, *Ann.
Bot. Lond. N.S.*, vol. 16, 1952.

other measure of abundance is determined for each basic unit. Then the basic units are combined into oblong pairs of two units. In turn the pairs are combined into blocks of four, the blocks of four into oblong blocks of eight, and so forth until all the units have been combined into a single block. The total sum of squares with a correction for the mean of the individuals in each unit is

$$\sum x_i^2 - \frac{(\sum x_i)^2}{n}$$

where x_i is the number of individuals in the ith unit of n units. This total sum of squares corrected for the mean can be divided into components based on the difference between each pair of single units in a two-unit block, on the difference between the two-unit blocks in the four-unit block, and so on. If the sum of squares for each difference is divided by the number of differences, a quantity termed the mean square results, which is analogous to the variance/mean ratio of Sec. 5-1.

It is impossible to justify the procedure that follows within the limitations of this book. However, we can look at the results intuitively. If the size of the block is small relative to the size of the clusters of individuals, it is likely that both blocks being combined will be within the cluster or outside the cluster. Therefore, the difference between the two blocks in number of individuals will be small, as will be the mean square which is a measure of these differences. As the size of the block increases relative to the size of the clusters, however, it becomes more and more likely that one block of the two combined will be in the patch and the other outside the patch. Consequently, when the two blocks combined are about the same size as the patch of individuals, the difference between the two will be great, and so will the mean square. Finally, if the block size is larger than the patch size, both blocks will include patches. If the patches are themselves randomly dispersed, the difference between blocks drops, as does the mean square. However, if the clusters of individuals are aggregated, the mean squares of increasing block size will increase again and will be greatest at the block size equal to the size of the clusters of clusters. Therefore, the presence and size of aggregations and aggregations of aggregations can be detected by plotting the mean square against block size. Peaks in the graph indicate aggregations are present and their approximate size.

The computational procedure is comparatively simple, and can be illustrated by referring to Table 5-4. For each block size count the number of individuals in each block, sum the squared numbers, and divide by the block size; that is, $\sum x_i^2$/block size. Then subtract from this the equivalent quantity for the next highest block size. This quantity is divided by the number of differences, degrees of freedom of the table, resulting in the mean square. For block size 1, the mean square representing the difference between the individual units combined into a block of two units is: (1) $\sum x_i^2$/block size $= 483,867.450$; (2) sum of squares $= 483,867.450 - 399,157.635 = 84,709.815$; and (3) with 512 differences between 1,024 basic units, the mean square equals $\frac{84,709.815}{512} = 165.449$. Similarly for block size 2, $(399,157.635 - 269,418.992)/256 = 506.792$. These mean squares are plotted against block size, as in Fig. 5-6. In Kershaw's (1964) example there are two sizes of aggregations indicated, an aggregation of individuals with an area equivalent to a block of 4 units, and aggregations of patches with an area equal to a block size of about 64 units.

Some deficiencies of the mean-square method are:

1 The area must be large enough to be broken into a sufficiently large number of initial basic quadrats, but small enough so that counting all the individuals does not become overbearing.

2 Block sizes are double the size below, so that the peak may be at block size 80 rather than at 64. The peak in the example indicates that the size of the aggregations of aggregations is between block size 32 and 128.

Table 5-4 **AN EXAMPLE OF THE USE OF PATTERN ANALYSIS TO DETECT SCALES OF AGGREGATION. KERSHAW (1964) CONSTRUCTED AN ARTIFICIAL POPULATION WITH TWO SCALES OF AGGREGATION, CLUSTERS OF INDIVIDUALS WITH AN AREA OF TWO SQUARE UNITS, AND CLUSTERS OF CLUSTERS OF EIGHT SQUARE UNITS**

Block size	$\sum x_i^2$/block size	Sum of squares	Degrees of freedom	Mean square
1	483,867.450	84,709.815	512	165.449
2	399,157.635	129,738.643	256	506.792
4	269,418.992	88,065.841	128	688.014
8	181,353.151	40,492.551	64	632.696
16	140,860.600	21,728.739	32	679.023
32	119,131.861	21,025.536	16	1,314.096
64	98,106.324	15,293.160	8	1,911.645
128	82,813.165	4,337.279	4	1,084.320
256	78,475.886	1,537.425	2	768.712
512	76,938.461	93.431	1	93.431
1024	76,845.030			

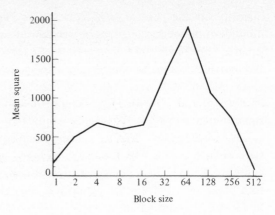

FIGURE 5-6

A plot of the mean square against block size for the example in the text to delimit the scale of aggregation and its presence. (*After Kershaw, 1964.*)

3   Because each successively larger block is formed of combinations of the original grid blocks, the mean-square estimates, which are estimates of variances, are not independent of each other. Therefore, the mean squares cannot be tested to see if they are significantly different from each other. Judgments are, therefore, subjective.

For a more complete discussion of these and other weaknesses of the mean-square method, see Pielou (1969).

5-2.3   Morisita's Index of Aggregation

Because of the effect of quadrat size on the measurement of aggregation it is desirable to have a measure of dispersion independent of the size of the sampling unit. This is possible under some rather special conditions. If it is possible to assume that the population consists of clumps or patches of individuals of different densities, and that within a clump individuals are randomly spaced, an index derived by Morisita (1959) can be used. Although relatively independent of quadrat size, the quadrat must not be so large as to intersect more than one clump of individuals. Morisita's index I_δ is

$$I_\delta = \frac{\sum_{i=1}^{N} n_i(n_i - 1)}{n(n - 1)} N \tag{5-3}$$

where N is the number of samples, n_i is the number of individuals in the ith sample, and n is the total number of individuals in all the samples. If I_δ

equals 1, the dispersion of individuals is at random; if greater than 1, the individuals are aggregated; and if less than 1, the population has a regular pattern. Using a hypothetical series of samples

$$2, 7, 9, 4, 1, 0, 9, 10, 13, 6, 1, 7, 8, 9, 1, 2, 5$$

$n = 94$, $N = 17$, and $\sum n_i(n_i - 1) = 655$, and

$$I_\delta = \frac{655}{(94)(93)} \quad (17)$$
$$= 1.2737$$

The value of I_δ is greater than unity, so the population appears to be aggregated. The significance of the deviation from 1 is tested by the statistic F

$$F = \frac{I_\delta(n - 1) + N - n}{N - 1}$$

or

$$F = \frac{(1.2737)(93) + 17 - 94}{16}$$
$$= 2.59$$

The value of F is compared to a table of the F distribution with $N - 1$ degrees of freedom for a numerator, and infinity for a denominator. In the example the value of F with 16 degrees of freedom as a numerator and infinity as a denominator is about 2.01, allowing for a 5 percent chance that a value of F as large as 2.01 will have arisen by chance. The calculated value of F is 2.59, and it is concluded that there is less than a 5 percent chance of an F as large as 2.59 arising by chance. It is concluded that I_δ is not equal to unity, and that the population has an aggregated dispersion pattern.

Despite the simplicity of Eq. (5-3) and its advantages, the assumptions are rather strict and should be considered before I_δ is used.

5-3 A NEAREST-NEIGHBOR METHOD

Because of the dependence of many of the indices of aggregation on the size of the sampling unit, Clark and Evans (1954) have proposed the mean distance between an individual and its nearest neighbor as a test for deviations from randomness. For their test and other similar tests the density of the population must be known. Nearest-neighbor methods also require the individuals to stay put while the measurements are being taken.

The procedure with plants is usually to tag each plant in the habitat, select an individual at random by using a table of random numbers, and measure the distance between it and its nearest neighbor. If N is the number of observations, the observed mean distance between a plant and its nearest neighbor is

$$\bar{r} = \frac{\sum r}{N}$$

The expected or mean value of the average distance between a randomly selected individual and its nearest neighbor if the dispersion of individuals is at random is

$$E(r) = \frac{1}{2p^{1/2}}$$

where p is the density of the population expressed as the number of individuals per unit area, where the units are those used in measuring r. If the dispersion pattern of individuals is at random, the ratio of the expected and observed average distance will be 1. The ratio between the two can be called R

$$R = \frac{\bar{r}}{E(r)}$$

If the population is aggregated, the value of R will be between zero and 1, and the more aggregated, the closer to zero will be its value. If the population has a regular dispersion pattern, the value of R will lie between 1 and 2.1496.

To test the significance of a deviation from the expected value of R, Clark and Evans (1954) suggest the use of a standardized normal variate, i.e., a variable from the standardized normal distribution

$$z = \frac{\bar{r} - E(r)}{SE(r)}$$

where $SE(r)$, the standard error of the expected value of r, is

$$SE(r) = \frac{.26136}{(Np)^{1/2}}$$

The value of z is compared with a table of the standardized normal distribution. The standardized normal distribution was discussed in Chap. 2. If the value of z is either larger than 1.96 or smaller than -1.96, the probability that the true value of z is 0 is less than 5 percent if we are interested only in whether the ratio is different from 1.   If we are interested in whether the population is aggregated, only values of z less than zero are considered.   If z is less than -1.96, the probability of a value of z this small is less than 2.5 percent. The choice between the one-tailed test or the two-tailed test must be made before the calculations are carried out.

Miller and Stephen (1966) applied Clark and Evans technique to aerial photographs of flocks of sandhill cranes.   They were interested in whether the individuals of a flock were regularly dispersed within the flock.   For flock number 29, $N = 252$, $p = .0094$ individuals, $\bar{r} = 5.82$, and $E(r) = 5.16$.   The ratio R is $5.82 \div 5.16 = 1.13$, and $z = (5.82 - 5.16)/.1698 = 3.89$.   The test is one-tailed because we are testing to see if the ratio is significantly larger than 1, not whether it is different from 1.   By using the one-tailed test the probability of a value of z this large arising by chance is less than .00001, and it is concluded that the individuals of the flock were dispersed in a regular manner.

To use Clark and Evans's technique the density of the population must be known.   Theoretically, it should be known exactly, entailing the counting of all the individuals in the population.   If the area is too large, an estimate of density can be made by taking a series of random quadrats in the area, estimating the density and average nearest-neighbor distance in each quadrat, and averaging the separate measurements from each quadrat.   Estimating density in this manner should be avoided if possible.   As a general rule the smaller the quadrat the better.   Usually as aggregation increases the variability of the nearest-neighbor distances also increases, increasing the number of quadrats needed to determine accurately the mean nearest-neighbor distance.

Other nearest-neighbor methods have been proposed.   Some of these are discussed in Pielou (1969).

5-4   TWO-PHASE MOSAICS AND THE CONCEPT OF PATCHES AND GAPS

All the indices of aggregation discussed in earlier sections of the chapter measure the dispersion patterns of individuals, with the possible exception of the mean-square technique.   It is also possible in some cases to measure the dispersion pattern of patches of individuals separated by areas where individuals of the species do not occur.   This concept of patches and gaps is most applicable to vegetatively reproducing plants forming patches of vegetation in which it is difficult to separate individuals into separate plants.   Many animals, such as barnacles, although they do not reproduce vegetatively, also occur in discrete

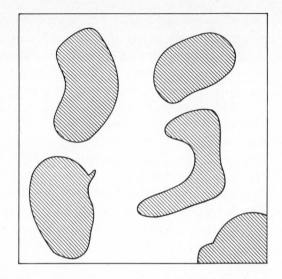

FIGURE 5-7
An artificial example of a vegetatively reproducing plant species occurring as patches of growth surrounded by gaps of open ground. The shaded areas are patches.

patches. Even if patches of individuals are not perfectly defined, it may be possible to approximate them by arbitrarily picking a low density at which the individuals are considered to be absent.

Figure 5-7 illustrates a hypothetical species population consisting of patches of "individuals" separated by empty gaps. Pielou (1964) calls this type of pattern a two-phase mosaic and has developed a method for testing the question, "Can the pattern of patches and gaps be considered to be randomly mingled?" The mathematical treatment of patches and gaps can be found in Pielou (1969). To sample a mosaic of patches and gaps a decision must first be made as to how wide an area constitutes a gap, because even in a patch areas of ground occur between individuals or parts of individuals. If this distance is r, any point on the ground a distance r or greater from the nearest individual is in a gap, but if less than r is considered to be in a patch. Circular quadrats of radius r are created, and if such a quadrat is randomly placed and any part of an individual occurs in the quadrat, the quadrat is said to have made a "hit." If the quadrat is empty, the quadrat has made a "miss." Now suppose two of these quadrats are contiguous and one randomly placed in the area being studied. The distance between them is $d = 2r$. If the first quadrat scores a hit and the second a miss, the combination is HM. Also, hit-hit, miss-miss, and miss-hit are possible. The field is sampled with this pair of contiguous quadrats for n pairs of observations. If the first quadrat is a hit, the second quadrat can be a miss or a hit. If the first quadrat is a hit, there is a probability that the second quadrat is a hit, $p_{1,1}$, and a probability that the second quadrat is a miss, $p_{1,2}$. The reasoning is the same if the first quadrat is a miss. These probabilities can be listed as a matrix P.

$$
\begin{array}{cc}
 & \text{Second quadrat} \\
 & \begin{array}{cc} \text{H} & \text{M} \end{array}
\end{array}
$$

$$
\begin{array}{cc}
\text{First} \quad \text{H} \\
\text{quadrat} \ \text{M}
\end{array}
\begin{bmatrix}
p_{1,1} & p_{1,2} \\
p_{2,1} & p_{2,2}
\end{bmatrix}
$$

Note that $p_{1,1} + p_{1,2} = 1$ and $p_{2,1} + p_{2,2} = 1$. The matrix is read as, "Given a hit in the first quadrat, the probability of a hit in the second quadrat is $p_{1,1}$," and so forth.

If the patches and gaps are randomly mingled, the matrix P will be a *realization of a Markov chain*. To test for random mingling of patches and gaps the procedure is to determine the probabilities of the matrix P, given that the sequence of hits and misses is a realization of a Markov chain, and to compare them to the actual observed probabilities from the sampling.

After sampling with contiguous quadrats with centers separated by a distance d, the quadrats are spaced with the distance between the centers equal to $2d$, and the sampling carried out with these two quadrats. The sampling is carried out for pairs of quadrats separated by 1, 2, 3, ..., t units, although one usually stops at lengths of four to five units.

The values of $p_{1,2}$ and $p_{2,1}$, based on the hypothesis that the series of patches and gaps are randomly mingled, are then estimated. Once $p_{1,2}$ and $p_{2,1}$ are known, $p_{1,1}$ and $p_{2,2}$ are automatically determined. The estimation equations are

$$
\left(\frac{n_2 + n_3 + m_2 + m_3}{p_{1,2}} \right) - \left(\frac{n_1}{p_{1,1}} \right) + \left[\frac{m_1(p_{2,1} - 2p_{1,1})}{p_{1,1}^2 + p_{1,2}p_{2,1}} \right] - \left(\frac{m_2}{p_{1,1} + p_{2,2}} \right)
$$
$$
+ \left(\frac{m_3 p_{2,1}}{p_{2,2}^2 + p_{1,2}p_{2,1}} \right) - \left(\frac{N + M}{p_{1,2} + p_{2,1}} \right) = 0 \quad (5\text{-}4)
$$

and

$$
\left(\frac{n_1 + n_2 + m_1 + m_2}{p_{2,1}} \right) - \left(\frac{n_3}{p_{2,2}} \right) + \left[\frac{m_3(p_{1,2} - 2p_{2,2})}{p_{2,2}^2 + p_{1,2}p_{2,1}} \right] - \left(\frac{m_2}{p_{1,1} + p_{2,2}} \right)
$$
$$
+ \left(\frac{m_1 p_{1,2}}{p_{1,1}^2 + p_{1,2}p_{2,1}} \right) - \left(\frac{N + M}{p_{1,2} + p_{2,1}} \right) = 0 \quad (5\text{-}5)
$$

The estimates of $p_{1,2}$ and $p_{2,1}$ are found by iteration of Eqs. (5-4) and (5-5). In these equations N is the number of quadrat pair observations made using the pair with unit length; M is the number of quadrat pair observations made using the pair with two-unit length; n_1, n_2, and n_3 are the frequencies of HH, HM + MH, and MM, respectively, in the first pair of quadrats; and m_1, m_2, m_3 are the frequencies found with the two-unit-long pair. To begin iteration initial estimates of $p_{1,2}$ and $p_{2,1}$ are $p_{1,2} = n_2/(2n_1 + n_2)$ and $p_{2,1} = n_2/(2n_3 + n_2)$. Pielou (1964) has also derived equations for determining the variances of the estimates.

Pielou (1964) has applied the method to a population of *Antennaria umbrinella* in a 2-square-meter area. Each quadrat pair was 2 centimeters in diameter ($d = 2$ cm), and quadrat pairs of unit lengths of 1, 2, 3, 4, 5, and 6 were used. With each of these six quadrat pairs 200 samples were taken. The observed frequencies of HH, HM + MH, and MM are listed in Table 5-5 for each of the six quadrat pairs.

For the first two quadrat pairs, $N = 200$, $M = 200$, $n_1 = 45$, $n_2 = 20$, $n_3 = 135$, $m_1 = 39$, $m_2 = 28$, and $m_3 = 133$. The rough estimates of $p_{1,2}$ and $p_{2,1}$ are $p_{1,2} = 20/(90 + 20) = .1818$ and $p_{2,1} = 20/(270 + 20) = .0690$. On substituting these two estimates into Eqs. (5-4) and (5-5) the deviations are $-.4085$ and -86.6156. Note that even though the deviation of the estimate of $p_{1,2}$ from the first equation is low, that of $p_{2,1}$ is high. Because both $p_{1,2}$ and $p_{2,1}$ are used in both equations, they are not independent of each other. Final estimates of $p_{1,2}$ and $p_{2,1}$ are .159 and .059, respectively, with deviations 2.9614 and -4.4600. These estimates could be improved, of course, by further iteration.

The expected frequencies of HH, HM + MH, and MM based on the hypothesis of a random mingling of the phases are calculated as

$$HH_n = \frac{p_{2,1}(p_{2,1} + k^n p_{1,2})}{(p_{1,2} + p_{2,1})^2}$$

$$MM_n = \frac{p_{1,2}(p_{1,2} + k^n p_{2,1})}{(p_{1,2} + p_{2,1})^2}$$

and

$$(HM + MH)_n = \frac{2p_{1,2}p_{2,1}(1 - k^n)}{(p_{1,2} + p_{2,1})^2}$$

where n is the number of units length between the centers of the pair of quadrats, and $k = 1 - p_{1,2} - p_{2,1}$. In the example $k = 1 - .159 - .059 = .782$, and for $n = 2$

$$HH_n = \frac{.059[.059 + (.782)^2(.159)]}{(.159 + .059)^2}$$

$$= .1939$$

Of 200 samples sampled with the pair of quadrats separated by two units, 19.39 percent, or 38.79, of them will consist of HH based on the hypothesis of a random mingling of phases. The expected frequencies of HH, HM + MH, and MM are listed in Table 5-5 along with the observed frequencies for all six pairs of quadrats. A chi-square goodness-of-fit test results in a value of $\chi^2 = 8.330$. This value of chi-square is not significant, and there is no reason to believe that the observed probabilities are not a realization of a Markov chain. It is concluded that the patches and gaps of *A. umbrinella* form a random mosaic.

It might also be useful to have estimates of the mean linear dimensions of the patches and the gaps. The linear dimensions of both gaps and

Table 5-5 **THE OBSERVED DISTRIBUTION AND THE EXPECTED DISTRIBUTION BASED ON A HYPOTHESIS OF RANDOMLY MINGLING OF PHASES IN PIELOU'S TWO-PHASE MOSAIC STUDY OF A POPULATION OF *A. umbrinella***

Length of quadrat pair, units of 2 cm	Event	Frequency	
		Observed	Expected
1	HH	45	45.5
	HM, MH	20	17.2
	MM	135	137.3
2	HH	39	38.8
	HM, MH	43	30.7
	MM	133	130.5
3	HH	33	33.5
	HM, MH	43	41.2
	MM	124	125.3
4	HH	27	29.4
	HM, MH	59	49.4
	MM	114	121.2
5	HH	27	26.2
	HM, MH	44	55.9
	MM	129	117.9
6	HH	27	23.7
	HM, MH	54	60.9
	MM	119	115.4

$\chi^2 = 8.330$

SOURCE: From Pielou, E. C., The Spatial Pattern of Two-Phase Patchworks of Vegetation, *Biometrics*, vol. 20, 1964.

patches are exponentially distributed if the pattern is random. The mean linear dimension of a patch is $1/\lambda_p$, and of a gap $1/\lambda_g$, where

$$\lambda_p = \frac{-p_{1,2}}{p_{1,2} + p_{2,1}} \left[\ln(p_{1,1} - p_{1,2})\right]$$

$$\lambda_g = \frac{-p_{2,1}}{p_{1,2} + p_{2,1}} \left[\ln(p_{1,1} - p_{2,1})\right]$$

and the measurement taken in units, 2 centimeters in the example. In this case

$$\lambda_p = \frac{-.159}{.159 + .059} \ln(.841 - .059)$$

$$= .1793$$

$$\frac{1}{\lambda_p} = 5.5757$$

Because the unit is 2 centimeters, the mean dimension of the patches is 11.15 centimeters. The mean dimension of the gaps is 30.05 centimeters. The total area occupied by patches is estimated as

$$a = \frac{\sum (f_{HH} + f_{HM}/2)}{N}$$

where for each quadrat pair f_{HH} is the number of HH events and f_{HM} is the number of HM events. In the example $a = \frac{322}{1200} = .2683$. Roughly 27 percent of the area is covered by patches of *A. umbrinella*.

For a pattern to be truly random, the patch lengths and gap lengths must be exponentially distributed; that is, $1/\lambda_p$ is the mean of the distribution of patch lengths. It is possible for the patches to be randomly dispersed but for the gaps to be aggregated, and vice versa. Aggregation of patches might be due not only to clumping of patches, but also to great variability in patch size due either to variable environmental conditions or to some intrinsic growth threshold of the group of individuals. Perhaps the population grows very large if it reaches a certain minimum size. It is also possible that the patches may have a tendency to be longer in one direction than in another, such as patches of plants following a ridge or valley. If patches or gaps are longer in one direction, the estimates of their mean dimensions are not valid.

5-5   CAUSES OF OBSERVED SPATIAL DISPERSION PATTERNS

A multitude of ecological phenomena gives rise to aggregated spatial patterns. In general regular and random dispersion patterns tend to be much rarer than aggregated patterns. Regular dispersion patterns may sometimes be observed if there is strong territoriality among the individuals of the population and each territory is of about equal size. The most obvious cause of aggregation is probably habitat heterogeneity. It is not the only cause. For example, if our hypothetical female insect of Sec. 5-1 lays egg clusters at random, and the number of larvae hatching from each cluster is also random, the dispersion pattern of the larvae among units is not random, but contagious (aggregated in terms of a continuous habitat). Nor is the pattern random if the number of larvae hatching has a logarithmic distribution or a binomial distribution. Each of these sets of assumptions has been used to derive a contagious distribution to fit observed frequency distributions of individuals among units. It should be emphasized, however, that it is impossible to prove the assumptions behind the distribution true by successfully fitting the distribution to a set of data. Sometimes two different distributions derived under different assumptions will both adequately match the observed frequency distribution. In fact, the negative binomial distribution can be derived under different and sometimes contradictory sets of assumptions (Bailey, 1964).

The reader can also verify, if he wishes, that many of the stochastic models of population growth of the first three chapters and the next two, if applied to a series of discrete units, ultimately result in nonrandom dispersion patterns.

Aggregation may also be an intrinsic part of the behavior of a species. Some bird species form winter feeding flocks, chimpanzees travel in groups, and wood lice congregate under boards and stones. Rohlf (1969), in a theoretical study of the probability of a female finding a mate, found that the advantage to a species with a sparse population of having an aggregated dispersion pattern is quite large. An aggregated dispersion pattern is only slightly disadvantageous if the density is great. Hairston (1959), in work with the soil arthropods in an old field in Michigan, found a decided tendency for rarer species to have more aggregated dispersion patterns than commoner ones. Therefore, there may be a selective pressure on species, particularly rare species, to aggregate.

Aggregated dispersion patterns, therefore, can arise in a multitude of ways. The degree and pattern of aggregation can be measured, but determining the biological reasons behind the observed pattern is not always easy. Inevitably the dispersion pattern depends on the biology of each species and the history of the population.

PART TWO

6

COMPETITION BETWEEN TWO SPECIES

A common definition of competition is: "the utilization of a limited, common resource by two or more species." Unfortunately in many cases a situation originally viewed as a competitive interaction between two species for a limited resource turns out to be not that at all. For example, the famous *Tribolium* competition experiments discussed in the next section were originally viewed as competition for a limited flour supply. It was later found that the different rates of survival of the two species, *Tribolium confusum* and *T. castaneum*, were due to an interaction between differential rates of increase, rates of cannibalism, and accumulation of metabolic wastes. It is difficult to view these factors as competition for a limited resource. In this chapter competition is defined as the "effect of the population density of one species on the rate of increase of a second except if one species serves as food for the other." Competition viewed in this way is rather generalized, but does allow for the multitude of factors that go into competitive interactions. Although competition for a limited resource may be the ultimate factor, competition is usually manifested as various forms of interference, i.e., proximate factors. The models in this chapter measure the effect of one population on another, thus the necessity of this definition.

The chapter begins with a discussion of the famous *Tribolium* experiments carried out by Park and his associates from the forties to the present.   These experiments illustrate some of the empirical findings when two similar species are both introduced into a limited amount of a food resource common to both species.   Following will be a short discussion of some deterministic models and a section on a stochastic model of the competitive process.

6-1   COMPETITION BETWEEN TWO SPECIES OF *TRIBOLIUM*

The experiments of Park and others on competition between *T. confusum* and *T. castaneum* represent a long-term and well-planned series of experiments rightfully well known among ecologists.   The basic experiment consisted of introducing known numbers of each of the two species into differing volumes of whole wheat flour with 5 percent by weight brewer's yeast.   Every 30 days the population in each experiment was completely censused by passing the flour through sieves of bolting cloth.   After censusing, all individuals of all stages were placed in a renewed media of the same volume.   Each experiment was generally replicated, and each replicate followed until one of the two species became extinct.   These experiments were performed with different initial proportions of individuals, with different volumes of media, at different temperatures, at different relative humidities, with different strains of the two species, and with or without the protozoan parasite *Adelina tribolii*.

In all the two-species experiments one species or the other always became extinct (Park, 1948).   Under a given set of conditions it was not always the same species, however.   In one experiment when four individuals of each species were introduced into 8 grams of flour, *T. castaneum* became extinct in 11 of the 15 replicates, and *T. confusum* in the remaining 4.   When the amount of flour was increased to 40 grams and the number of adults of each species introduced was increased to 20, *T. castaneum* became extinct in all 9 replicates.   These two experiments were carried out at 29.5°C and 70 percent relative humidity.

Park (1954) studied the effect of different environmental conditions on the outcome of competition between the two species.   In a series of experiments started with four adults of each species in 8 grams of flour at different temperatures and relative humidities, the probability of one species or the other becoming extinct was dependent on the conditions of the experiment. *Tribolium confusum* was favored at low temperatures and high humidity, and *T. castaneum* at high temperatures and low humidity.   At the lowest temperatures and humidities used, *T. castaneum* reached its physiological limit and did not persist in the single-species controls.   Therefore, its elimination in the two-species experiments at 24°C and 30 percent relative humidity was not entirely due to competition from *T. confusum*.   However, competition from *T. confusum* drastically reduced the mean duration of the species in the two-species cultures.   In the control cultures *T. castaneum* persisted at 24°C and 30 percent relative humidity for an average of 350 days, but when in competition with *T. confusum* survived only 27 days.

The following generalizations can be made about the *Tribolium* experiments:

1   One species or the other will always become extinct, given enough time.
2   Under some sets of conditions one species of the two will always become extinct.
3   Under other sets of conditions the outcome of competition depends on the initial number of each species introduced into the flour.
4   In some cases the outcome of competition is completely probabilistic; one species or the other becomes extinct, with probabilities dependent on the environmental conditions and the initial number of the two species.

A large percentage of the *Tribolium* experiments fall into category *4*, including the first one described.   Because deterministic models cannot account for this observation, their treatment will be cursory, followed by a more complete discussion of a stochastic model.

6-2   DETERMINISTIC MODELS OF TWO-SPECIES COMPETITION

The logistic equation used to represent the growth of a population in a limited environment was $dN/dt = N(r - bN)$.   The parameter b represented the effect of density on the rate of increase of the population.   The logistic equation can be extended to include the effect of not only a species' own density, but also the density of another species.

$$\frac{dN_1}{dt} = N_1(r_1 - b_{11}N_1 - b_{12}N_2)$$

$$\frac{dN_2}{dt} = N_2(r_2 - b_{21}N_1 - b_{22}N_2)$$

The parameters b_{11} and b_{22} represent the effect of the density of a species on its own rate of increase, and b_{12} and b_{21} the effect of the density of the other species on a species' rate of increase. If b_{12} and b_{21} are zero, the density of neither species affects the other, and the equations reduce to simple logistic equations. Essentially the competition equations are merely obvious extensions of the logistic equation and suffer from the same assumptions. The outcome of competition depends on the relative sizes of the parameters b_{11}, b_{22}, b_{12}, and b_{21}, as well as the intrinsic rates of increase r_1 and r_2.

There are four possible outcomes of competition allowed by the competition model:

Case *1*: $r_2 b_{12} > r_1 b_{22}$ and $r_2 b_{11} > r_1 b_{21}$: The limiting value of N_1 is zero and of N_2, K, the carrying capacity of the environment for the second species. Thus the first species will always become extinct.

Case *2*: $r_1 b_{22} > r_2 b_{12}$ and $r_1 b_{21} > r_2 b_{11}$: The limiting values are $N_1 = K$ and $N_2 = 0$. Species 2 will always become extinct.

Case *3*: $r_1 b_{22} > r_2 b_{12}$ and $r_2 b_{11} > r_1 b_{21}$: In this case a stable equilibrium between the two species results, with stable densities

$$N_1 = \frac{r_1 b_{22} - r_2 b_{12}}{b_{11} b_{22} - b_{12} b_{21}}$$

and

$$N_2 = \frac{r_2 b_{11} - r_1 b_{21}}{b_{11} b_{22} - b_{12} b_{21}}$$

Case *4*: $r_2 b_{12} > r_1 b_{22}$ and $r_1 b_{21} > r_2 b_{11}$: In this case there is an unstable equilibrium, and one species or the other will become extinct, depending on the initial densities of each species.

There are no explicit solutions of these differential equations, as there were for the logistic equation. Therefore, it does not appear possible to fit these models to real data easily (although see Vandermeer, 1969). However, the process can be approximated by assuming population growth to consist of discrete jumps in density, by relating the density of each population at time $t + 1$ to its density at time t. The matrix models of Chap. 1 were discrete approximations to a continuously changing population density. With this approximation the density of a population has only discrete values, rather than continuously changing with time. However, as the time interval used becomes smaller, the discrete model becomes a better and better approximation to the true continuous time situation. By using single intervals of time equal to 1 the differential equations become *difference equations*. The difference equation for the logistic process is (Leslie, 1958)

$$N(t + 1) = \frac{\lambda N(t)}{1 + \alpha N(t)}$$

where $N(t)$ is the density at time t, and $N(t + 1)$ is the density at time $t + 1$. The parameter α represents the effect of population density at time t on the theoretically unlimited rate of growth $\lambda N(t)$. Because λ, the finite rate of increase, is the multiplication of the population in one interval of time, and if density does not affect the rate of increase, $N(t + 1) = \lambda N(t)$. However, if density does affect the rate of increase, $\alpha > 0$, then the multiplication of the population is reduced by the factor $1 + \alpha N(t)$. The parameter is easily estimated as $\alpha = (\lambda - 1)/K$. Therefore, by making time consist of discrete units of time the growth of the population can be simulated by calculating $N(t + 1)$, using $N(t + 1)$ as $N(t)$, and repeating the process.

6-2.1   The Leslie-Gower Model of Competition

An extension of Leslie's difference equation model of logistic growth is the two-competing-species difference equation model of Leslie and Gower (1958).

$$N_1(t + 1) = \frac{\lambda_1 N_1(t)}{q_1(t)}$$

$$\alpha_1, \alpha_2, \beta_1, \beta_2 > 0 \tag{6-1}$$

$$N_2(t + 1) = \frac{\lambda_2 N_2(t)}{q_2(t)}$$

where $q_1(t) = 1 + \alpha_1 N_1(t) + \beta_1 N_2(t)$, and $q_2(t) = 1 + \alpha_2 N_2(t) + \beta_2 N_1(t)$. The parameters α_1 and α_2 represent the effect of the density of a species on the rate of increase of the species, but β_1 is the effect of the density of species 2 on species 1, and β_2 is the reverse. The parameters λ_1 and λ_2 are the respective finite rates of increase of the two species, and $N_1(t)$ and $N_2(t)$ are the densities of each species at time t. The two populations will reach equilibrium when $q_1 = \lambda_1$ and $q_2 = \lambda_2$, or when

$$L_1 = \frac{\alpha_2(\lambda_1 - 1) - \beta_1(\lambda_2 - 1)}{\alpha_1\alpha_2 - \beta_1\beta_2}$$

$$L_2 = \frac{\alpha_1(\lambda_2 - 1) - \beta_2(\lambda_1 - 1)}{\alpha_1\alpha_2 - \beta_1\beta_2}$$

The equilibrium points L_1 and L_2 of N_1 and N_2 may represent either a stable equilibrium or an unstable equilibrium.   If the ratios x and y are defined as

$$y = \frac{\beta_1(\lambda_2 - 1)}{\alpha_2(\lambda_1 - 1)} \qquad x = \frac{\beta_1\beta_2}{\alpha_1\alpha_2}$$

when $1 < y < x$, the stationary state is unstable and one species or the other will become extinct, depending on the initial number of the two species. On the other hand, when $x < y < 1$, the stationary point is stable and both species will persist. In addition, if $y < 1$, $x > y$, either species 2 always becomes extinct or, when $y > 1$, $x < y$, the first species always disappears.

The parameters λ_1 and λ_2 are estimated in single-species controls by the cohort method. Then provided the logistic model is correct for each single-species population, $\alpha_1 = (\lambda_1 - 1)/K_1$ and $\alpha_2 = (\lambda_2 - 1)/K_2$. The two remaining parameters β_1 and β_2 can be found by substituting observed values of $N_1(t)$, $N_2(t)$, $N_1(t + 1)$, and $N_2(t + 1)$ into Eq. (5-1) and solving for β_1 and β_2. If $\lambda_1 = 1.3234$, $\lambda_2 = 1.4714$, $\alpha_1 = .002812$, $\alpha_2 = .004029$, $N_1(t) = 33$, $N_2(t) = 9$, $N_1(t + 1) = 39$, and $N_2(t + 1) = 11$, then for β_1

$$39 = \frac{(1.3234)(33)}{1 + (.002812)(33) + \beta_1(9)}$$

$$\beta_1 = .00414$$

and similarly $\beta_2 = .00496$. Better estimates can be found by taking several sets of observed densities and averaging the calculated values of β_1 and β_2. The averaging process should also provide a rough test of the reality of the model, because if the model does mimic the real interaction, the values of β_1 and β_2 arrived at for each set of observed densities will be fairly consistent and will not be correlated with time. In other words the values of β_1 and β_2 will neither increase nor decrease as a consistent trend with time because they are assumed to be constants.

An example of the use of the model is included in the next section.

6-3 THE STOCHASTIC VERSION OF THE LESLIE-GOWER MODEL

Stochastic events can have an important influence on the outcome of a two-species competitive interaction. Leslie and Gower (1958) have derived a stochastic version of the discrete time model of competition. The expected value of the population densities of each species at time $t + 1$ is indicated as $E[N_1(t + 1)|N_1(t), N_2(t)]$ and $E[N_2(t + 1)|N_1(t), N_2(t)]$ read as the expectation of $N_i(t + 1)$, given that the densities of N_1 and N_2 at time t are $N_1(t)$ and $N_2(t)$. These expected densities are the same as the deterministic equations in Eq. (6-1). In order to derive the variance of each expected density at time $t + 1$ it is first necessary to assume that either the birthrate remains constant or that the death rate remains constant. If the birthrate remains constant

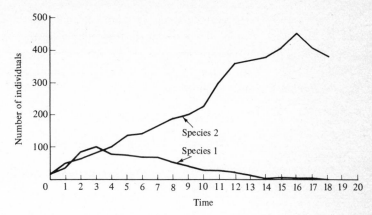

FIGURE 6-1
A Monte Carlo simulation of competition between two hypothetical species using the parameters $\lambda_1 = 2.5$, $\lambda_2 = 2.0$, $\alpha_1 = .0030$, $\alpha_2 = .0025$, $\beta_1 = .0105$, $\beta_2 = 0050$, and $N_1 = N_2 = 20$.

for both species, the variance of the expected density of each species at time $t + 1$ is approximately

$$\text{var}[N_i(t + 1)] = \left(\frac{2b_i}{r_i(t)} - 1\right)\left(\frac{\lambda_i}{q_i(t)} - 1\right)\lambda_i(t)N_i(t) \qquad \lambda_i(t) \neq 1$$

$$= 2b_i N_i(t + 1) \qquad\qquad\qquad \lambda_i(t) = 1 \qquad (6\text{-}2)$$

where i indicates either the first or the second species; i.e., Eq. (6-2) is the variance of the expected density at time $t + 1$ for both species. Also, $\lambda_i(t) = \lambda_i/q_i(t)$, and $r_i(t) = \ln \lambda_i(t) = b_i - d_i(t)$. If the finite rate of increase λ of a species is between 2.0 and 2.5, and the ratio b_i/d_i is between 3.2 and 9.0, the variance is approximately twice the expected value, that is, $2E[N_i(t + 1)]$.

A Monte Carlo simulation for the birthrate constant model is graphed in Fig. 6-1 with parameters $\lambda_1 = 2.5$, $\lambda_2 = 2.0$, $\alpha_1 = .0030$, $\alpha_2 = .0025$, $\beta_1 = .0105$, $\beta_2 = .0050$, and $N_1(0) = N_2(0) = 20$. Assuming that the distribution of possible densities at time $t + 1$ is approximately normal, random normal deviates can be used to simulate stochastic events as follows:

$$E[N_1(t + 1)] = \frac{(2.5)(20)}{1 + (.0030)(20) + (.0105)(20)}$$

$$= 39.37$$

$$E[N_2(t + 1)] = \frac{(2.0)(20)}{1 + (.0025)(20) + (.0050)(20)}$$

$$= 34.78$$

Because λ_1 and λ_2 are both between 2.0 and 2.5, and if we assume for both species that the ratio of the intrinsic birthrate and death rate is between 3.2 and 9.0, then an approximate variance for both species is twice the expected value. If for either species or both these assumptions are not true, Eq. (6-2) must be used. Therefore,

$$\text{var}[N_1(t + 1)] = (2)(39.37) = 78.74$$
$$\text{var}[N_2(t + 1)] = (2)(34.78) = 69.57$$

By drawing two random normal deviates, .016 and $-.266$, two possible densities of the species at time $t + 1$ are

$$N_1(t + 1) = 39.37 + (78.74)^{1/2}(.016)$$
$$= 39.51 \text{ individuals}$$

and

$$N_2(t + 1) = 34.78 + (69.57)^{1/2}(-.266)$$
$$= 32.56 \text{ individuals}$$

The course of the interaction is simulated by using $N_1(t + 1)$ and $N_2(t + 1)$ as $N_1(t)$ and $N_2(t)$, and repeating the process. In Fig. 6-1 the simulation has been carried out up to $t = 20$.

The deterministic stationary states for these two species in the above example are

$$L_1 = \frac{(.0025)(2.5 - 1) - (.0105)(2.0 - 1)}{(.0025)(.0030) - (.0105)(.0050)}$$
$$= 150 \text{ individuals}$$

and $L_2 = 100$. The equilibrium, however, is unstable, because $y = 2.8$ and $x = 7.0$ or $7.0 < 2.8 < 1$. Because the equilibrium is unstable one species or the other will become extinct, depending on the initial number of each.

Leslie and Gower (1958) have also derived the variance of the expected densities at time $t + 1$ of both species if the death rates rather than the birthrates are assumed to be constant. The appropriate equations may be found in their article. If the model is to be applied to a real competitive interaction, the choice between the constant birthrate and the constant death rate versions depends on which model makes more sense biologically, even though neither the birthrate nor the death rate may be constant.

6-3.1   Unstable Equilibrium

If there is an unstable stationary point, the extinction of one species or the other depends not only on the initial number of each species but also on stochastic fluctuations. Therefore, with a starting density combination of the two species there is a probability p that one species will become extinct, and a probability $1 - p$ that it will be the other species. For some initial combinations of the two species p may be very close to 0 or 1.

Leslie et al. (1968) studied the effect of the initial density of each species on the outcome of competition between *T. confusum* and *T. castaneum*. Thirty-five cultures with different initial numbers of the two species were created. In each culture the species to become extinct and the number of days to extinction were noted. At the same time a stochastic simulation of competition between the two species was carried out to compare the observed results with those predicted by the model. The parameters used were

	Tribolium castaneum	Tribolium confusum
λ	1.7507	1.6854
α	.008435	.004129
β	.009734	.042983
d	.228	.170

The constant death rate model was used in computing the variance of the expected values. Sixty replications of the simulation for each initial combination of individuals were made. In some of the replications *T. castaneum* persisted, but in others *T. confusum*. The probabilities of *T. castaneum* persisting for each initial density combination as determined from 60 replicate simulations are shown in Table 6-1. In Table 6-2 the species to become extinct for each initial experimental combination is listed, as well as the number of days to extinction. The observed results are in agreement with the predicted probabilities.

Returning to the numerical example, Leslie and Gower determined the probabilities of extinction using the constant birthrate model of each of their two hypothetical species for 56 initial combinations of the two species, N_1 varying from 10 to 350 individuals, and N_2 from 20 to 175 individuals. Again the probabilities were determined from a set of replicates of each density combination. The initial combinations of individuals giving equal probabilities of a given species becoming extinct are shown in Fig. 6-2. The straight lines were fitted to the data points by a form of regression known as probits. Probits are discussed in most elementary statistics texts. Note that the unstable equilibrium point 100–150 lies on the 50 percent line.

Table 6-1   ESTIMATED PROBABILITIES OF *T. CASTANEUM* PERSISTING OVER *T. CONFUSUM* FOR DIFFERENT INITIAL DENSITIES OF THE TWO SPECIES. THE PROBABILITIES WERE EACH DETERMINED FROM 60 REPLICATE SIMULATIONS USING THE PARAMETERS LISTED IN THE TEXT

Initial density of *T. confusum*	Initial density of *T. castaneum*			
	4	8	16	32
8	.900	.983	1.000	1.000
16	.667	.967	1.000	1.000
32	.333	.917	1.000	1.000
64	.050	.317	1.000	1.000

SOURCE: From Leslie, P. H., Park, T., and Mertz, D. B., The Effect of Varying the Initial Numbers on the Outcome of Competition between Two *Tribolium* Species, *J. Anim. Ecol.*, vol. 37, 1968.

Table 6-2   THE SPECIES c, *T. CASTANEUM*, OR b, *T. CONFUSUM*, WHICH WAS ELIMINATED IN 25 EXPERIMENTS STARTED WITH DIFFERENT INITIAL DENSITIES. THE DAYS TO EXTINCTION OF THE SPECIES BECOMING EXTINCT ARE LISTED IN PARENTHESES

Initial density of *T. confusum*	Initial density of *T. castaneum*				
	4	8	16	32	64
4	b(720)	b(510)	b(360)	b(540)	b(480)
8	b(690)	b(480)	b(510)	b(480)	b(480)
16	b(900)	b(660)	b(600)	b(480)	b(630)
32	Missing	b(810)	b(570)	b(480)	b(510)
64	c(510)	c(510)	b(570)	b(720)	b(870)

SOURCE: From Leslie, P. H., Park, T., and Mertz, D. B., The Effect of Varying the Initial Numbers on the Outcome of Competition between Two *Tribolium* Species, *J. Anim. Ecol.*, vol. 37, 1968.

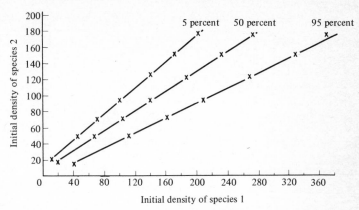

FIGURE 6-2
Lines of equal probability giving a 5-, 50-, and 95-percent chance of species 1 surviving in competition with species 2 for different initial densities of the two species. (*After Leslie and Gower, 1958.*)

Any initial density combination on the 50 percent line gives equal odds to either species of being the species to become extinct. Any density combination on the 95 percent line gives species 2 a 5-percent chance of surviving, or viewed another way a 95 percent chance of extinction. Any point to the right of the line almost surely will lead to the extinction of species 2 under this set of environmental conditions.

These simulations have also been made with the constant death rate model. A graph similar to Fig. 6-2 for the constant death rate model may be found in Leslie and Gower (1958). For relatively small initial densities the probability of extinction or survival of the two species appears to be much the same in both models. However, as N_2 increases in magnitude, the slopes of the lines for the constant death rate model become steadily greater than those for the constant birthrate model, leading to a closer grouping of the lines. The calculated 50 percent line for both models was the same up to about $N_2 = 125$, but above 125 individuals the constant death rate model gave a significantly steeper line than did the constant birthrate model, resulting in an upward curve to the line. However, for initial numbers of N_2 below 125 the predictions of the two models are in close agreement.

6-3.2 Stable Equilibrium

Leslie and Gower also studied an example of a stable equilibrium between two competing species. I might reiterate that no stable equilibrium has ever been found in the *Tribolium* experiments of Park. If the parameters are

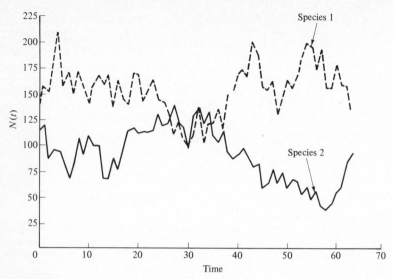

FIGURE 6-3

The fluctuations in density of two hypothetical species around a stable equilibrium in a Monte Carlo simulation using the parameters $\lambda_1 = 2.5$, $\lambda_2 = 2.0$, $\alpha_1 = .00800$, $\alpha_2 = .00625$, $\beta_1 = .0030$, and $\beta_2 = .0025$. (*From Leslie and Gower, 1958.*)

$\lambda_1 = 2.5$, $\lambda_2 = 2.0$, $\alpha_1 = .00800$, $\alpha_2 = .00625$, $\beta_1 = .0030$, and $\beta_2 = .0025$, then $L_1 = 150$ and $L_2 = 100$. Also, $x = .15$, $y = .2667$, and, therefore, there is a stable equilibrium.

In a simulation of this interaction using the constant birthrate model both species rapidly reached their equilibrium densities and began to fluctuate around them. A simulation, ignoring the first six time intervals, is shown in Fig. 6-3. There is a tendency for both species to drift away from their equilibrium points. From $t = 20$ to $t = 35$ the second species started to drift toward zero, but later increased again and by $t = 70$ had again reached its equilibrium point. At the same time the density of the first species increased, drifting in the opposite direction.

6-3.3 One Species Always Becomes Extinct

Only those cases in which one species invariably becomes extinct according to the deterministic model remain to be discussed. If $y < 1$ and $x > y$, the second species will always become extinct if stochastic fluctuations are ignored. This is not necessarily true when the effects of stochastic events are included. Using the parameters $\lambda_1 = 2.5$, $\lambda_2 = 2.0$, $\alpha_1 = .0030$, $\alpha_2 = .0025$,

$\beta_1 = .00375$, and $\beta_2 = .00500$, and the constant birthrate model, 60 replicate simulations were carried out for different initial density combinations of the two species. The probabilities of species 2 becoming extinct for each combination are listed in Table 6-3. Even though the deterministic model predicts species 2 will always become extinct ($y = 1$, $x = 2.5$), stochastic fluctuations have made the outcome probabilistic. Technically under all conceivable conditions the outcome of a competitive interaction is probabilistic, because p can never be exactly 1 or zero, although it may approach 1 or zero so closely that for all practical purposes the outcome is deterministic.

6-3.4 Some Comments

Because the Leslie-Gower model is basically an extension of the logistic equation, the assumptions underlying the model are essentially the same. The effect of density on the rate of increase is linear, there are no time lags, and all the individuals of each population are treated as equivalent.

Leslie (1962) formulated a discrete time stochastic model dividing each species of *Tribolium* into two age groups: immatures and adults. Unfortunately the predicted results of the calculations using this model were not in agreement with the empirical results from experiments of Park. In some cases it is possible to derive a model for each individual situation. Taylor (1968) has created a model of the competitive interaction between *T. confusum* and *T. castaneum*. The model is specific to these two species, but is far more realistic than the generalized Leslie-Gower model. The general models, such

Table 6-3 **PROBABILITIES OF SPECIES 2 BECOMING EXTINCT FOR DIFFERENT INITIAL DENSITIES OF THE TWO HYPOTHETICAL SPECIES. SIXTY REPLICATE SIMULATIONS OF EACH INITIAL COMBINATION WERE CALCULATED**

Density of species 1	Density of species 2			
	150	200	350	400
25	98.33	95.00	81.67	60.00
50	100.00	98.33	93.33	90.00
100	100.00	100.00	100.00	100.00

SOURCE: From Leslie, P. H., and Gower, J. C., The Properties of a Stochastic Model for Two Competing Species, *Biometrika*, vol. 45, 1958.

as the Leslie-Gower model, are most useful in studying the theoretical properties of a competitive interaction. However, because of their generality the models are not particularly good as accurate pictures of a specific situation. To be truly realistic a model should be derived for each individual situation and take into account the multitude of vagaries of the biology of the two species. The methods of deriving specific models are beyond the scope of this book.

A continuous time model of a competitive interaction was derived by Bartlett (1960) and used by Barnett (1962) to simulate competition between *T. castaneum* and *T. confusum*. The results of the simulations are similar to those of the Leslie-Gower discrete time model.

6-4 COMMENTARY

All the competition models discussed in this chapter contain a parameter measuring the influence of the population density of one species on the rate of increase of the second. This parameter is only an empirical measure and does not explain how one species affects another. Usually, the parameter breaks down into several different biological interactions. A good case in point is the competitive interaction between *Tribolium confusum* and *T. castaneum*.

The probability of elimination of one species or the other in the *Tribolium* experiments turns out to be a balance, or a lack of balance, between the intrinsic rates of increase of the two species and differential mortalities because of cannibalism. The interactions in a single-species culture of the c-IV-a strain of *T. castaneum* may be used as an example. In a population started with all adults egg laying is heaviest during the first 4 days, tapering off later as the first larvae hatch. The larvae cannibalize eggs, their voracity increasing with age. As the larvae grow older and increase in density a point is reached where almost no eggs survive. When this group of larvae reaches pupation, eggs can again survive to hatch. The adults also cannibalize eggs, although less so than the larvae.

This type of life history produces waves of pupae, because it creates cohorts of larvae. These waves are about 35 to 40 days apart. However, adult density tends to be much less variable because the adults cannibalize the pupae; the commoner the adults and pupae, the more pupae eaten. As the number of adults increases, the number of pupae surviving to adulthood decreases. This life history pattern varies among strains in each of the two species.

When this strain of *T. castaneum* is in competition with the b-I strain of *T. confusum*, all these factors come into play, not only within species, but between species; i.e., members of *T. castaneum* cannibalize stages of *T. confusum*, and vice versa.   This strain of *T. castaneum* almost always wins in competition over the b-I strain of *T. confusum*.   Park et al. (1968) cite the following deciding factors:

1   *Tribolium castaneum* has a higher rate of increase *r* and is able to make up its losses faster.

2   *Tribolium castaneum* has a more rapid development rate than *T. confusum*.   The more rapid development of the *T. castaneum* larvae means that cannibalism will be strongest toward the eggs of *T. confusum* and also toward the pupae of *T. confusum*.

3   The adults of *T. castaneum* preferentially cannibalize the pupae of *T. confusum* over their own pupae.

All these interactions change with changing strains, environmental conditions, and initial densities of the two species.   The great complexity of even this relatively simple experiment is quite apparent.

One of the most general terms used in ecology is *niche*.   The niche of a species may be defined as the position of the species' population in the community.   The position is defined by the species' relation to the full array of resources and interactions affecting it.   The *competitive exclusion principle* holds that two species with the same ecological niche cannot persist together. Because of the variability of nature, however, two species by the very fact of being different cannot have exactly the same niche.   In fact, even within a population each individual's requirements and interactions are different. Usually it is impossible to define and measure the full array of resources and interactions of a species' population, although in some cases an approximate measure of a niche can be based on one or two of the most important and measurable factors influencing the species.   The competitive exclusion principle is weakened by the undefinability of a niche, and leads to the circular reasoning that two species which survive together must have different niches.   It is worth remembering that the competition equations do allow for a stable equilibrium under the right sets of condition, and that the absence of competitive exclusion does not mean the two species are not competing. Although the theoretical niche construct has been used very successfully to study the coevolution of two competing species, in my opinion it is of limited value in ecology because it is rarely possible to quantify a niche in a realistic way.   The concept has simplicity, but simplicity is not necessarily a virtue if the interaction is inherently complex.

Perhaps the best experimental confirmation of competitive displacement in the field has been provided by Connell (1961), using the two barnacles *Chthamalus stellatus* and *Balanus balanoides* in the intertidal zone of the Isle of Cumbe off the coast of Scotland. These two species are markedly stratified, *C. stellatus* occurring at the high-tide line and *B. balanoides* in a band below the high-tide line. *Balanus balanoides* could not persist in the band where *C. stellatus* lived, because it was not able to survive the desiccation and higher temperatures there. However, *stellatus* could survive in the *B. balanoides* zone, and in fact did survive when they were planted in an open area protected from the encroaching *B. balanoides*. If they were not protected, the larger *B. balanoides* overgrew or undercut individuals of *C. stellatus*. *Chthamalus stellatus* was limited to those places where *B. balanoides* could not survive, even though it could conceivably live elsewhere except for competition from *B. balanoides*.

7

PREDATOR-PREY AND HOST-PARASITE INTERACTIONS

Host-parasite and predator-prey relationships are taken up simultaneously, as many of the observations and models of this chapter are relevant to both types of interactions. In fact, a parasite might be considered to be a very specialized predator. If a model is not relevant to one topic or the other, it will be stated.

The discussion begins with deterministic models, one developed by Lotka and Volterra, and a second created in conjunction with a stochastic model by Leslie. Finally the Nicholson-Bailey model will be presented. Following the deterministic models a stochastic model of the same situation will be given. A discussion of the influence and importance of the spatial patterns of the parasites and hosts follows the sections on deterministic and stochastic models. The final section discusses the functional response of predators to prey density.

7-1 DETERMINISTIC MODELS

Utida (1957) studied the numerical changes in the Azuki bean weevil, *Callosobruchus chinensis*, and its larval parasite, *Heterospilus prosopidis*, in six experimental populations. The initial fluctuations of both species in all six

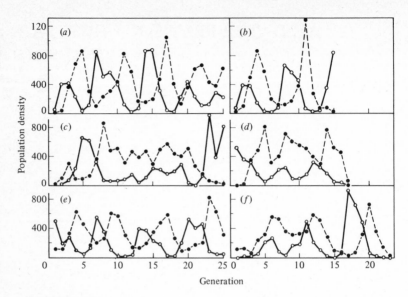

FIGURE 7-1
Fluctuations in the population density of the Azuki bean weevil, *C. chinensis* (○), and its larval parasite, *H. prosopidus* (●), in six replicate cultures. Figure 7-2 is a continuation of the upper left replicate. (*From Utida, 1957.*)

replicates were great. In five of the replicates either the parasite or the host population became extinct in the relatively short time of 15 to 25 generations (Fig. 7-1). In the sixth replicate at the top left of the figure the interaction continued on until the experiment was terminated at the 110th generation (Fig. 7-2). An examination of Figs. 7-1 and 7-2, particularly Fig. 7-2, leads to some tentative conclusions:

1 The initial fluctuations of both species are great.

2 The peaks and low points in the density of the parasite and host alternate. If the host increases in density, the parasite follows suit shortly after. A decline in the density of the host is followed by an inevitable decrease in the parasite population. After the decrease in the parasite density, the host population again increases in density.

3 Figure 7-2 seems to indicate that the severe oscillations of the populations at the beginning of the experiment tend to damp with time, although it also appears that there are strong tendencies for the damping process to be interrupted and the oscillations reestablished.

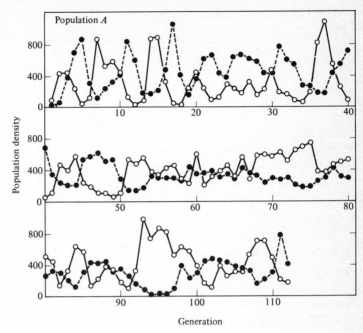

FIGURE 7-2

Fluctuations in density of the Azuki bean weevil, *C. chinensis* (◯), and its larval parasite, *H. prosopidus* (●), in the upper left replicate of Fig. 7-1. (*From Utida, 1957.*)

4 The probability of the extinction of the predator population or the predator and prey populations is great.

Gause (1934) conducted an experiment using a predator, *Didinium nasutum*, and a prey species, *Paramecium caudatum*. When these two species were cultured together in an infusion of oats with no place for *P. caudatum* to hide, the prey population became extinct in all cases, the predators dying of starvation shortly afterward. These results can be interpreted in two ways: (*1*) the predators overexploited their prey, or (*2*) the *Paramecium* population became extinct because of chance fluctuations. Usually the first explanation is advanced, although, as will be seen later, the second possibility is equally tenable.

Gause (1934) performed an experiment similar to the first, but this time provided a sediment for the *Paramecium* to hide in. The prey were exterminated in the clean medium, leading to the extinction of the predator. Later the surviving *Paramecium* from the sediment repopulated the culture.

7-1.1   The Lotka-Volterra Model

The Lotka-Volterra model of the predator-prey relationship is

$$\frac{dN_1}{dt} = (r_1 - b_1 N_2)N_1$$

$$\frac{dN_2}{dt} = (-d + b_2 N_1)N_2$$

where N_1 is the density of the prey population, N_2 is the density of the predator population, r_1 is the intrinsic rate of increase of the prey population, d is the intrinsic death rate of the predator population, and b_1 and b_2 are constants expressing the effect of the density of one species on the rate of growth of the other.   The first equation reads, "The rate of change in density of the prey population with time is a function of the intrinsic rate of increase of the prey minus losses due to the density of the predator population." Similarly, the second equation states, "The rate of change in the density of the predators is equal to a gain due to the density of the prey minus the intrinsic rate of death."

The assumptions implicit in the model are:

1   Neither the prey nor the predator population inhibits its own rate of growth.
2   The environment is completely closed and homogeneous.
3   Every prey or host has an equal probability of being attacked.
4   The predators or parasites have an unlimited rate of increase.
5   Prey density has no effect upon the probability of a prey or host individual being eaten.
6   Predator density has no effect on the probability of a predator or parasite catching prey.

In the Lotka-Volterra model the populations of the predator and prey undergo oscillations of constant amplitude and period.   The predator population is always one-quarter out of phase with the prey population. There is no damping of the oscillations.   If stochastic fluctuations occur, the predator or the predator and prey populations inevitably become extinct. If stochastic events push the two populations away from the deterministic trajectory, both populations begin drifting further and further from the predicted values.   Finally, at the low point of the cycle the population of the predator, and sometimes also the prey, have drifted to such low densities that the chance loss of a few individuals pushes the population over the zero axis, and the population becomes extinct.   If the prey population becomes extinct first, the predator population must follow. The assumptions of the model are not very realistic, and, therefore, the model will not be considered further.

7-1.2   The Leslie Model

Unlike the Lotka-Volterra model the Leslie model assumes that the rate of increase of a predator population has an upper limit, and that the growth of the prey population in the absence of the predator is density dependent. At least initially the remaining assumptions are the same as those of the Lotka-Volterra model. The two differential equations are

$$\frac{dN_1}{dt} = (r_1 - a_1 N_1 - b_1 N_2)N_1$$

$$\frac{dN_2}{dt} = \left(r_2 - \frac{a_2 N_2}{N_1}\right)N_2$$

Note the similarity of this model to the competition equations of Chap. 6 and the logistic process of Chap. 3. The parameter a_1 is the influence of the prey species on its own growth, and b_1 is the effect of the density of the predator population on the population growth of the prey. The growth of the predator population is limited by the function $a_2(N_2/N_1)$, which decreases the rate of increase of the predator population as N_2 increases. As in the competition models the explicit solution of the equations does not exist, and it is necessary to resort to a discrete time approximation.

Working with discrete intervals of time an analogous set of difference equations is

$$N_1(t + 1) = \frac{\lambda_1 N_1(t)}{1 + \alpha_1 N_1(t) + \gamma_1 N_2(t)}$$

$$N_2(t + 1) = \frac{\lambda_2 N_2(t)}{1 + \alpha_2[N_2(t)/N_1(t)]}$$

(7-1)

The values of λ_1 and α_1 are estimated from single-species control experiments, and the finite rate of increase of the predator population, λ_2, may be determined by presenting the predators with an unlimited supply of prey. The values of α_2 and γ_1 are estimated by using Eq. (7-1) and observed values of $N_i(t)$ and $N_i(t + 1)$, as was done in the similar competition equation in Chap. 6.

In the Leslie model both populations approach a stable equilibrium by a series of damped oscillations. In contrast, the Lotka-Volterra model does

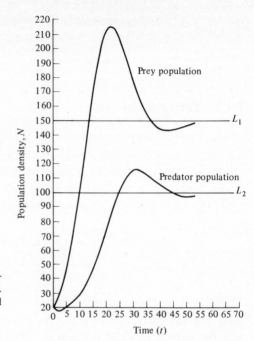

FIGURE 7-3
A simulation of the predator-prey relationship using Leslie's deterministic model.
The parameters of the simulation are listed
in the text.

not damp to an equilibrium.   The predator population density is one-quarter
cycle behind the prey population density.   The stable equilibrium point is

$$L_1 = \frac{\alpha_2(\lambda_1 - 1)}{\gamma_1(\lambda_2 - 1) + \alpha_1\alpha_2}$$

$$L_2 = \frac{(\lambda_1 - 1)(\lambda_2 - 1)}{\gamma_1(\lambda_2 - 1) + \alpha_1\alpha_2}$$

To illustrate the use of this model and some of its properties a numerical
example with parameters $\lambda_1 = 1.2574$, $\lambda_2 = 1.1892$, $\alpha_1 = .0005148$, $\alpha_2 = .2838$,
$\gamma_1 = .0018018$, and $N_1(0) = N_2(0) = 20$ is given in Fig. 7-3.   The stable
stationary state of the prey population is $L_1 = 150$, and of the predators
$L_2 = 100$.

Leslie and Gower (1960) modified Leslie's deterministic model to protect
a percentage of the individuals from the predators, as Gause provided
sediment to protect a proportion of the *Paramecium* from the *Didinium*.
If k percent of the population is exposed to predation $(0 \leq k \leq 1)$, the
population density of the prey population after one interval of time is

$$N_1(t + 1) = \left[\frac{(1 - k)\lambda_1}{1 + \alpha_1 N_1(t)} + \frac{k\lambda_1}{1 + \alpha_1 N_1(t) + \gamma_1 N_2(t)}\right] N_1(t)$$

and of the predator population

$$N_2(t + 1) = \frac{\lambda_2 N_2(t)}{1 + (\alpha_2/k)[N_2(t)/N_1(t)]}$$

The equilibrium density of the predator population is

$$L_2 = \frac{k(\lambda_2 - 1)}{\alpha_2} L_1$$

and the positive root of

$$L_1{}^2 + \left[\frac{1 - (1 - k)\lambda_1}{\alpha_1} + \frac{1 - k\lambda_1}{B} \right] L_1 - \frac{\lambda_1 - 1}{\alpha_1 B} = 0$$

where

$$B = \alpha_1 + \frac{k\beta_1(\lambda_2 - 1)}{\alpha_2}$$

is the stationary state of the prey population. Taking as an example $\lambda_1 = \lambda_2 = 1.25$, $\alpha_1 = .001$, $\alpha_2 = .250$, $\gamma_1 = .050$, and $k = .5$, the stationary state of the prey population is

$$B = .001 + \frac{(.5)(.05)(1.25 - 1)}{.25}$$

$$= .026$$

$$L_1{}^2 + \left[\frac{1 - (1 - .5)(1.25)}{.001} + \frac{1 - (.5)(1.25)}{.026} \right] L_1 - \frac{1.25 - 1}{(.001)(.026)} = 0$$

$$= L_1{}^2 + 389.42L_1 - 9615.38 = 0$$

The positive root of this equation is found from the familiar algebraic expression

$$X = \frac{-b + (b^2 - 4ac)^{1/2}}{2a}$$

where a is the coefficient of the square term, here 1, b is the coefficient of the second term, $L_1(389.42)$, and c is 9615.38. The positive root of the equation above is 23.3, the equilibrium density of the prey population. The equilibrium density of the predator population is

$$L_2 = \frac{.5(1.25 - 1)}{.250} 23.3$$

$$= 11.6 \text{ predators}$$

Some equilibrium densities of the system with different values of k and the above parameters are listed in Table 7-1. If a proportion of the prey is protected, the predator population becomes extinct only if the number of protected prey is 100 percent.

The Leslie model does not mimic the results of Gause's experiments, although there are some resemblances to Utida's. The model may be correct in predicting a series of damped oscillations, as shown at the beginning of this section.

7-1.3   Time Lags

One of the possible reasons for the departure of experimental results from the changes predicted by the Lotka-Volterra and Leslie models is a time lag on the part of the predator in reacting to the increasing or decreasing density of the prey species. The model, because it is closely related to the logistic equation, assumes that the reaction time of one species to a change

Table 7-1 STATIONARY STATES FOR PREDATOR AND PREY POPULATIONS WHEN k PERCENT OF THE PREY POPULATION IS EXPOSED TO PREDATION. THE PARAMETERS FOR THE CALCULATIONS ARE LISTED IN THE TEXT

k	L_1	L_2
.00	250.0	0.0
.05	230.1	11.5
.10	194.0	19.4
.15	155.7	23.4
.20	125.0	25.0
.25	90.7	22.7
.35	50.4	17.6
.50	23.3	11.6
.65	12.8	8.3
.80	8.1	6.5
1.00	4.9	4.9

SOURCE: From Leslie, P. H., and Gower, J. C., The Properties of a Stochastic Model for the Predator-Prey Type of Interaction between Two Species, *Biometrika*, vol. 47, 1960.

in the density of the other is instantaneous. Wangersky and Cunningham (1957) introduced a parameter representing the time lag of the predator population in reacting to a change in the density of the prey population. The differential equation for the prey population is similar to the Leslie model

$$\frac{dN_1}{dt} = [r_1 - a_1 N_1(t) - b_1 N_2(t)]N_1(t)$$

but the differential equation expressing the change in the predator population is

$$\frac{dN_2}{dt} = [-d_2 + b_2 N_1(t - \tau)]N_2(t - \tau)$$

where b_1 is the effect of the predator on prey density, b_2 is the effect of the prey population on the predator, τ is the time lag, and d_2 is the intrinsic death rate of the predator. In contrast to the Leslie model the predator population does not inhibit its own growth.

After some exceedingly complex calculations, Wangersky and Cunningham show that, in general, as the time lag increases the tendency to oscillate becomes greater, until finally for a long time lag the same continuous oscillations as predicted by the Lotka-Volterra model are found.

7-1.4   The Nicholson-Bailey Model

One of the more widely known of the deterministic models of the predator-prey/host-parasite relationship is the Nicholson-Bailey model. Because the model was originally based specifically on a host-parasite system, the terms *host* and *parasite* will be used. Parts of the model are also applicable to a predator-prey situation. Nicholson and Bailey (1935) base their model on the following assumptions:

1   Searching for hosts on the part of the parasite population is completely at random.

2   The hosts are uniformly distributed over an area, and the area is uniform in respect to food and other environmental characteristics.

3   The model has been strictly formulated for an internal parasite, and the host is eaten not at the time of contact but during the next interval of time. Each interval of time is taken as one generation, and the generation times of both the host and the parasite are exactly the same.

4   A parasite can recognize a previously parasitized host. No host ever receives more than one parasite egg.

5   The number of eggs per parasite is unlimited.

6   In terms of predators, the predators are insatiable.

7   The searching area of the parasite is independent of the density of the prey.

8   The parasite finds a constant proportion of the prey in its search area from generation to generation.

If F is the fecundity of the hosts in one generation, R_0 of Chap. 1, and a is the proportion of hosts in the parasite's search area found by the parasite in its lifetime, then there will be F time the density of host individuals in the next generation, of which the proportion $e^{-a(\text{number of parasites})}$ remains unparasitized and survives.   Therefore, if the density of the host or prey at time t is $h(t)$, and the number of parasites is $p(t)$, the density of the host population after one generation is

$$h(t + 1) = Fh(t)e^{-ap(t)} \tag{7-2}$$

and the number of parasites

$$p(t + 1) = Fh(t)[1 - e^{-ap(t)}] \tag{7-3}$$

To test the Nicholson-Bailey model, Debach and Smith (1941) performed an experiment using the housefly, *Musca domestica*, and its pupal parasite, *Nasonia vitripennis*.   The experiment was set up to mimic the assumptions of the model, and so it is important to outline the experimental procedure. Debach and Smith first arbitrarily set the generation time equal to 24 hours and the fecundity of the hosts equal to 2.0.   All forms of mortality to the hosts were eliminated, except the parasites.   To begin, 36 fly pupae, equivalent to two times 18 adults, were distributed in 2 quarts of barley, and 18 parasites were added.   After the parasites had searched for 24 hours the number of parasitized pupae was counted.   After 24 hours 20.6 hosts on the average had been parasitized, and 15.4 had not been parasitized.   To begin the next generation 31 fly pupae were introduced into the barley, two times 15.4 adults; all the original parasites were assumed to have died, leaving the 21 parasites that had emerged from the parasitized hosts.   This procedure was continued for seven artificial generations.   From these data the value of a was found from the first generation to be

$$a = \frac{-\ln[h(t + 1)/h(t)F]}{p(t)}$$

or $a = -\ln(15.36)/18 = .045$.   With this value of a, the number of parasites and hosts after one generation is predicted from Eqs. (7-2) and (7-3) as $h(t + 1) = (2)(18)e^{-(.045)(18)} = 16.015$ flies, and from Eq. (7-3) $p(t + 1) = 19.985$ parasites.   The average observed values in the experiment were $p(t + 1) = 21$ parasites and $h(t + 1) = 15.4$ hosts.   The expected and observed densities

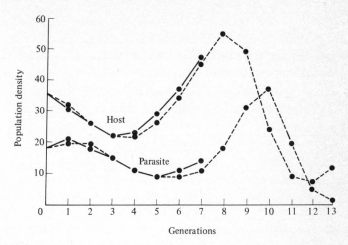

FIGURE 7-4

Results of the interaction between populations of the parasite *N, vitripennis* and the host *M. domestica* in seven successive generations. Solid lines, experiment curves; dotted lines, estimated densities from the Nicholson-Bailey model. (*From Debach and Smith, 1941.*)

of the host and parasite populations after each generation are shown in Fig. 7-4. The following criticisms can be made of this experiment:

1 The generation time was constant and arbitrary.
2 The rate of increase of the host was arbitrarily set.
3 The sources of mortality to the host, other than the parasites, were eliminated.
4 The parameter *a* was estimated from the data and not set a priori.

The Nicholson-Bailey model results in a series of oscillations of ever-increasing amplitude. The increase in the amplitude of the oscillations leads to the eventual extinction of the predator or the predator and the prey populations.

7-1.5 General Comments

The three deterministic models of the predator-prey interaction lead to three different predictions of the probable course of the interaction. The Lotka-Volterra model predicts continuous oscillations, the Leslie model a series of damped oscillations, and the Nicholson-Bailey model a series of increasing oscillations. Although the Nicholson-Bailey model was successfully fit to Debach and Smith's very artificial experiment, the experiments of Utida did not exhibit ever-increasing amplitudes in the fluctuations of the parasites and hosts. The Lotka-Volterra model may be rejected because of its many unrealistic assumptions, although it must be admitted that if stochastic events are taken into consideration the model can mimic some of the properties of the experiments of Gause and Utida. The similarities between the model and the experiments may be real or only fortuitous. Perhaps of all the models the Leslie model was derived from the most realistic assumptions, although it does not predict the extinction of either the predator or the prey population. The stochastic version of the Leslie model presented in Sec. 7-2 can lead in certain circumstances to the extinction of the predator or prey population and may, therefore, be the most reasonable approach. The expansion of the functional response models of predation and parasitism discussed shortly to include numerical population changes may eventually prove to be the most realistic way of mimicking the predator-prey interaction.

7-2 A STOCHASTIC MODEL

A possible explanation for the failure of either the Leslie or the Lotka-Volterra deterministic models to mimic the results of Gause's experiments is the possibility of the chance extinction of the predator and prey populations. To investigate the effects of chance fluctuations the stochastic discrete time version of the Leslie model will be discussed.

The expected population density of the prey and predator populations after one interval of time is

$$E[N_1(t + 1)] = \frac{\lambda_1}{q_1(t)} N_1(t)$$

$$E[N_2(t + 1)] = \frac{\lambda_2}{q_2(t)} N_2(t)$$

where

$$q_1(t) = 1 + \alpha_1 N_1(t) + \gamma_1 N_2(t)$$

$$q_2(t) = 1 + \alpha_2 \frac{N_2(t)}{N_1(t)}$$

which is just another way of writing Eq. (7-2). The variance of the expected values for a constant birthrate model are

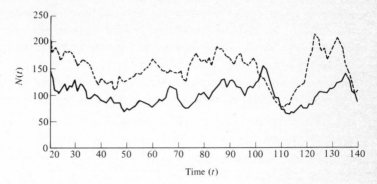

FIGURE 7-5
A simulation of the predator-prey interaction between two hypothetical species using the Leslie-Gower discrete time stochastic model. The simulation is shown from $t = 20$ to $t = 140$. Broken line, density of prey; solid line, density of predators. (*From Leslie and Gower, 1960.*)

$$\text{var}[N_1(t + 1)] = \left|\frac{2b_1}{r_1(t)} - 1\right|\left|\frac{\lambda_1}{q_1(t)} - 1\right|E[N_1(t)] \qquad r_1(t) \neq 0$$

$$= 2b_1 E[N_1(t)] \qquad r_1(t) = 0$$

$$\text{var}[N_2(t + 1)] = \left|\frac{2b_2}{r_2(t)} - 1\right|\left|\frac{\lambda_2}{q_2(t)} - 1\right|E[N_2(t)] \qquad r_2(t) \neq 0$$

$$= 2b_2 E[N_2(t)] \qquad r_2(t) = 0$$

where $r_i(t) = \ln[\lambda_i/q_i(t)]$. The parameters b_i are the intrinsic birthrates and are considered to be constant. Leslie and Gower (1960) have also derived the variances for a death rate constant model. Using the example of the Leslie model in Sec. 7-1, with $\lambda_1 = 1.2574$, $\lambda_2 = 1.1892$, $b_1 = .2577$, $b_2 = .2521$, $\alpha_1 = .0005148$, $\alpha_2 = .2838$, and $\gamma_1 = .0018018$, and steady states of $L_1 = 150$ prey and $L_2 = 100$ predators, a Monte Carlo simulation of the fluctuations of the prey and predator populations is shown in Fig. 7-5 for $t = 20$ to $t = 140$. Both populations fluctuate irregularly around their steady states, and the fluctuations away from the equilibrium can be severe. The fluctuations of both species tend to be positively correlated.

The mean value of the prey population for 800 time intervals in the simulation was 149.74, and for the predator population 97.09. The observed variances and covariances were: variance $N_1 = 635.6$, variance $N_2 = 214.0$, and covariance $N_1, N_2 = 63.6$.

The extinction of either or both of these two hypothetical populations is unlikely under these conditions. Leslie and Gower, however, investigated the effect of changing the parameters γ_1 and α_2, the effect of each population on the other. To be more specific γ_1 may be the voracity of the predator, and α_2 the dependence of the predator population on the density of the prey. With the single population parameters roughly as before, that is, $\lambda_1 = \lambda_2 = 1.25$, $b_1 = b_2 = .25$, and $\alpha_1 = .001$, Leslie and Gower tried three combinations of the parameters α_2 and γ_1, of which the second and third combinations were

Second combination: $\qquad \gamma_1 = .025 \qquad \alpha_2 = .5 \qquad L_1 = 18.5 \qquad L_2 = 9.3$
Third combination: $\qquad \gamma_1 = .050 \qquad \alpha_2 = .250 \qquad L_1 = 4.9 \qquad L_2 = 4.9$

Because of the high values of α_2 and γ_1 the equilibrium densities of both species are low, making chance extinction of either the predator or the prey population a very real possibility. In five simulations of the second combination of parameters starting with $N_1 = 200$ and $N_2 = 15$, the predator became extinct in all cases and did so rapidly. Even though the predator became extinct before the prey population in all five simulations, there is still no guarantee that the prey population will persist. The possibility of the chance extinction of the prey population following the extinction of the predators is still substantial, because of the population's low density at the time of the extinction of the predator.

In the third combination, also started with $N_1 = 200$ and $N_2 = 15$, the predator population became extinct before the prey 9 times, and the prey population before the predator 10 times, for a total of 19 simulations. In 5 of the 10 times the predator became extinct first, the prey population

Table 7-2 **MEAN TIME TO EXTINCTION FOR PREDATOR AND PREY POPULATIONS WHEN A PROPORTION OF THE PREY k IS EXPOSED TO PREDATION**
($\lambda_1 = \lambda_2 = 1.25$; $b_1 = b_2 = .25$; $\alpha_1 = .001$; $\alpha_2 = .250$; $\gamma_1 = .050$)

k	Mean time to extinction of prey	Mean time to extinction of predator
.0	5.8×10^{40}	...
.05	9.8×10^{34}	6.2×10^2
.10	6.0×10^{25}	1.3×10^4
.15	7.4×10^{17}	4.8×10^4
.20	1.2×10^{13}	6.5×10^4
.25	1.5×10^8	1.7×10^4
.35	2.4×10^4	1.5×10^3
.50	3.3×10^2	1.6×10^2
.65	76	58
.80	37	34
1.00	20	21

SOURCE: From Leslie, P. H., and Gower, J. C., The Properties of a Stochastic Model for the Predator-Prey Type of Interaction between Two Species, *Biometrika*, vol. 47, 1960.

had been reduced to such low numbers that the probability of its chance extinction, even though the predators were gone, was great.

If the voracity of the predator is great, the chance of the predator population surviving is small, as was found by Gause. The extinction of the predator or parasite population limited to a given prey species in a small homogeneous habitat appears to be almost certain unless, perhaps, territorial behavior limits the number of predator individuals in a given area.

Leslie and Gower (1960) also considered protecting a segment of the prey population from predation. The mean time to extinction for the predator and prey populations for different values of k, and the parameters $\lambda_1 = \lambda_2 = 1.25$, $\alpha_1 = .001$, $\gamma_1 = .050$, $b_1 = b_2 = .25$, and $\alpha_2 = .250$ are given in Table 7-2. As the proportion of the prey population exposed to predation becomes greater, the time to extinction of both populations decreases. This observation leads to the rather surprising conclusion that the predator population survives longest if the majority of the prey population is protected from predation.

7-3   SPATIAL PATTERNS, REFUGIA, AND THE PREDATOR-PREY INTERACTION

In considering the results of Gause's experiments with *Paramecium caudatum* and *Didinium nasutum* it at first seems surprising that there are any predators and parasites at all. The Leslie-Gower model, if modified to protect part of the host or prey population, indicated that protecting part of the prey population increased the expected lifetimes of both the prey and the predator populations. The partial protection of the host population from predation probably holds the key to the continuance of predator and parasite populations in nature.

The classic observation on the influence of spatial dispersion on the fluctuations of predator and prey populations is the eradication of a species of the genus *Opuntia* in Australia by the moth *Cactoblastis cactorum*. The larvae of the moth feed internally in the cactus, and the situation is analogous to the host-parasite interaction, although the interaction is, strictly speaking, between a herbivore and a plant. The cactus was a serious pest of agricultural land in the early twentieth century and in 1925 covered 30 million acres of land. In 1925 the moth was introduced from Argentina. By 1934 the moth had almost completely wiped out the cactus. Small patches of cactus remained and grew, only to be eventually discovered by moths and eradicated. In the meantime, however, the cactus managed to disperse and form other small isolated patches. At present both species are rare; small patches of cacti grow, are discovered by moths, and are eradicated, but not before dispersal of seeds from this patch has formed other small, isolated pockets. The cactus

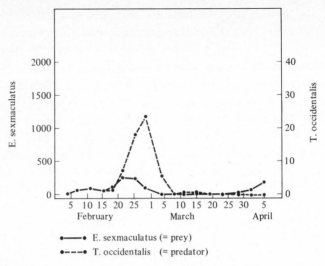

FIGURE 7-6
Densities per orange area of the prey, *E. sexmaculatus*, and the predator, *T. occidentalis*, with four large areas of food for the prey grouped at adjacent, joined positions. (*From Huffaker, 1958.*)

seems to be continually one step ahead of the moth, and as a consequence both the moth and the cactus persist.

To see if a similar situation was true for a predator-prey relationship, Huffaker (1958) performed a series of unique experiments utilizing a predatory mite *Typhlodromus occidentalis* and a prey species the six spotted mite, *Eotetranychus sexmaculatus*. Huffaker created artificial universes of partially exposed oranges. The reader should refer to Huffaker's article for details of the experimental design. First, he introduced 20 prey individuals onto four oranges placed in a square of four, and 11 days later added two female *T. occidentalis*. The predator population quickly increased, almost wiping out the prey population. As a consequence the predators became extinct in about $1\frac{1}{2}$ months (Fig. 7-6). Huffaker then placed four oranges at random in a tray

FIGURE 7-7
Three oscillations in density of a predator-prey relationship in which the predatory mite *T. occidentalis* preyed upon the six-spotted mite *E. sexmaculatus* in a complex environment of petroleum jelly barriers, 120 oranges, and 6 orange feeding areas. The graphic record shows the sequence of densities per orange area, while the pictorial record, charts A to R, shows both densities and positions within the universe. The horizontal line next to each letter shows the period on the time scale represented by each chart. Prey: 0 to 5, nil density, white; 6 to 25, low density, light stipple, 26 to 75, medium density, horizontal lines; 76 or over, high density, solid black. Predator: 1 to 8, one white circle. (*From Huffaker, 1958.*)

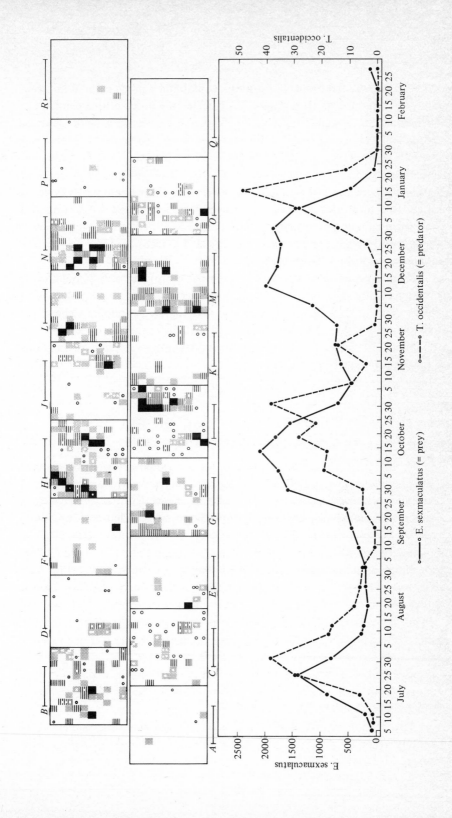

with 40 positions, the other 36 positions containing rubber balls of approximately the same size as the oranges. Again the predator became extinct, but both populations reached considerably higher densities than before. These experiments were continued for different positions, numbers of oranges, and placement of oranges, invariably leading to the extinction of the predator and sometimes also the prey populations. Finally, Huffaker devised a complex experiment involving 120 positions filled with 120 almost completely covered oranges (the total area of exposed orange was equal to six whole oranges) separated by a complex system of partial barriers of petroleum jelly. In this system both populations persisted for almost 8 months (Fig. 7-7) before the predator population became extinct.

It appears that a complex dispersion pattern of the prey population is at least partially responsible for the persistence of predator populations in nature. The prey population may also live in areas, refugia, where the predators cannot go. The sediment in Gause's experiment is an example of a refugium. Refugia serve as sources of individuals to refound populations of prey exterminated by predators.

7-4 THE FUNCTIONAL RESPONSE OF PREDATORS AND PARASITES TO PREY DENSITY

A functional response is the response of an individual predator or parasite to the presence of prey. Do predators attack faster if more prey are present? Are predators more successful in capturing prey at low densities? In contrast, a numerical response measures the change in the population density of predators because of changes in the density of the prey, and vice versa. The effect of prey density on the number of prey eaten by a single predator, the functional response of a predator to prey density, can be of three types (Fig. 7-8). Type 1 is a linear rise in the consumption of prey to a plateau. At the plateau the predator is eating as much as he can or wants. The predator does not have to search for prey. The type-1 curve is most characteristic of a filter-feeding predator. The type-2 curve is a negatively accelerated rise to a plateau and is representative of most invertebrate predators and some fish. The type-3 functional response is a sigmoid curve rising to a plateau. The s shape of the curve is due to learning on the part of the predator and is characteristic of vertebrates.

This section begins with a discussion of Holling's disk model of the functional response of a predator to prey density. A similar model has been developed by Watt (1959).

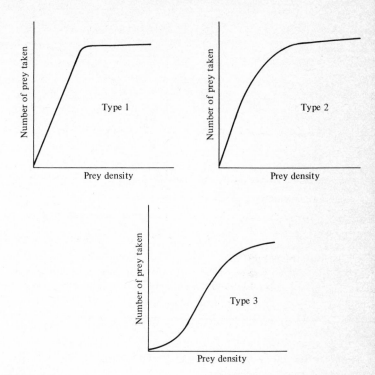

FIGURE 7-8
Diagrammatic representation of the three types of functional responses of a predator
to differences in prey density.

7-4.1   Holling's Disk Model

In deriving his disk model of the functional response of a predator to prey
density, Holling (1959) created an artificial situation in which "prey" were
represented by sandpaper disks 4 centimeters in diameter thumbtacked to a
3-foot-square table.   A blindfolded subject, the "predator" stood in front of
the table and searched for the disks for 1 minute by tapping with her
finger.   As each disk was found, it was removed, set to one side, and the
searching continued.   Each experiment was replicated eight times at densities
of disks ranging from 4 to 256 per 9 square feet.   The results of one such
experiment are shown in Fig. 7-9.   The curve representing the functional
response is type 2.

The number of disks found in a fixed interval of time T_t is a function
of at least two parameters: the rate of searching of the predator, a, and the
time taken to remove a prey once it has been found, b.   The more time it

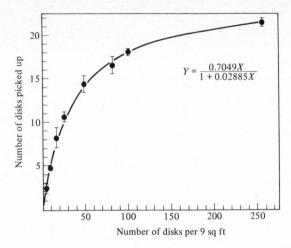

FIGURE 7-9
Functional response of a subject searching for sandpaper disks by touch. Each data point is the average of eight replicates plus 2 SE. (*From Holling, 1959.*)

takes to handle a prey, capturing and eating it, the less time available to the predator for searching.

Holling found that the functional response curve, type 2, could be represented by the equation

$$y = \frac{T_t a x}{1 + abx} \tag{7-4}$$

where y is the number of disks removed, and x is the density of the disks. The constant a, the rate of searching multiplied by the probability of finding a given disk, he called the instantaneous rate of discovery. The parameters a and b may be estimated from a series of data points by transforming Eq. (7-4) into the linear relationship

$$\frac{y}{x} = -aby + T_t a$$

Regressing y/x on y (see Chap. 9) the slope of the regression line, the regression coefficient, is $-ab$, and the intercept is $T_t a$. Because T_t is known, the division of the intercept by T_t yields a, and division of $-ab$ by $-a$ the value of b.

This is the functional response in its simplest form, including only the time consumed in searching and handling prey. Holling termed this the basic functional response. In two later articles Holling (1965, 1966) expanded his model to include time spent eating, digesting, and learning. If a learning

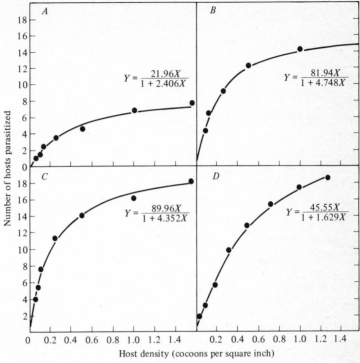

FIGURE 7-10

Functional responses of *D. fulginosus* searching for *N. sertifer* cocoons in the laboratory, based on data of Burnett (1951). (*A*) to (*C*) Experiments conducted at 16, 20, and 24°C, respectively, with different host densities achieved by changing the cage size. (*D*) Experiment conducted at 24°C with size constant. (*From Holling, 1959.*)

component is added to the model, an *s*-shaped curve similar to the type-3 functional response results.

Burnett (1951), in a series of experiments, determined the number of *Neodiprion sertifer* cocoons found by the parasite *Dahlbominus fulginosus* at a given host density. Holling fit his model to four such experiments conducted at different temperatures (Fig. 7-10). The fits of the curves to the data are quite good. The effect of temperature was exerted through the constant *a*, the instantaneous rate of searching, and *b*, the time spent in handling the prey. The parameter *A* increased from .9 to 3.4 to 3.7 as the temperature increased, and *B* decreased from 2.6 to 1.4 to 1.2. Increasing temperature, therefore, increased the searching rate and decreased the amount of time it took the predator to handle the prey. Changing the area of search (Fig. 7-10*D*) modified the response by changing *A* from 3.7 to 1.9. The value of *B* was little changed.

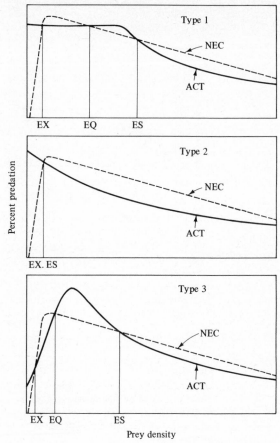

FIGURE 7-11

Population regulatory effects of the three types of functional response to prey density. ACT = actual percent predation; NEC = percent mortality necessary to stabilize populations; EX = threshold density for population extinction; EQ = equilibrium density; ES = threshold density for population escape. (*From Holling, 1965.*)

7-4.2   The Functional Response and the Stability of Prey Populations

In Fig. 7-11 the three functional response types have been plotted as the percentage or proportion of prey individuals destroyed at a given prey density. In the type-1 response a constant percentage of prey individuals is destroyed, until the plateau is reached and the predator cannot cope with the increasing numbers.   In the type-2 response the percent of predation declines continuously because of the negatively accelerated feature of the functional response curve.

In the type-3 response the proportion destroyed rises as prey density increases, because of the initial acceleration of the *s*-shaped response. Following the point of inflection the curve has a negative acceleration, and the proportion destroyed declines continually afterward.

The percent mortality inflicted by the predator is labeled *ACT*. There is also a percent mortality necessary to keep the population from increasing, *NEC*. A possible form of the *NEC* curve for different prey densities is drawn in Fig. 7-11. The *NEC* curve reflects the assumption that a smaller percent mortality from predation is needed at low densities of prey than at inter- mediate densities to keep the population from increasing. At high densities the effect of food shortage increases the mortality to the population, decreasing the necessary mortality due to predation. If *ACT* is greater than *NEC*, prey density will decline, and if *ACT* is less than *NEC*, prey density will increase. If *ACT* equals *NEC*, the density of prey will remain the same from generation to generation. Those points where the *NEC* and *ACT* curves cross are the points, equilibria, where *ACT* = *NEC*. These equilibria may be either stable, i.e., once the prey density reaches that point it does not change; or unstable; i.e., the prey density will move off in one direction or the other, depending on chance.

If the slope of the *ACT* curve is greater at the point of intersection than the slope of the *NEC* curve, the equilibrium point is stable; and if the density of prey individuals reaches this point it should, at least theoretically, stay there. If the slope of the *NEC* curve is greater than the slope of the *ACT* curve at the intersection, the point is unstable and the density of the prey population will move either to a stable equilibrium, to extinction, or to catastrophic increase.

In the type-1 response there are three equilibria, one stable and two unstable, the unstable ones labeled extinction, *EX*, and escape, *ES*. With a type-1-response predator, prey population stability can be achieved if the density of the prey species reaches the *EQ* point. If the prey density passes to the left of *EX*, the extinction of the prey population is assured; and if the prey population density passes *ES*, the population is doomed to a release of sudden growth. Response type 2 has no stable equilibrium point. Apparently, invertebrate predators and parasites with type-2 responses are not able to stabilize prey density.

Response type 3 is virtually the same as the type-1 response, but with an important exception. Although a stable equilibrium point is a theoretical possibility for type-1 and -3 responses, in actual situations it can never be reached and held, because time lags in the response of the predators to changes in prey density, stochastic fluctuations, and density-independent mortality are continually pushing prey density away from the equilibrium point and setting up a series of oscillations. However, some vertebrates, notably birds, may switch from several prey species to a single species if that species becomes unusually common (Tinbergen, 1960). In this way they are more responsive and are better able to damp nascent oscillations in prey density than are type-1 predators.

7-4.3 The Functional Response and Numerical Changes

Holling's disk model and its extension mimic the functional response of a single predator or parasite to prey density. The model has been enlarged to include competition between several parasite individuals in looking for hosts, and competition between parasite progeny if only one individual can survive among several eggs laid on the same host (Griffiths and Holling, 1969). The model is strictly applicable to the parasite-host relationship. Griffiths and Holling based the development of the model on experiments with the parasitic ichneumonid *Pleolophus basizonus* and its host the European sawfly *N. sertifer*. The cocooned larvae of the sawfly are attacked by the parasite.

To study the effect of intraspecific competition in the form of inter-ference between adult parasites on the number of eggs laid per female at different host and parasite densities, experiments were conducted in 4×8 foot cages. Hosts at six densities from 9 to 126 cocoons per cage were presented to parasite females at densities of 2, 4, 7, 10, or 15 per cage. The results of these experiments are shown in Fig. 7-12. There are no significant slopes, indicating that competition among the parasites was not significant even at the highest parasite densities used.

From several similar experiments in the literature Griffiths and Holling concluded that competition among parasites in the form of interference was usually not a significant factor, except at some unrealistically high parasite densities. If P is the number of parasites per square meter, and A is the number of attacks per predator or parasite per unit of time as determined by the Holling disk model, the number of hosts attacked, N_A, per square meter during time T_t is

$$N_A = AT_t P \qquad (7\text{-}5)$$

if there is no reduction in attack or number of eggs laid per parasite by interference from other parasites during the attack process. In the functional response model T_t was the time taken for one complete attack cycle at a given density of hosts. The value of A is the factor which multiplied

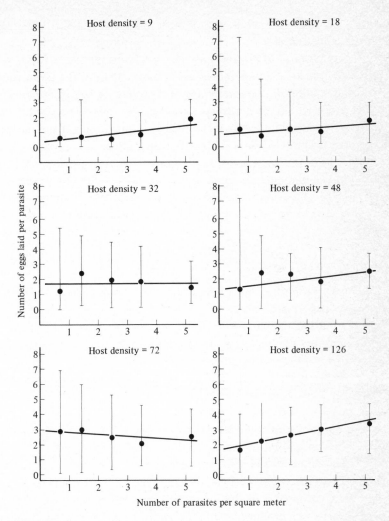

FIGURE 7-12
Relation between number of eggs laid per parasite and parasite density at six host densities, *P. basizonus* attacking cocoons of *N. sertifer*. Each point represents the average of from 15 to 50 replicates, and the vertical lines demonstrate the ranges. (*From Griffiths and Holling, 1969.*)

times T_t equals 1, that is, the number of hosts attacked per unit of time at some host density.

Equation (7-5) is valid only if a host can be attacked only once. If a parasite attacks, the prey individual is removed from the population. Some parasites exhibit *discrimination* and will not lay an egg on a parasitized host. This is not true of *P. basizonus*, however. In this species a female parasite may lay an egg on an individual already parasitized. If the number of eggs per host is compiled as a frequency distribution, i.e., no eggs per host, one egg per host, two eggs per host, etc., this frequency distribution is best fit by the negative binomial distribution (see Sec. 5-1). If the frequency distribution of eggs per host individual is fit by the negative binomial, Eq. (7-5) can be modified to include the effects of imperfect discrimination as

$$N_{HA} = N_0 \left[1 - \left(1 + \frac{N_A}{N_0 k} \right)^{-k} \right] \tag{7-6}$$

where N_A is the number of eggs laid per square meter, N_0 is the number of hosts per square meter, N_{HA} is the number of hosts attacked per square meter, and k is the parameter of the negative binomial. The value of k may be found by the methods listed in Chap. 5, or by alternative methods used by Griffiths and Holling.

The negative binomial gives a good fit to the frequency distribution of eggs per host, because the distribution of eggs in each square meter was not random. Because each female laid one egg per host, the contagious distribution of eggs could be due to the contagious dispersion pattern of the hosts or nonrandom searching on the part of the parasites.

The number of parasites produced can be determined once the number of hosts attacked has been predicted. The effects of superparasitism must also be considered. In the interaction between *P. basizonus* and *N. sertifer* one parasite was produced per parasitized cocoon, regardless of the number of eggs laid on the cocoon. This is a form of contest competition. The first instar parasite larvae will kill each other off until only one parasite remains. The number of parasites produced is simply

$$P = S_P N_{HA} \tag{7-7}$$

where P is the number of parasites, and S_P is a constant representing the survival rate of parasite progeny. In this example 95 percent of the progeny survived to emerge as adult parasites, so $S_P = .95$.

If discrimination is shown by a parasite species, and individuals once attacked cannot be attacked again, the effective host density is reduced by 1 every time a host becomes parasitized. In correcting for the effects of discrimination it is necessary to generate attacks in a series of steps, allowing only enough time in each step so that the resulting reduction in unattacked prey does not significantly affect the attack rate. By making this small decrease in host or prey density equal to $.01N_{0X}$, where N_{0X} is the starting prey density before any attack occurs, the necessary small interval of time T_i is that value during which $N_{HA} \leq .01N_{0X}$. Hence

$$.01N_{0X} \geq N_0 - N_0\left(\frac{N_0 k + AT_i P}{N_0 k}\right)^{-k}$$

$$T_i \leq \frac{N_0 k}{AP}\left[\left(\frac{N_0}{N_0 - .01N_{0X}}\right)^{1/k} - 1\right]$$

(7-8)

Values of $.01N_{0X}$ and T_i satisfying these inequalities are suitable for the computations. In the computations the value of A is substituted into Eq. (7-8), together with k and the initial value of N_0, to yield the number of units of time allowed for the first round of attack. This time is used with Eq. (7-5) to yield the number of prey attacked, N_A, during the interval. Because the time allowed for attack is short, only a small proportion of those attacked can be multiple ones, and the number of hosts attacked, N_{HA}, is equal to N_A. The next round of attack is initiated by subtracting the density of attacked hosts N_A from the starting host density, because with perfect discrimination these prey cannot be attacked again. Setting the new host density equal to N_0, A, T_i, N_A, and N_{HA} are recalculated, and the process continued until the time allotted for attack is used up. Because discrimination is perfect, the only two classes in the frequency distribution of eggs per host are no eggs and one egg. Therefore, the parameter k of the negative binomial cannot be estimated. Instead the number of times a host is visited by a female is recorded, the frequency distribution compiled, and the parameter k estimated from this frequency distribution rather than the frequency distribution of eggs per host.

If competition between larvae in superparasitized hosts is scramble rather than contest, the number of progeny parasites produced is

$$P = S_P N_0\left[\sum_{x=1}^{x=b} xP_x + (b - \alpha)\sum_{x=b+1}^{\infty} P_x\right]$$

(7-9)

where:

> P = number of parasite adults emerging
> b = capacity of each host for yielding adults
> P_x = probability of a host having x eggs as predicted by the negative binomial distribution with mean N_A/N_0 which in this case is

$$P_x = \left| \frac{(k + x - 1)N_A}{x(N_0 k + N_A)} \right| P_{x-1}$$

$$P_0 = \left(1 + \frac{N_A}{N_0 k}\right)^{-k}$$

(7-10)

> α = degree of scramble $(0 \leq \alpha \leq b)$
> S_P = progeny survival rate.

With $b = 1$ and $\alpha = 0$, Eq. (7-9) reduces to Eq. (7-7). Equation (7-10) is a recurrence relation in which consecutively higher values of P_x are calculated from P_{x-1}, starting with $P_{x-1} = P_0$. In scramble competition all parasite larvae develop in the host, even though the host may not be able to support them all. Griffiths and Holling have carried out simulations of the number of parasites produced at different parasite and host densities, using Holling's disk model and the appropriate equations. The steps in the computations are:

1 Determine the parameters.
2 Calculate A for N_0.
3 Is discrimination present?
   Yes = calculate N_{HA} from Eq. (7-5) via Eq. (7-8).
   No = calculate N_{HA} from Eq. (7-6).
4 If discrimination is absent, is the competition between larvae?
   Scramble = calculate P from Eq. (7-9).
   Contest = calculate P from Eq. (7-7).

The results of some simulations are shown in Fig. 7-13. The number of progeny produced per unit area with scramble competition is lessened by competition among the larvae. This is not the case in contest competition. The value of k also has a profound effect on the number of progeny produced. Low values of k indicate a contagious distribution of eggs, while k approaches infinity as the distribution of eggs approaches randomness. In contest competition the more random the search of the parasites the more progeny are produced. With scramble competition, however, the value of k with the highest selective advantage depends very much on the population densities of predator and prey.

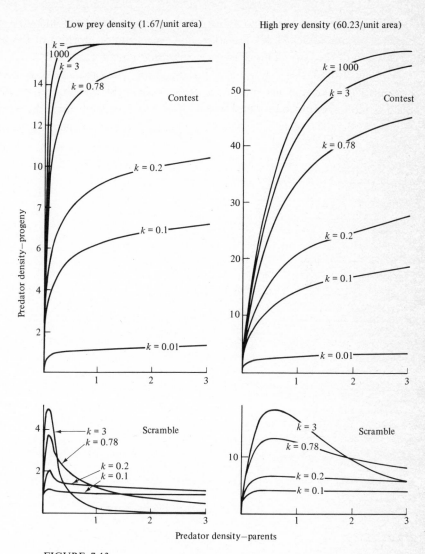

FIGURE 7-13
Progeny production at different predator densities for different degrees of contagion of attack under conditions of contest and scramble competition among progeny. (*From Griffiths and Holling, 1969.*)

In general, the ability to avoid previously attacked hosts increases the number of prey attacked per predator, and hence the proportion of the host population destroyed.   This difference is considerable if k is small.

Conceivably, this extension of the functional response model could be expanded to take into account the rate of increase of the prey population. Such a model would undoubtedly prove far more realistic than the simple models of Secs. 7-1 and 7-2, particularly if the effects of chance fluctuations could be included.

8

EPIDEMICS AND DISEASE

In 1936 a polyhedrosis virus of the sawfly *Neodiprion hercyniae* was accidentally introduced into Canada from Europe. The sawfly itself had originally been introduced from Europe and had become a moderately serious pest in Canada. By 1938 numerous diseased *N. hercyniae* larvae were found in the spruce forests of New Brunswick. The disease spread from 1939 to 1942, and by 1942 occurred almost everywhere the sawfly did. Larval mortality due to the disease varied from 50 to 99 percent, 99 percent being the more common figure (Nielson and Morris, 1964). By 1945 *N. hercyniae* had been reduced far below its original numbers in Canada and has remained there since.

Despite the obvious importance of disease organisms as mortality factors, few really quantitative studies of epidemics in plant and animal populations outside of man have been made. This is possibly due to the rather formidable mathematics involved. The complexity cannot be appreciated from the short account given here. A comprehensive account can be found in an excellent book by Bailey (1957).

Because of the mathematical difficulties involved in deriving models of epidemics these models tend to be in many cases unrealistic, although perhaps no more unrealistic than the similar models of the previous chapters. However, they are realistic enough to warrant more use in ecology than they have had. Out of necessity almost all the examples in this chapter are either numerical, or are taken from human epidemics. The chapter begins with a discussion of the so-called simple epidemic and then moves on to a more general type of epidemic. The third section discusses the modeling of recurrent epidemics and endemic diseases. Concluding the chapter is a discussion of the effect of dispersion patterns on the spread of an epidemic.

8-1 SIMPLE EPIDEMICS

In a simple epidemic a group of individuals completely mingle with each other in a homogeneous area. The disease is introduced by one or more *infectives*, and the other individuals of the population are termed *susceptibles*. The model is concerned only with susceptibles; individuals who for some reason or other are immune are excluded.

The development of a disease can be divided into two periods: a *latent period* during which the disease is not infectious, and an *infectious period* during which the disease can be passed on to others. In a simple epidemic the latent period is nonexistent, and an individual is assumed to be infectious as soon as he is infected. The time elapsing between the exposure to infection and the appearance of symptoms is called the *incubation period*.

In a simple epidemic all susceptibles eventually become infected, and there is no removal of infected individuals from the population. In a human population a mild cold epidemic in a school classroom is an example, although the assumptions are rather restrictive. In insect populations there are several diseases that cause the diseased or dead animal to attach itself to a leaf or blade of grass, such as the common fungal disease of flies, creating a situation roughly analogous to no removal. Dead or diseased plants are rarely removed, and if dead plants are infectious the assumption of no removal may be fulfilled. Because dead or diseased individuals in natural plant and animal populations are often not removed from the population, the assumption of no removal may not be as limiting as it is for human populations.

8-1.1 A Deterministic Model

In the deterministic model of the simple epidemic a single infective individual is introduced into a population of size n. If x is the number of susceptibles and y is the number of infectives at any time, then obviously

$x + y = n + 1$.   First define the parameter β as the rate of contact between individuals in the population.   Then during some interval of time Δt the number of new cases of the disease during Δt is $\beta xy \, \Delta t$.   The number of new cases causes a decrease in the number of susceptible remaining. Therefore, the change in x during Δt is $\Delta x = -\beta xy \, \Delta t$.   If the time interval Δt becomes infinitely small, we eventually arrive at a differential equation

$$\frac{dx}{dt} = -\beta x(n - x + 1) = -\beta xy$$

If the time scale is changed to $\tau = \beta t$, the above differential equation has the solution (Bailey, 1957)

$$x = \frac{n(n + 1)}{n + e^{(n+1)\tau}}$$

If $\beta = .1$, $t = 1, 2, \ldots, 10$, and $n = 20$, then at $t = 3$, $\tau = (.1)(.3) = .3$, and the number of susceptibles left at $t = 3$ is

$$x = \frac{(20)(21)}{20 + e^{(21)(.3)}}$$

$$= .74 \text{ susceptibles}$$

The number of susceptible individuals left for increasing amounts of time τ is shown in Fig. 8-1.

Often the data gathered are in the form of the number of new cases per time period, referred to as the epidemic curve.   The rate at which new cases arise, $-dx/d\tau$, is

$$-\frac{dx}{d\tau} = \frac{n(n + 1)^2 e^{(n+1)\tau}}{[n + e^{(n+1)\tau}]^2}$$

The rate $dx/d\tau$ is negative because it represents the change in x, and x, the number of susceptibles, is decreasing as the epidemic develops.   As the density of the population increases, the development of the epidemic becomes more rapid.   The epidemic will be complete in a much shorter time for a dense population than for a sparse one.   At time $\tau = .2$, if $n = 20$, the rate of appearance of new cases is

$$-\frac{dx}{d\tau} = \frac{(20)(21)^2 e^{(21)(.2)}}{[20 + e^{(21)(.2)}]^2}$$

$$= 78.27$$

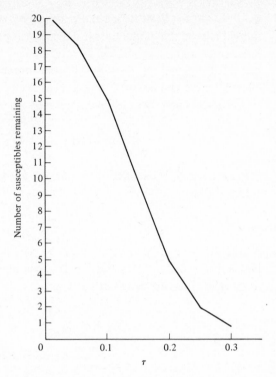

FIGURE 8-1
Number of susceptibles remaining in a simple epidemic for increasing periods of time τ. The original number of susceptibles is 20.

The epidemic curve for $n = 20$ is given in Fig. 8-2. This curve is at its maximum, i.e., the rate of appearance of new cases is greatest, when $\tau = (\log n)/(n + 1)$, $x = 1/(2n + 1)$, and

$$-\frac{dx}{d\tau} = \frac{1}{4(n + 1)^2}$$

or in the example when $\tau = .1426$, $x = 15.5$, and $-dx/d\tau = 110.2$.

In Fig. 8-2 the same epidemic curve as derived by stochastic methods is also plotted. Although the points of the maxima of the two curves are about the same, the stochastic curve is more asymmetrical and tapers off more slowly. In cases in which more than an approximation of a simple epidemic is needed the deterministic model is inadequate. However, when compared to the complexity of the stochastic model, the deterministic model

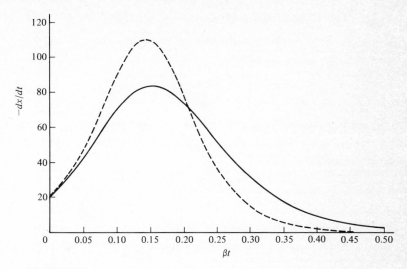

FIGURE 8-2

Comparison of deterministic and stochastic epidemic curves for a simple epidemic with $n = 20$. Broken line, deterministic curve; solid line, stochastic curve. (*From Bailey, 1957.*)

becomes exceedingly attractive. Because of the complexity of the stochastic model it will not be discussed. The reader is referred to Bailey (1957) for a discussion of the stochastic simple epidemic.

8-2 THE GENERAL EPIDEMIC

In a general epidemic infectives are removed from the population either by their death, isolation, or recovery with a subsequent immunity to the disease. A population of size n is considered to consist at any one time of x susceptibles, y infectives, and z individuals who have been removed by death, isolation, or recovery. The value of z at the beginning of the epidemic is zero, and x and y have their own values, except it is preferable that y be initially small. As in the simple epidemic the rate of contact between individuals is β. In addition, individuals are removed at a rate γ. A relative removal rate ρ is defined as $\rho = \gamma/\beta$, or the rate of removal relative to the rate of contact. If at the beginning of the epidemic the number of susceptibles is x_0, an epidemic cannot start unless $x_0 > \gamma/\beta$. Thus $\rho = x_0$ represents a *threshold density* of susceptibles necessary for an epidemic to build up.

8-2.1 The Deterministic Model

As in the simple epidemic the change in the number of susceptibles in the population is postulated to be

$$\frac{dx}{dt} = -\beta xy$$

The rate of change in y, the number of infectives, is equal to the number of new infected individuals being added to the population minus those removed, or

$$\frac{dy}{dt} = \beta xy - \gamma y$$

and the change in the number of individuals removed from the population is equal to the removal rate times the number of infectives in the population, or

$$\frac{dz}{dt} = \gamma y$$

Kermack and McKendrick (1927), by solving the differential equations listed above, have shown that the number of susceptibles in the population is given by the equation

$$x = x_0 e^{-z/\rho}$$

If $\rho = 10$, $x_0 = 100$, and $z = 20$, $x = (100)(e^{-20/10}) = 13.53$ susceptibles left by the time z individuals have been removed from the population. The number z depends on t, time. An approximate expression for the change in number of individuals removed with the change in time is

$$\frac{dz}{dt} = \gamma \left[n - x_0 + \left(\frac{x_0}{\rho} - 1 \right) z - \left(\frac{x_0 z^2}{2\rho^2} \right) \right]$$

which differs from the original expression for dz/dt because the initial value of x, x_0, has been stipulated. If this differential equation is solved (see Bailey, 1957), we will find that the number of individuals removed by some time t is

$$z = \frac{\rho^2}{x_0} \left[\frac{x_0}{\rho} - 1 + \alpha \tanh \left(\frac{1}{2} \alpha \gamma t - \phi \right) \right]$$

where

$$\alpha = \left[\left(\frac{x_0}{\rho} - 1 \right)^2 + \frac{2 x_0 y_0}{\rho^2} \right]^{1/2}$$

and

$$\phi = \tanh^{-1}\left[\frac{1}{\alpha}\left(\frac{x_0}{\rho} - 1\right)\right]$$

Tanh(x) is

$$\tanh(x) = \frac{e^x - e^{-x}}{e^x + e^{-x}}$$

and $\tanh^{-1}(x)$ can be approximated by the series

$$\tanh^{-1}(x) = x + \frac{x^3}{3} + \frac{x^5}{5} + \frac{x^7}{7} + \cdots$$

Both $\tanh(x)$ and $\tanh^{-1}(x)$ are tabulated in many compendiums of tables. If $x_0 = 100$, $y_0 = 1$, $\rho = 20$, and $\gamma = 1.00$, at $t = 5$

$$\alpha = \left[\left(\frac{100}{200} - 1\right)^2 + \frac{(2)(100)(1)}{(20)^2}\right]^{1/2}$$

$$= 4.06$$

$$\phi = \tanh^{-1}\left[\frac{1}{406}\left(\frac{100}{20} - 1\right)\right]$$

$$= 1.82$$

$$z = \frac{(20)^2}{100}\left\{\frac{100}{20} - 1 + (4.06)\tanh[(.5)(4.06)(1)(5) - 1.82]\right\}$$

$$= 32.24 \text{ individuals removed by } t = 5$$

The epidemic curve is given by

$$\frac{dz}{dt} = \frac{\gamma\alpha^2\rho^2}{2x_0}\,\text{sech}^2\left(\frac{1}{2}\alpha\gamma t - \phi\right)$$

where sech(x) is equal to $1/\cosh(x)$, and $\cosh(x)$ is $(e^x + e^{-x})/2$.

The epidemic curve using these approximate equations is a smooth, bell-shaped curve. The equations listed give the number of individuals removed, because it is rarely possible to determine exactly when an individual is first infected, but it is usually known when he has been removed, at least in human populations.

The total size of the epidemic, the total number of individuals removed by the end of the epidemic, is

$$w = \frac{\rho^2}{x_0}\left(\frac{x_0}{\rho} - 1 + \alpha\right)$$

or in the example

$$w = \frac{400}{100}\left(\frac{100}{20} - 1 + 4.06\right)$$

$$= 32.24 \text{ individuals}$$

Of the 100 individuals who were originally susceptible, 32.24 of them will have caught the disease and been removed either by death, isolation, or recovery.

The Kermack and McKendrick approximation to the epidemic curve consistently underestimates the infection rate, and consequently the total size of the epidemic (Kendall, 1956). In order to estimate the exact total size of the epidemic, a shift in outlook is needed. In the deterministic model the epidemic was considered as having started at the beginning and allowed to go to completion. For an exact solution the peak of the epidemic curve is taken as zero, and the number of individuals removed before the peak of the curve and after the peak of the curve are symbolized as ζ_1 and ζ_2. Therefore, the total size of the epidemic is $\zeta_1 + \zeta_2$. The values of these two statistics are the positive and negative roots, $-\zeta_1$ and ζ_2, of the equation

$$y_0 - \zeta + \rho(1 - e^{-\zeta/\rho}) = 0$$

ζ_1 is negative because we are going backward from zero. If $y_0 = 1$ and $\rho = 8$, the two roots satisfying this equation are $\zeta_1 = -12.3$ and $\zeta_2 = 13.0$,

Table 8-1   SOME CHARACTERISTICS OF A GENERAL DETERMINISTIC EPIDEMIC

Intensity of epidemic	Ratio of population to threshold, n/ρ	Percentage of population infectious at central epoch, y_0/n	Percentage of removals occurring before central epoch, $\zeta_1/(\zeta_1 + \zeta_2)$
.00	1.000	.00	50.0
.10	1.054	.13	49.5
.20	1.116	.56	49.1
.30	1.189	1.33	48.5
.40	1.277	2.55	47.9
.50	1.386	4.30	47.1
.60	1.527	6.79	46.2
.70	1.720	10.33	45.0
.80	2.012	15.55	43.4
.90	2.558	24.20	40.8
.95	3.153	31.87	38.3
.98	3.992	40.27	35.4

SOURCE: From Bailey, N. J. T., "The Mathematical Theory of Epidemics," Hafner Publishing Company, New York, 1957.

and the total size of the epidemic is 25.3 individuals.   Note that the distribution is skew.   The intensity of the epidemic i can be defined as the percentage of the total original population of susceptibles n infected and removed during the course of the epidemic, or $i = (-\zeta_1 + \zeta_2)/n$.   In this case $i = (12.3 + 13.0)/100 = 25.3$ percent.   The percentage of the total removals before the peak of the epidemic curve is $-\zeta_1/(-\zeta_1 + \zeta_2)$.   The relationship between the ratio of the original population and the threshold for increase of the epidemic n/ρ and the intensity of the epidemic is shown in Table 8-1.   The percent total removal before the peak of the epidemic curve, and the percentage of the population infectious at the peak of the curve given by

$$y_0(\text{peak}) = n - \rho - \rho \log\left(\frac{n}{\rho}\right)$$

are also listed.

As the size of the original population of susceptibles becomes large relative to the removal rate, the intensity of the epidemic becomes greater and greater.   As the ratio n/ρ increases, the skewness of the epidemic curve also increases.   The greater this skewness the less satisfactory are the approximate equations of Kermack and McKendrick.   In fact, this can be illustrated by using the numerical example calculated to illustrate the Kermack-McKendrick equations.   In that example $n = 100$ and $\rho = 20$, giving a ratio of 5.00.   From Table 8-1 this corresponds to an intensity of almost 100 percent, or almost all of the 100 individuals will become infected even though the approximate equation predicts only 32.24 would.

Because of the asymmetry of the curve when the intensity of the epidemic is great, the curve peaks very rapidly, and the majority of the removals occurs after the number of new removals reported has peaked.   This type of distribution of removals corresponds to the type of epidemic curve often found for notifications of new cases of infectious diseases in human populations.

8-2.2   Parameter Estimation

The estimation of the parameters ρ, β, and γ used in the models of the general epidemic serves first of all to make clear the greatest stumbling block in applying these models to natural populations of organisms other than man. Normally data for man are collected in the form of the number of new cases of a disease per time interval in a household.   Therefore, there is usually an implicit assumption that n is small.   The maximum likelihood estimators discussed below utilize N households of n individuals, n usually varying between 1 and 5. It may be possible to generalize these estimators to animal or plant populations, if the individuals occur in discrete groups, such as larvae in pine cones or mites on a leaf. The use of laboratory experiments eliminates this problem of sampling, but the experiments may have little relation to the real situation in nature.

Data for the estimation of the parameters can be gathered in two ways. The first and simplest is the total number of cases in a household. With these data it is possible to estimate the relative removal rate ρ, but not β or γ. Normally the estimation of ρ will have to suffice, because the estimation of β and γ require that the removal times also be known, a difficult set of data to gather in most natural populations other than man.

To estimate the relative removal rate if the total number of cases in each discrete unit is known, take a group of N units, twigs or leaves, for example, each with exactly n susceptibles in addition to the original infective. To keep things simple a, the original infective, will always be 1. For each discrete unit of n individuals the number of new cases in addition to the first is represented as a_w. In Table 8-2 data on the observed number of new cases of scarlet fever in households of three ($n = 2$ plus the one infective, $a = 1$), from Wilson et al. (1939), are given. The total number of households with three individuals ($n = 2$) is 235. The number of secondary cases a_w, where the possibilities are no new cases, one new case, and two new cases, are listed in the table. The *maximum likelihood score* for ρ, $S(\rho)$, is

$$S(\rho) = \sum_{w=0}^{n} a_w S_w(\rho)$$

Table 8-2 OBSERVED AND EXPECTED NUMBERS OF SECONDARY CASES a_w FOR EPIDEMICS OF SCARLET FEVER IN HOUSEHOLDS OF THREE ($n = 2$)

Number of secondary cases, w	Number of households observed, a_w	Number of households expected, NP_w
0	172	169.5
1	42	46.0
2	21	19.5
Total (N)	235	235.0

SOURCE: From Bailey, N. J. T., "The Mathematical Theory of Epidemics," Hafner Publishing Company, New York, 1957. Based on data from Wilson, E. B., Bennett, C., Allen, M., and Worchester, J., Measles and Scarlet Fever in Providence, R.I., 1929–34 with Respect to Age and Size of Family, *Proc. Am. Phil. Soc.*, vol. 80, 1939.

Bailey (1957) has given a table of values of $S_w(\rho)$ for $a = 1$ and $n = 2, 3, 4$, and 5. In this example $n = 2$; turn to the table for $n = 2$ and find the value of ρ in the table setting $S(\rho)$ equal to zero, or

$$S(\rho) = a_0[S_0(\rho)] + a_1[S_1(\rho)] + a_2[S_2(\rho)] = 0$$

The values of a_0, a_1, and a_2 are 172, 42, and 21, and the values for ρ of $S_0(\rho)$, $S_1(\rho)$, and $S_2(\rho)$ are given in Bailey's table. For this example the score for $\rho = 5.1$ is

$$S(5.1) = (172)(.055233) + (42)(-.076557) + (21)(-.290142)$$

$$= .1917$$

and $S(5.2) = -.0464$. Therefore, the maximum likelihood value of the relative removal rate of this epidemic of scarlet fever is between 5.1 and 5.2, and by interpolation is 5.18. Because this is an estimate of ρ, it is necessary to determine the standard error of the estimate. The variance of the estimate of ρ can be determined by using an *information function*. The information function $I(\rho)$ is approximately

or

$$I(\rho) = \frac{S(\rho_1) - S(\rho_2)}{\rho_2 - \rho_1}$$

$$I(\rho) = \frac{(.1917) - (-.0464)}{5.2 - 5.1}$$

$$= 2.381$$

The reciprocal of $I(\rho)$ is the variance of the estimate; that is, $1/2.381 = .42$. By taking the square root of the variance the standard error of the estimate of ρ is $\pm.66$, and the estimate of ρ is $\rho = 5.18 \pm .66$.

If there are N discrete units of n susceptibles in addition to the original infective, and the number of additional cases besides the first infective are known, and in addition the time interval between the last two observed removals in each discrete unit is known, it is possible to estimate β and γ as well as ρ. The relative removal rate is estimated as before, and then $\hat{\gamma} = (N - a_0)/T$ and $\hat{\beta} = \hat{\gamma}/\hat{\rho}$, where T is the sum of the time intervals between the last two observed removals in each of the families, t; that is, $T = \sum t$. The variances of the estimates of β and γ are

$$\text{var}(\hat{\gamma}) = \frac{\hat{\gamma}^2(\hat{\rho} + n)}{Nn}$$

$$\text{var}(\hat{\beta}) = \frac{\hat{\beta}^2(\hat{\rho} + n)}{Nn} + \frac{\hat{\beta}^2}{\hat{\rho}^2 I(\rho)}$$

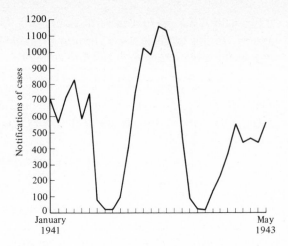

FIGURE 8-3
Monthly notifications of chicken pox in Philadelphia, Pa., from January 1941 to
May 1943.

8-3 RECURRENT EPIDEMICS

In a general epidemic the disease is introduced into a population and rapidly
increases to a peak, later slowly dying away. Eventually all the susceptibles
become infected, or by chance some are not infected. In either case the
epidemic ends with the extinction of the disease. Some human diseases, such
as plague, operate in this manner, but others, as illustrated by monthly notifications of cases of chicken pox in Philadelphia from January 1941 to May 1943
in Fig. 8-3, appear to continually reoccur.

In 1947 an outbreak of the European spruce sawfly, *Neodiprion hercyniae*,
occurred near Sault Ste. Marie, Ontario. This sawfly, originally a pest in
eastern Canada in spruce forests, is now largely controlled by an introduced
viral disease of the larvae and by various predators and parasites. However,
the outbreak at Sault Ste. Marie was free of the disease. In 1950 Bird and
Burk (1961) sprayed a plot of 20 spruce trees in the area with the virus,
and the number of infected and uninfected larvae followed from July 21, 1950,
until the end of 1959. The results of this experiment are shown in Fig. 8-4.
The viral disease of the sawfly larvae clearly has the form of a series of
recurring epidemics, although the cyclic nature of the disease may be due in
part to yearly changes in the density of the larvae.

The explanation for these recurrent epidemics is the addition of
susceptibles to the population, either by birth or by immigration. As a wave

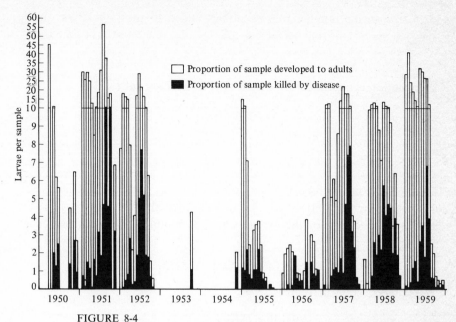

FIGURE 8-4

Larval populations of the spruce sawfly and the prevalence of disease on 20 selected spruce trees in a plot at Sault Ste. Marie, Ontario, where there were no introduced parasites. (*From Bird and Burk, 1961.*)

of the epidemic builds up, the number of susceptibles is reduced below the threshold value.   Later, after the extinction of the disease, the population of susceptibles again rises above the threshold value, and if the disease is reintroduced from outside, or managed to survive within the population as spores, the disease will reoccur and there will be another outbreak.   This series of events tends to be cyclic, the disease continually reoccurring.   Such a disease is termed *endemic*.

8-3.1   Stochastic Simulation of a Recurrent Epidemic

Bartlett (1956) has developed a discrete time model of recurrent epidemics used to mimic with notable success recurrent epidemics, not only in a homogeneous population but also in spatially dispersed populations.   The discrete time model is restricted to diseases having a definite latent period between the original infection and the period of infectiousness of the individual.   If s_t is the number of susceptibles at time t, I_t the number of infectives at time t, and m the number of susceptibles added to the population during each time period, the number of susceptibles after one interval of time is

$$s_{t+1} = s_t + m - I_{t+1}$$

This equation is dependent on the number of infectives at $t + 1$

$$I_{t+1} = s_t P(I_t) + z$$

where

$$P(I_t) = 1 - (1 - p)^i \qquad i = I_t$$

where p is the probability of a contact between any two specified individual members of the population during one time interval sufficient to spread the disease, and z stands for the Monte Carlo deviation added with mean zero and a variance given by

$$\text{var}(I_t) = s_t P(I_t)[1 - P(I_t)]$$

The sequence of calculations is:

1   Calculate I_{t+1} equal to $s_t P(I_t)$.
2   Calculate the variance of I_{t+1}.
3   Multiply the random normal deviate times the standard deviation and add it to I_{t+1}.
4   Calculate s_{t+1}.

For example, with $s_0 = 50$, $I_0 = 2$, $m = 5$, and $p = .05$: (1) $E(I_{t+1}) = (50)(.0795) = 3.97$, (2) $\text{var}(I_{t+1}) = (50)(.0795)(.9205) = 3.66$, (3) $I_{t+1} = 3.97 + (3.66)^{1/2}(.21)$, and if .21 is the random normal deviate, $= 4.36$, and (4) $s_{t+1} = 50 + 5 - 4.36 = 50.64$ susceptibles. A Monte Carlo simulation using these numerical values is shown in Fig. 8-5. The value of m is set rather high in order to lengthen the mean time to extinction of the disease. With $m = 5$ the epidemic went through two cycles, and after 27 periods of time became extinct. The length of time in one interval is equal to the incubation period of the disease. After the demise of this disease, susceptibles will build up at the rate of five per time period, and if infectives are reintroduced and $\rho < n$, the process will reoccur. If there is a small, constant immigration of infectives, the recurrent cycles of the disease will continue indefinitely. Apparently the number of susceptibles added per period was not sufficient to keep the process going indefinitely. Bartlett (1960) has shown that there is a critical point where an increase in the number of susceptibles added (or the size of the population) per time period causes the mean time to the extinction of the disease to become very long. For measles the critical

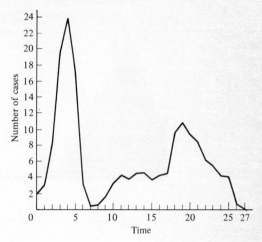

FIGURE 8-5
A stochastic simulation of a recurrent epidemic. The curve represents the number of new cases reported per time period.

community size to insure the perpetuation of the disease indefinitely in the community without reintroduction from outside is about 200,000 individuals.

Bartlett (1956) has used this model to mimic the behavior of the disease ectomelia in a population of mice studied experimentally by Greenwood et al. (1936). Using an incubation period of 4 days, the simulation ran for the equivalent of almost 3 years, the disease becoming extinct after 30 cycles. The results of the simulation mimic closely the actual history of the disease in the mouse population. Bailey (1957) discusses methods of estimating the parameter p from data.

8-4 DISPERSION PATTERNS AND THE SPREAD OF A DISEASE

The mathematical modeling of epidemics involving spatial pattern is complex and has so far proven elusive. In lieu of a continuous time stochastic model, a Monte Carlo simulation of a recurrent measles epidemic performed by Bartlett (1957) will be discussed. The simulations were carried out using a 6×6 grid of squares, and the dispersal of infectives between the squares simulated. In the first simulation the 36 squares were grouped into four blocks of nine squares each labeled A, B, C, and D. In quarters A and D each block was started with 75 susceptibles and 3 infectives, and in quarters B and C there were 75 susceptibles and no infectives. Over a long period

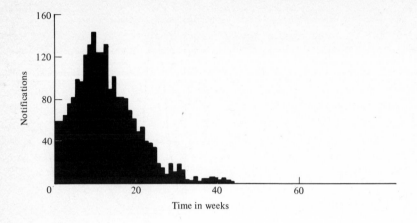

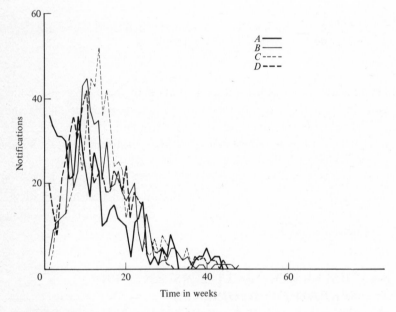

FIGURE 8-6

Simulation of an epidemic in a 6 × 6 grid. The average number of susceptibles for the entire grid is 1,800. The upper figure shows the total notifications per week until the disappearance of the disease; the lower figure shows the weekly notifications for each quarter of the grid. (*From Bartlett, 1957.*)

of time the mean number of susceptibles for the whole grid was 1,800, and the mean number of infectives 54. Details of the model used can be found in Bartlett (1956). The time unit used was 1 week, and the parameters used were: (*1*) rate of infection per week per infected person per susceptible person in one cell (.01), (*2*) rate of entry in one cell per week of new susceptibles (.75), (*3*) rate of removal or recovery of infectives per week per infected person (.5), and (*4*) rate of migration per week per infected person to any one neighboring cell with a common border (.25). In the first simulation with an average total population of 1,800 susceptibles (Fig. 8-6) the epidemic went through one cycle and died out in about 50 weeks. In a second simulation (Fig. 8-7) the equivalent population size was increased to 200,000, and in this case the simulated epidemic persisted for 75 weeks. Bartlett speculates that the critical population size given these parameters is about 250,000 susceptibles, if the disease is to persist indefinitely without reintroduction. The consideration of spatial effects does not appear to alter the behavior of the epidemic much from that expected in a confined population.

8-4.1 Spread of a Disease in a Plant Population

Diseased plants in a plant population are often found to be patchily dispersed, patches of diseased plants being intermixed with gaps of healthy plants. Within each patch a certain proportion of the plants will be diseased, and another proportion healthy. The disease can spread in two ways: (*1*) increase in the proportion π of the total area of the stand occupied by patches, and (*2*) increase in the proportion v of within-patch trees that are diseased. Pielou (1965) presents a method of estimating these two parameters.

To estimate the parameters π and v take one or several transects through the study area, noting the linear sequence of diseased and healthy trees. An example might be

$$\text{D } \underline{\text{D D}} \text{ } \underline{\text{H H H}} \text{ } \underline{\text{D}} \text{ } \underline{\text{H H H}} \text{ } \underline{\text{D D D}} \text{ H}$$

Each sequence of one or more of the same letter is termed a run, and the runs for this particular sequence are underlined. The two end letters of the sequence are ignored. The mean length of each letter is determined. The mean run length of diseased trees is $\bar{d} = (2 + 1 + 3)/3 = 2$. To determine π and v it is necessary to make two assumptions:

1 If a tree is in a patch, the probability that it is diseased is independent of its locality or the condition of the trees next to it. More succinctly diseased and healthy trees within a patch are not segregated.

2 The patches and gaps are randomly mingled (see Chap. 5 for a further discussion of this topic).

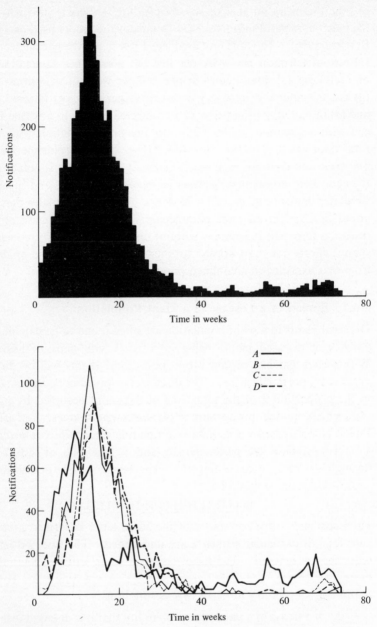

FIGURE 8-7
Simulation of an epidemic in a 6 × 6 grid.   The average number of susceptibles for the entire grid is 200,000.   The upper figure shows the total notifications per week until the disappearance of the disease; the lower figure shows the weekly notifications for each quarter of the grid.   (*From Bartlett, 1957.*)

Assumption *1* can be tested by estimating the product av, where a is the probability that a patch tree is succeeded by another patch tree in the linear transect.   This product is estimated by using the mean run length of diseased trees, and is

$$av = \frac{\bar{d} - 1}{\bar{d}}$$

or if $\bar{d} = 2$, then $av = (2 - 1)/2 = .5$.   The probability of encountering a run of r diseased trees is

$$d(r) = (av)^{r-1}(1 - av)$$

or for $r = 1$, $d(1) = (1 - .5) = .5$.   If there were 100 runs of D, 50 of them would be expected to consist of a single D.   Similarly the expected frequency of runs, 2, 3, 4, ..., n diseased trees can be calculated and compared to the observed frequency distribution by using a chi-square goodness-of-fit test.   If the observed and expected frequency distributions are not significantly different, assumption *1* will be fulfilled.

Pielou (1965) studied the spread of the root rot fungus *Armillaria mellea* in a douglas fir (*Pseudotsuga menziesii*) plantation in British Columbia.   The investigation was carried out in 1959 and 1962 to study the spread of the disease.   The observed and expected frequencies of runs of diseased trees are listed in Table 8-3.   The chi-square tests in both cases are insignificant, and it is concluded that assumption *1* is true for this douglas fir population.

Table 8-3   **OBSERVED AND EXPECTED FREQUENCY DIS-TRIBUTIONS OF RUN LENGTHS OF DISEASED TREES IN A DOUGLAS FIR PLANTATION**

	1959		1962	
Run length	Observed	Expected	Observed	Expected
1	67	59.8	73	70.3
2	18	22.5	32	35.7
3	4	8.5	18	18.2
4	2	...	9	9.2
5	1	...	5	4.7
6	2	5.2	3	4.9
7	2	...	3	
	96		143	

$(a\hat{v}) = .3766$
$\chi^2 = 4.828$
$P(\chi^2, 2 \text{ df}) = .089$

$(a\hat{v}) = .5086$
$\chi^2 = .791$
$P(\chi^2, 4 \text{ df}) = .940$

SOURCE: From Pielou, E. C., The Spread of Disease in Patchily-Infected Forest Stands, *Forest Sci.*, vol. 11, 1965.

The testing of the second assumption is more complex. First, the three parameters a, v, and y must be estimated. The parameter y is the probability that a tree in a gap is succeeded by a tree in a patch in the linear transect. Also, define $b = 1 - a$, $x = 1 - y$, and $u = 1 - v$. These parameters are estimated from the three equations

$$\bar{d} = \frac{1}{1 - av} \tag{8-1}$$

$$\bar{h} = \frac{uy + b}{vy(1 - av)} \tag{8-2}$$

$$h(1) = \frac{bvy + a^2uv}{1 - av} \tag{8-3}$$

where $\bar{h}$ is the mean run length of healthy trees, and $h(1)$ the expected number of runs of a single healthy tree. First, choose a trial value of v between $(\bar{d} - 1)/\bar{d}$ and $2(\bar{d} - 1)/\bar{d}$; substitute this value of v into Eq. (8-1) and solve for a. Second, substitute these values of a and v into Eq. (8-2) and solve for y. Finally, substitute these values of a, v, and y into Eq. (8-3) and compare the result with the observed value of $h(1)$. By adjusting v, values of a and y are found which together with v satisfy Eq. (8-3). With a, v, y, and also b, u, and x estimated, the expected frequency distribution of runs of healthy trees for $r = 1, 2, 3, 4,$ and 5 can be calculated using the equations in Table 8-4. The observed and expected frequency distributions of runs of

Table 8-4   THE PROBABILITY $h(r)$ OF A RUN OF r HEALTHY TREES

r	$h(r)$
1	$\dfrac{1}{1 - av}[bvy + a^2uv]$
2	$\dfrac{1}{1 - av}[bvxy + a^3u^2v + 2abuvy]$
3	$\dfrac{1}{1 - av}[bvx^2y + a^4u^3v + 2abuvy(x + au) + a^2bu^2vy + b^2uvy^2]$
4	$\dfrac{1}{1 - av}[bvx^3y + a^5u^4v + 2abuvy(x + au)^2 + a^2bu^2vy(x + 2au) + b^2uvy^2(2x + au)$ $+ 2ab^2u^2vy^2]$
5	$\dfrac{1}{1 - av}[bvx^4y + a^6u^5v + 2abuvy(a^3u^3 + xa^2u^2 + x^2au + x^3) + a^2bu^2vy(3a^2u^2 + 2aux + x^2)$ $+ b^2uvy^2(3x^2 + 2aux + a^2u^2) + 4ab^2u^2vy^2(x + au) + b^2u^2vy^2(a^2u + by)]$

SOURCE: From Pielou, E. C., The Spread of Disease in Patchily-Infected Forest Stands, *Forest Sci.*, vol. 11, 1965.

healthy trees for 1959 and 1962 in the douglas fir plantation are given in Table 8-5. Neither chi-square value is significant, and it is concluded that the patches and gaps are randomly mingled.

The parameter π can be estimated as

$$\hat{\pi} = \frac{1 - x}{2 - x - a}$$

The estimated values of a, v, x, and π for the study area in both 1959 and 1962 are listed in Table 8-6. Both v and π have increased, indicating that the disease has spread by both an increase in patch area and by an increase in the proportion of infected trees within a patch.

Table 8-5 OBSERVED AND EXPECTED DISTRIBUTIONS OF THE RUN LENGTHS OF HEALTHY TREES IN A DOUGLAS FIR PLANTATION

Run length	1959 Observed	1959 Expected	1962 Observed	1962 Expected
1	23	23.0	54	54.0
2	14	13.1	22	25.3
3	9	9.4	11	14.7
4	8	8.2	17	11.1
5	8	6.1	6	7.4
≥ 6	28	30.3	27	24.6
	90		137	

$\bar{h} = 5.3778$
$\chi^2 = .885$
$P(\chi^2, 3 \text{ df}) = .829$

$\bar{h} = 3.4453$
$\chi^2 = 4.971$
$P(\chi^2, 3 \text{ df}) = .174$

SOURCE: From Pielou, E. C., The Spread of Disease in Patchily-Infected Forest Stands, *Forest Sci.*, vol. 11, 1965.

Table 8-6 ESTIMATES OF THE PARAMETERS a, x, v, AND π

	1959	1962
a	.708	.842
x	.778	.748
v	.532	.604
π	.432	.615

SOURCE: From Pielou, E. C., The Spread of Disease in Patchily-Infected Forest Stands, *Forest Sci.*, vol. 11, 1965.

PART THREE

ANALYSIS AND MODELING OF
NATURAL POPULATIONS

One of the principal goals of ecology is to create models of population change in order to predict the future density of the population. Models may also be formulated to serve as a summary of the set of actions and interactions within a population. Therefore, the purpose of models is not always predictive. Models vary in their complexity, usefulness, methods of derivation, and cost of collecting the data required to use them. The use of a complicated model instead of a simple one for prediction may not be justified if the cost of gathering the data outweighs the added precision gained from its use. Complicated models are almost invariably deterministic, because of the incredible complexity of formulating stochastic models in even the simplest of situations. Stochastic models, however, sometimes give different predictions than deterministic models based on the same assumptions, as has already been seen in earlier chapters of this book.

Watt (1968) proposes the following classification of models:

1 Models that attempt to explain changes in the size of a population on the basis of the relationship between the size of the reproductive segment of the population and the size of the resultant offspring population.

2   Models that use regression methods to relate the numbers in each age group for each year to the numbers in one or more age groups for the previous year.

3   Models that attempt to explain changes in a population in terms of factors intrinsic to the population. Factors extrinsic to the population such as weather or changing temperature are assumed to remain constant. The model considers such factors as change in growth of individuals and mortality to each age group.

4   Models that are complicated but not steady-state models. This category is "open-ended" in that there is no limit to the degree of complexity that can be built into the model. As many environmental factors as required can be included in the model, such as temperature, humidity, competitor species, parasites, diseases, dispersal, and the results of strategies imposed by man.

Section 9-1 is an introduction to the formulation of the fourth type of model. Section 9-2 is a brief discussion of the analysis of variance and its uses in detecting important environmental variables. Section 9-3 deals with regression and correlation. In some cases regression equations can be used to formulate relatively simple models of some particular population process, such as the number of eggs laid with changing temperature. In order to avoid making this chapter a textbook of statistical methods, the discussion of regression and the analysis of variance is kept brief, with emphasis on what each method tests or models, the assumptions made, the computations involved, and the analysis of an example.

Inductive regression techniques cannot always be used to model a situation effectively. Section 9-4 briefly discusses the formulation of models using a mixture of inductive and deductive methods. Because the formulation of models using methods other than regression often entails the use of calculus, only inductive modeling methods are discussed in detail. The derivation of deductive models is extensively discussed in Watt (1968).

9-1   TYPE-4 MODELS

Suppose there is an insect population, a pest perhaps, that annually goes through a series of life history stages, including several larval stages, an egg stage, a pupal stage, and an adult stage. Two sets of data are available, the number of eggs and the probable values of the environmental factors determining how many of the eggs survive to become adults later in the year.

If the environmental factors are designated as variables $X_1, X_2, \ldots, X_n$, the number of adults descended from these eggs will be a function of the environmental variables

$$\text{Number of adults} = f(X_1, X_2, \ldots, X_n) \tag{9-1}$$

Constructing a model of this function directly would be a complicated if not impossible task. The derivation of the model depends on simplifying its construction by somehow breaking the whole process, the survival of eggs to adults, into smaller components, modeling each component, and reassembling the components to make the completed model.

There are two basic ways in which the function represented by Eq. (9-1) can be broken down: (1) a series of components which are added, and (2) a series of components which are multiplied. A model may consist of both additive and multiplicative components. Each of these components is a function of several variables. Given an imaginary situation in which the variable to be predicted is a function of 10 other variables

$$Y = f(X_1, X_2, \ldots, X_{10})$$

the division of the process into components might make the model

$$Y = f(X_1, X_2, X_3) + f(X_4, X_5, X_6, X_7) + f(X_8) + f(X_9, X_{10})$$

The functions represented by each of these components are more amenable to modeling than the single 10-variable function.

Holling's component model of the functional response of a predator to prey density (Holling, 1966) is based on dividing the time period between the eating of one prey to the eating of the next into four components: (1) time taken in a digestive pause after a prey is eaten, TD, (2) time spent searching for a prey, TS, (3) time spent pursuing prey, TP, and (4) time spent eating prey, TE. Each of the four components is modeled separately, and the four components added to give the total time taken by a predator from the eating of one prey to the eating of the next; that is, $T = \text{TD} + \text{TS} + \text{TP} + \text{TE}$.

Changes in population density can often be represented by a series of multiplicative components. In our hypothetical insect population the development of eggs to adults may be broken down into three components: (1) the proportion of eggs surviving to become larvae, (2) the proportion of larvae surviving to become pupae, and (3) the proportion of pupae surviving to become adults. These three components are equivalent to: (1) the probability E of an egg surviving to become a larva, (2) the probability L of a larva surviving to become a pupa, and (3) the probability P of a pupa surviving to become an adult.   Because the probability of two independent probabilistic events both occurring is equal to the product of their probabilities, the probability of an egg surviving to become an adult is $E \times L \times P$.   Therefore, the number of adults A resulting from 100 eggs is $A = 100 \; ELP$.   Each of the components E, L, and P is a function of one or several environmental variables, so that the model might be

$$A = 100[f(X_1, X_2, X_3)][f(X_4, X_5)][f(X_6, X_7, X_8, X_9)]$$

A type-4 model is open-ended.   If the model as formulated is not accurate, it is possible to add more variables and reformulate the components they affect.   Because of the complexity of the computations, and because of the several advantages of computer programming languages, cyclical operations and either/or possibilities to name two, type-4 models are almost always intimately tied in with computer programming and simulation.

The construction of a type-4 model of population change can be illustrated by a simplified, hypothetical example of the way one might go about making such a model and programming it for a computer.   The purpose of the following discussion is to acquaint the reader with the general principles involved, and not to teach computer programming.   Therefore, the program steps are made as simple as possible by using NCE Fortran, a simplified version of Fortran IV, the most commonly used computer programming language.   Imagine an animal population divided into three life history stages: egg, larva, and adult.   The stages are not overlapping, and reproduction occurs only in the adults.   Each stage is also affected by a source of mortality.   The increase in the density of the population in one generation, using the simple deterministic model of Chap. 1, is $N_{t+1} = e^r N_t$.   Note that no distinction is made between the three stages of the animal, and also that r represents the intrinsic rate of increase but says nothing about why animals die or when and how they are born.   In contrast, the type-4 model breaks this process down into several components such as: (1) mortality to eggs, (2) mortality to larvae, (3) mortality of adults before laying eggs, and (4) number of eggs laid per adult.   Say that at the beginning of the generation there are 100 eggs, and that 45 percent of the eggs are destroyed every generation by an egg parasite.   If we represent the variable eggs as EGG (in Fortran a variable can be more than a single letter), and parasitism by PARA, the model of egg survival is

```
1    READ,EGG
2    PARA = .45
3    SECS = EGG – (PARA*EGG)
```

The first *statement*, statement 1, instructs the computer to read the value of EGG, the number of eggs, from some source such as a punched card. The second statement sets the value of parasitism at a constant 45 percent, and the third statement says that the number of larvae, SECS, is equal to the number of eggs minus the number of eggs lost to parasitism. The asterisk indicates multiplication. If 30 percent of the larvae are lost to a predator every generation, PRD, the number of adults is

4   ADLT = SECS − (PRD*SECS)

If 10 adults are lost to a disease every generation

5   ADLT = ADLT − 10.0

Unlike a normal equation, statement 5 is legal in NCE Fortran. The meaning is, "Take the value of ADLT, subtract 10.0, and call this new value ADLT."

The possibility now arises that because of all this mortality the population will become extinct and the computations must stop. If the population has not reached extinction, the process continues. This contingency is provided for by using an IF statement.

6   IF(ADLT)9,9,7

This statement reads, "If the number of adults is less than 0 or equal to 0, go to statement 9." The first two numbers represent less than 0 and equal to 0. Statement 9 is

9   STOP

and the computations stop. The population has become extinct. If the population does not become extinct, and each adult lays 10 eggs, ADLT is greater than 0 and the program goes to statement 7 which is

7   EGG = 10.0*ADLT
8   PRINT,EGG,ADLT
       GO TO 3

Statement 8 prints the number of adults and eggs, and the next statement starts the whole process over again with a new value for the number of eggs as computed by the program. The computations and cycles will continue until either the population becomes extinct or the computer is told to stop.

The calculations in real situations often become long and complicated. In order to follow the general flow of the computations the program steps are usually illustrated as a flow chart. The flow chart for this hypothetical

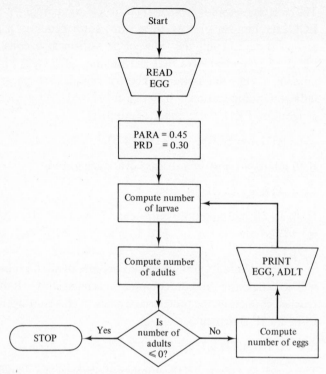

FIGURE 9-1
A flow chart for the programmed model of the hypothetical insect population in the text.

example is shown in Fig. 9-1. The shape of the boxes depends on the type of program step or steps involved, e.g., IF statements are in diamonds, READ and PRINT statements in trapezoids, computations in rectangles, and start and stop instructions in ovals.

This example is meant only for illustration and is grossly unrealistic. Formulating a type-4 model, in summary, consists of four steps:

1 Break the process down into its component steps.
2 Determine the variables significantly affecting the different components.
3 Model the components and reassemble the model, including programming of the model for a computer if necessary.
4 Test the model with real or artificial data to determine if the model accurately mimics the process we originally set out to model. Application of the model to data will suggest improvements in the component models.

9-2   THE ANALYSIS OF VARIANCE

The usefulness of the analysis of variance can best be illustrated by an example. Suppose an experimenter is interested in determining the effect of different temperatures on the number of seeds produced in a plant species. In his experiment he grows five individual plants at each of five different temperatures. The five temperatures are called *treatments*, t, and there are $t = 5$ treatments. At each temperature there are five *replicates*, r, and $r = 5$. The total number of observations, the number of seeds produced by a plant, is $rt = 25$. The 25 plants are allocated randomly among the treatments, five to each treatment. Each individual observation is designated by X_{ij}, where there are $i = 1, 2, ..., t$ treatments, and $j = 1, 2, ..., r$ replicates. The total number of observations is $N = 25$ in the example. One possible outcome of the experiment might be

	Temperature, °C						
	$20°$	$30°$	$40°$	$50°$	$60°$	$\bar{x}_{.j}$	
Seeds per	68	57	58	59	62	60.8	
plant	76	59	56	57	66	62.8	$t = 5$
	67	60	58	57	65	61.4	$r = 5$
	67	59	57	58	63	60.8	$N = 25$
	71	61	56	57	64	61.8	
$\bar{x}_{i.}$	69.8	59.2	57.0	57.6	64.0	$\bar{x}_{..} = 61.52$	

The question of primary interest to the experimenter is whether or not the mean number of seeds set per plant differs at different temperatures, i.e., whether or not changes in temperature affect seed production. Each individual observation X_{ij} is the sum of three components: the mean value over all the treatments, μ, a component due to the effect of the ith treatment temperature, a_i, and an error term for each observation, e_{ij}. The distribution of the error terms is the same for all the individuals and is normally distributed with a mean of zero and a variance σ_e^2. Symbolically the observation X_{ij} is

$$X_{ij} = \mu + a_i + e_{ij}$$

If the temperatures, the treatments, do not affect seed production differently, all the a_i terms are zero and the model reduces to $X_{ij} = \mu + e_{ij}$. The purpose of this *one-way analysis of variance* is to check the hypothesis that all the effects of the treatments are zero; that is, $H_0: a_1 = a_2 = a_3 = a_4 = a_5 = 0$. The notation H_0 : stands for *null hypothesis*. The *alternative hypothesis* is H_1: at least one nonzero a_i.

Now suppose that the experimenter is interested in the effects of these five temperatures and none other. In this case the effects a_i are fixed and constant. This is called a model-I analysis of variance. Often, however, the treatments are only a random sample from a large number of possible treatments. If the experimenter is interested in the effects of temperature on seed production in general and chose these five treatments to study, the a_i are then a random sample from a large number of possible a_i. Therefore, for this situation it is the variance of the a_i that is of interest, rather than specific treatment effects. This is a model-II analysis of variance. For convenience of notation let the mean of the observations for each treatment be $\bar{x}_{i.}$ and the mean of each replication $\bar{x}_{.j}$. The mean of all the observations is $\bar{x}_{..}$. The sum of the squared values of all the observations is

$$\text{SS total} = \sum_i^t \sum_j^r X_{ij}^2$$

We can also compute a sum of squares for the mean over all the observations as

$$\frac{\sum_i \sum_j (X_{ij})^2}{N}$$

and if this quantity is subtracted from the sum of squares total and divided by $N - 1$, the variance of the observations results (see Chap. 2). The degree of freedom is equal to the number of observations. There is a total of 25 degrees of freedom in the sample. However, because the mean is estimated before the variance is calculated, 1 degree of freedom is lost, and the variance is computed by dividing by $N - 1$ rather than N.

The total sum of squares may be broken down into components for the overall mean, a sum of squares for treatments, and an error sum of squares. The sums of squares are additive and add up to the total sum of squares. The same is true of the degrees of freedom. The following statements are proved in most elementary statistics texts. The sum of squares for the error terms is estimated as

$$\sum_i \sum_j (X_{ij} - \bar{x}_{i.})^2$$

When this quantity is divided by its degrees of freedom $t(r - 1)$, the result is termed the *mean square* of the errors. The mean square is an estimate of the variance of the error terms, σ_e^2. If the treatment effects are all zero, the variance of the error terms, that is, σ_e^2, can also be estimated by the sum of squares for treatments

$$r \sum_i (\bar{x}_{i.} - \bar{x}_{..})^2$$

This sum of squares has $t - 1$ degrees of freedom, and again a mean square is found by dividing the sum of squares by the degrees of freedom. If there are no treatment differences, the mean square is also an estimate of the error or *residual* variance. If, however, there are treatment differences, the mean square represents the error variance plus an additional component due to significant treatment effects. If the experiment is a model-I analysis of variance, the treatment mean square represents

$$\sigma_e^2 + \frac{r \sum a_i^2}{t - 1}$$

but if it is a model-II analysis

$$\sigma_e^2 + r\sigma_a^2$$

In the model-II analysis of variance the mean square consists of two variances, the error variance and the variance of the treatment effects times r. If the treatments do not affect the individual observations, the *variance component* $r\sigma_a^2$ is 0.

If there are no significant treatment effects, the mean square for treatments and the mean square for errors are equal. The equality of the two may be tested by an F *test*. An F test tests for the equality of two variances and is

$$F = \frac{\text{treatment mean square}}{\text{error mean square}}$$

and is not significantly different from 1, if the mean squares are equal. If there are significant treatment effects, the treatment mean square will be larger than the error mean square, and F will be greater than 1. To perform the test enter a table of the F distribution, with the number of degrees of freedom of the treatment sum of squares as a numerator, and the number of degrees of freedom of the error sum of squares as a denominator, and find the tabulated value of F for some given probability of significance. If one wishes to be 95 percent certain, then find the value of F for the given degrees of freedom under .05 in the table. If the observed value of F is larger than this number, the probability of an F value this large arising by chance is less than 5 percent. If the F value is larger than the tabulated F value, the null hypothesis is rejected, and it is concluded that at least one of the treatment effects is not zero, provided the model is model I.   If the analysis is model II, the conclusion is that the variance component $\sigma_a^2 r$ is significantly different from zero.

9-2.1   Computations in the One-Way Analysis of Variance

If the number of replicates for each treatment are equal, the computational equations for the treatment sum of squares (SS) is

$$\text{SS treatment} = \frac{\sum_i (\sum_j X_{ij})^2}{r} - \text{SS mean}$$

Because the sums of squares are additive, the sum of squares for the error is SS total − SS mean − SS treatments.   If sample sizes are unequal, the sum of squares for the mean is

$$\text{SS mean} = \frac{(\sum_i \sum_j X_{ij})^2}{\sum_i r_i}$$

and the treatment sum of squares is

$$\text{SS treatments} = \sum_i \frac{(\sum_j X_{ij})^2}{r_i} - \text{SS mean}$$

The double summation sign

$$\sum_{i=1}^{t} \sum_{j=1}^{r} X_{ij}$$

states that all values of X_{ij} in the matrix are summed.   More formally the observations in each column or replicates, $j = 1$ to r, are summed, and the sums of each column are added, $i = 1$ to t.   Similarly

$$\sum_{i=1}^{t} \left(\sum_{j=1}^{r} X_{ij} \right)^2$$

indicates that the squared column totals are added.

Evans (1951) measured the distance moved by an individual chiton, *Lepidochitona cinereus*, in 10 minutes under three different conditions of illumination.   The data are listed in Table 9-1.   The analysis of variance is listed as a table.

Source	Degrees of freedom	Sum of squares	Mean square	F
Total	45	1572.58		
Mean	1	1063.85		
Treatments	2	346.78	173.39	44.92*
Error	42	161.95	3.86	

* Significant at the 1 percent probability level.

With 2 and 42 degrees of freedom and a 99 percent probability level, the tabulated F value is about 5.18.   The observed value is much larger than this, and it is concluded that differences in lightings do have a significant effect on the distances moved by the chiton.   Because this is a model-I analysis of variance, the treatments chosen are the only ones of interest, it is concluded that at least one and perhaps all three of the treatment effects is significantly different than zero.

The indiscriminate use of the F statistic test in the analysis of variance can be very misleading.   Two statistical populations subjected to two different treatments of one sort or another will always have different means, although in many cases the difference between two means may be minuscule. In other words, no two things in the universe are identically equal.   It is always possible to arrive at a significant F statistic, provided we use enough

Table 9-1 **DISTANCES IN CENTI-METERS MOVED IN 10 MINUTES BY THE CHITON *L. CINEREUS* UNDER DIFFERENT CONDITIONS OF ILLUMINATION** $(t = 3, \ r = 15)$

	Treatments	
Sunlight	Artificial light	Darkness
10.6	5.7	0.0
5.4	3.6	0.0
8.0	3.6	4.0
8.5	6.3	2.6
6.3	5.0	0.0
11.0	4.6	0.0
4.7	4.1	2.0
10.3	7.3	0.9
8.6	6.0	0.5
8.0	4.3	0.0
4.3	2.0	0.0
12.3	6.2	4.5
10.6	8.7	2.0
5.4	4.2	1.1
8.6	3.9	3.1
$\sum^r$ 122.6	75.5	20.7

SOURCE: Evans, F. G. C., An Analysis of the Behavior of *Lepidochitona cinereus* in Response to Certain Physical Features of the Environment, *J. Anim. Ecol.*, vol. 11, 1953.

replicates, even if the means of the two treatments differ by a minute and unimportant amount. The important part of an analysis of variance is not the F statistic, but the relative sizes of the mean squares which estimate the magnitude of the various components and parameters of the analysis of variance model. The most significant question of the analysis of variance is not whether differences in treatments exist, but how large the differences are. Therefore, the real object of an analysis of variance is to estimate the differences between treatment means, and to place confidence intervals about the estimates so that we can say that the true difference between the means, which is unobservable and must be estimated by a sample, falls in a given interval with some predetermined probability.

In the chiton experiment the mean distances moved by the chitons under the three different light treatments are: (1) sunlight $= \bar{x}_1 = 8.1733$ centimeters, (2) artificial light $= \bar{x}_2 = 5.0333$ centimeters, and (3) darkness $= \bar{x}_3 = 1.3800$ centimeters. Designate the observed absolute difference between the means of two treatments as d_{ij}, for example, $|\bar{x}_1 - \bar{x}_2| = 3.1340$. Similarly $d_{1,3} = 6.7933$, and $d_{2,3} = 3.6533$. Estimates of the differences between means are simply the differences between the estimates of the means of any two treatments. In order to place a confidence interval about our estimate of the difference between means first calculate the *least significant difference*, the lsd

$$\text{lsd} = t_{\alpha,\,\text{df}} \sqrt{s_e^2 \left(\frac{1}{r_1} + \frac{1}{r_2}\right)}$$

where s_e^2 is an estimate of the error variance, the error mean-square term of the analysis of variance table, and $t_{\alpha,\,\text{df}}$ is found in a table of the t distribution. The degrees of freedom, df, is the number of degrees of freedom associated with the error mean square, and α is the probability that the confidence interval will not contain the true difference between the means, usually taken as 5 or 1 percent.

In the chiton example $s_e^2 = 3.86$, $r_1 = 15 = r_2 = r_3$, df $= 42$, and $t_{.05,\,42} = 2.020$. Therefore, the lsd is

$$\text{lsd} = 2.020\sqrt{3.86(\tfrac{1}{15} + \tfrac{1}{15})}$$
$$= 1.4492$$

A confidence about the estimated difference between means is $d_{ij} \pm \text{lsd}$ with probability $1 - \alpha$ of being true. Let δ_{ij} be the true difference between means estimated by d_{ij}. The confidence interval is

$$P(d_{ij} - \text{lsd} \le \delta_{ij} \le d_{ij} + \text{lsd}) = 1 - \alpha$$

The confidence interval about the difference in means of sunlight and artificial light is $P(3.1340 - 1.4492 \le \delta_{1,2} \le 3.1340 + 1.4492) = 95$ percent, or $P(1.6848$

$\leq \delta_{ij} \leq 4.5832) = .95$.   The confidence interval is read; the probability of the true difference falling in the interval 1.6336 to 4.5130 centimeters is 95 percent. The confidence intervals for all three differences in means are

$$P(1.6848 \leq \delta_{1,2} \leq 4.5832) = .95$$
$$P(5.3441 \leq \delta_{1,3} \leq 8.2425) = .95$$
$$P(2.2041 \leq \delta_{2,3} \leq 5.1025) = .95$$

The probability statement applies only to a single confidence interval.   The probability of all three differences falling in their respective intervals is not 95 percent, but some smaller probability.   If the differences in means are unrelated, the true probability will be the product of the probabilities; that is, $(.95)^3 = .85$, because the probability of three independent events all occurring is the product of their probabilities.   If the differences were unrelated in the chiton experiment, the probability of all three differences falling in their confidence intervals would be approximately 85 percent.   In this case, however, the events $d_{1,2}$ and $d_{1,3}$ are not independent, because treatment 1 occurs in both differences.   In fact, in the chiton experiment, there are no independent differences, and the probability of all three differences falling in their confidence intervals is not 85 percent, but some other probability.   There are, however, statistical methods analogous to the lsd which do not assume independence.   These methods are discussed in many standard statistical texts.

9-2.2   Nested One-Way Analysis of Variance

A special form of the one-way analysis of variance is the nested one-way analysis of variance.   Such an analysis might arise in an experiment in which the fecundity of females is being measured at several different temperatures, and at each temperature there are more than one group of females, for example, three culture bottles of flies at each temperature.   In this type of experiment differences in means may be caused either by the effects of the treatments or by differences among the cultures of flies.   In this *two-level* nested analysis of variance the analysis of variance table is

Source	Degrees of freedom	Sum of squares	Mean square	F
Total	N			
Mean	1			
Treatments	$t - 1$		$SS/(t - 1)$	
Subgroups	$t(s - 1)$		$SS/t(s - 1)$	
Within subgroups (error)	$ts(r - 1)$		$SS/st(r - 1)$	

The appropriate sums of squares are computed as

$$\text{SS total} = \sum_i^t \sum_j^s \sum_k^r X_{ijk}^2$$

where $i = 1, 2, \ldots, t$, $j = 1, 2, \ldots, s$ subgroups, and $k = 1, 2, \ldots, r$ replicates within each subgroup. Also

$$\text{SS mean} = \frac{(\sum_i \sum_j \sum_k X_{ijk})^2}{tsr}$$

$$\text{SS treatments} = \frac{\sum_i (\sum_j \sum_k X_{ijk})^2}{rs} - \text{SS mean}$$

$$\text{SS among subgroups} = \frac{\sum_i \sum_j (\sum_k X_{ijk})^2}{r} - \text{SS treatments} + \text{SS mean}$$

$$\text{SS within subgroups} = \text{SS total} - \text{SS among subgroups} + \text{SS treatments}$$
$$\text{(error)} \qquad\qquad\qquad\qquad\qquad\qquad\qquad\qquad - \text{SS mean}$$

where r is the number of replicates in each subgroup, and s is the number of subgroups in each treatment. A significant mean square among groups indicates significant treatment effects are present. A significant mean square among subgroups shows that the subgroups within each treatment group have different mean values. It is possible to have significant differences among subgroups, but not among treatments. This implies, for example, that there are differences among the cultures used, but these differences are not related to the particular treatments tested.

An important assumption of the two-level nested analysis of variance is that the subgroups are randomly allocated to the treatments. If there are 10 cultures of flies and the experiment calls for 2 cultures to be subjected to five different temperature regimes, the cultures must be allocated randomly among the treatments. Thus, the subgroups always form a model-II analysis of variance. The treatment variable among groups may be model I or model II. If the treatment variable is a fixed effect, the entire analysis of variance is *mixed*.

To test the mean squares among groups and subgroups, the mean square among subgroups is first tested against the mean square within subgroups. Second, the mean square among groups is tested over the mean square among subgroups. In some experiments there are sub-subgroups. This is termed a three-level nested analysis of variance. Examples of nested analyses of variance and a discussion of them may be found in most statistics texts.

9-2.3   Two-Way Classification

Suppose larval survival had been measured for all possible combinations of three temperatures and three humidities. Assuming only one measurement per combination, i.e., no replication, the data are tabulated as

	Temperature 1	Temperature 2	Temperature 3	$t = 3$
Humidity 1	86	63	71	
Humidity 2	92	68	69	
Humidity 3	87	68	73	
$s = 3$				

In a two-way classification each individual observation is due to: (1) an overall mean; (2) the effect due to treatment t_i, a_i; (3) the effect due to treatment s_j, b_j; (4) and the effect due to the interaction of the t_i and s_j treatments, $t_i s_j$; and (4) an error term, e_{ij} if there is no replication, and e_{ijk} if there are r replicates of each treatment combination. In terms of the analysis of variance model

$$X_{ij} = \mu + a_i + b_j + (a_i b_j) + e_{ij}$$

or with replication

$$X_{ijk} = \mu + a_i + b_j + (a_i b_j) + e_{ijk}$$

As in the one-way analysis of variance a sum of squares for each term of the model is calculated, and from the sum of squares a mean square estimated to be used in testing for the significance of each term. Whenever possible each treatment combination should be replicated with $k = 1, 2, \ldots, r$ replicates for each treatment combination. The reason for the replication will become painfully clear shortly. If there are r replications for each treatment combination, the computational equations for the sums of squares are

$$\text{SS total} = \sum_i^t \sum_j^s \sum_k^r X_{ijk}^2$$

$$\text{SS mean} = \frac{\left(\sum_i \sum_j \sum_k X_{ijk}\right)^2}{tsr}$$

$$\text{SS treatment } t = \frac{\sum_i \left(\sum_j \sum_k X_{ijk}\right)^2}{sr} - \text{SS mean}$$
(columns)

$$\text{SS treatment } s = \frac{\sum_j \left(\sum_i \sum_k X_{ijk}\right)^2}{tr} - \text{SS mean}$$
(rows)

$$\text{SS interaction} = \frac{\sum_i \sum_j \left(\sum_k X_{ijk}\right)^2}{r} - \text{SS mean} - \text{SS treatment } t$$
$$- \text{SS treatment } s$$

$$\text{SS error} = \text{SS total} - \text{SS mean} - \text{SS columns} - \text{SS rows} - \text{SS interaction}$$

In a two-way classification both treatments may be fixed by the experimenter, corresponding to a model-I analysis of variance; both treatments may be random variables, a model-II analysis of variance; or in some cases one treatment may be fixed and the other random, a mixed model. The F test for the significance of any of the three terms of the model depends upon the components of the mean squares for each of the three models. The expected mean squares for each model are listed in Table 9-2.

To test for the significance of any of the treatment or interaction effects, an F test is performed by dividing the mean square for the term by another mean square with the same components except the one of interest. In a model-I analysis of variance the mean squares for treatment t, treatment s, and the interaction consist of the error variance σ_e^2 plus an additional term representing the fixed treatment or interaction effect. In each case the appropriate mean square for testing any of the three other mean squares is the error mean square. A significant F for each mean square indicates that the additional component of the mean square is significantly larger than zero and that the treatment or interaction effect is present. For example, testing for treatment t a significant F in the F test, (mean square columns)/(mean square error), means that the effects of treatment t, the a_i, are significantly different from zero; that is, $(rs)/(t-1) \sum a_i^2$ is greater than 0. In a model-I analysis of variance all the mean squares are tested over the error mean square.

Table 9-2 **THE DEGREES OF FREEDOM AND EXPECTED MEAN SQUARES IN A TWO-WAY CLASSIFICATION ANALYSIS OF VARIANCE**

Term	Degrees of freedom	Expected mean square—model I
Columns	$t-1$	$\sigma_e^2 + rs/(t-1) \sum_i a_i^2$
Rows	$s-1$	$\sigma_e^2 + rt/(s-1) \sum_j b_j^2$
Interaction	$(t-1)(s-1)$	$\sigma_e^2 + r/(s-1)(t-1) \sum_i \sum_j (a_i b_j)^2$
Error	$ts(r-1)$	σ_e^2

Term	Model II	Mixed (t fixed, s random)
Columns	$\sigma_e^2 + r\sigma_{txs}^2 + rs\sigma_t^2$	$\sigma_e^2 + r\sigma_{txs}^2 + rs/(t-1) \sum_i a_i^2$
Rows	$\sigma_e^2 + r\sigma_{txs}^2 + rs\sigma_s^2$	$\sigma_e^2 + rt\sigma_s^2$
Interaction	$\sigma_e^2 + r\sigma_{txs}^2$	$\sigma_e^2 + r\sigma_{txs}^2$
Error	σ_e^2	σ_e^2

This is not true of a model-II analysis of variance. The variance of the interactions σ^2_{txs} can be tested against the error mean square. However, the treatment mean squares include both the error variance and the interaction variance. Treatment t includes the error variance, the interaction variance, and the variance of the treatment effects, the a_i. The appropriate mean square for testing the two treatment variances is the interaction mean square. A significant F test in this case indicates that the variance of the treatment effects, the a_i or b_j, is significant. In a model-II analysis of variance, therefore, the presence of significant interactions is tested for by dividing the interaction mean square by the error mean square. The two treatment mean squares are each tested by dividing the mean square by the interaction mean square.

In a mixed model the interaction mean square consists of the error variance and a variance due to the interactions. Test the interaction mean square by dividing by the error mean square. The mean square for the random treatment consists of only the error variance and the variance of the random treatment effects. The mean square for the random variable is also tested by dividing by the error mean square. The fixed treatment mean square consists of three components, the error variance, the interaction variance, and the fixed treatment effects, the a_i of treatment t in Table 9-2. Therefore, the appropriate mean square for testing the significance of the fixed treatment effects is the interaction mean square.

Guild (1952) counted the number of three species of earthworms in a series of 25 plots. Four samples, replicates, were collected in each plot. The data for five of these plots are given in Table 9-3. The t treatments $(t = 5)$ were plots. The s treatments $(s = 3)$ are the earthworm species. There were four replicates $(r = 4)$ per plot. The analysis of variance table is

Source	Degrees of freedom	Sum of squares	Mean square	F
Total	60	19,012.0000		
Mean	1	12,673.0667		
Columns (plots)	4	746.7666	186.6916	4.7384**
Rows (species)	2	3,107.2333	1553.6167	39.4319**
Interaction	8	711.9334	88.9917	2.2587*
Error	45	1,773.0000	39.4000	

* Significant at the 5 percent probability level.
** Significant at the 1 percent probability level.

Both the plots and the species were fixed by the experimenter, and so this is a model-I analysis of variance. If the plots and the species both represented random samples from a large number of possible plots and species, the situation would have been model II. If the situation had been model II, the treatment effects would have been tested for by dividing the treatment mean squares by the interaction mean square rather than the error mean square.

·   The analysis of variance shows that the number of worms depends on both the plots in which the counts are taken and on the species of the worms. There is also a significant interaction between plots and species. Each species tends to occur in a specific type of plot, although more field work is needed to determine the exact nature of the interaction.

9-2.4   Two-Way Classification without Replication
If each treatment combination is not replicated, the summation signs for the replicates drop out of the computational equations for the sums of squares.

Table 9-3   NUMBERS OF THREE SPECIES OF EARTH-WORMS IN FIVE PLOTS, FOUR REPLICATES PER PLOT, IN A FIELD NEAR MIDLOTHIAN, SCOTLAND

| | Plots ($t = 5$) | | | | | |
	1	2	3	4	5	$\sum\limits^{t}\sum\limits^{r}$
Allolobophora	16	43	32	21	16	
longa	41	48	22	14	25	487
	23	31	20	32	14	
	16	21	20	16	16	
	96	143	94	83	71	
Allolobophora	3	13	8	4	14	
caliginosa	7	18	5	9	21	
	17	19	12	3	18	239
	10	18	8	7	25	
	37	68	33	23	78	
Lumbricus	5	12	1	8	4	
terrestris	18	11	2	6	18	
	11	9	3	5	8	146
	5	4	7	1	8	
	39	36	13	20	38	
$\sum\limits^{s}\sum\limits^{r}$	172	247	140	126	187	

SOURCE: Data from Guild, W. J. M., Variation in Earthworm Numbers within Field Populations, *J. Ecol.*, vol. 21, 1952.

If replication is absent, the sum of squares within replicates will be the same as the total sum of squares.   Therefore, after the sums of squares for columns, rows, and mean are subtracted from the total sum of squares, we are left with a single sum of squares, which is equivalent to the interaction sum of squares in the analysis of variance with classification.   If there is no replication, this remaining sum of squares is termed the remainder sum of squares.   Without replication the presence of a significant interaction cannot be tested for.   If the analysis of variance is model II, the two treatment mean squares can still be tested over the remainder mean square, because it is equivalent to the interaction mean square of Table 9-2.   However, if the analysis of variance is model I, there is no appropriate mean square to test either of the two treatment mean squares unless it is possible to assume that there is no interaction.   If we can make this assumption, the σ_{txs}^2 term drops out of the remainder mean square, leaving only σ_e^2.   If there is no interaction, the remainder mean square is an estimate of the error variance, and can be used to test for significant treatment effects.   In a mixed model the fixed treatment can be tested against the remainder mean square, but the random variable cannot unless interaction is assumed to be absent.

The computational equations for the sums of squares are arrived at by dropping out the summation over replications.   The sum of squares for the remainder, equivalent to the interaction sum of squares of the replicated two-way classification, is

$$\text{SS remainder} = \text{SS total} - \text{SS mean} - \text{SS columns} - \text{SS rows}$$

There is no error sum of squares.

Williams (1951) counted the number of moths collected on five 3-day periods for each of three different mercury-vapor light traps at the Rothamstad Experiment Station in England.

| | Trap | | | | |
	1	2	3	$x_{i.}$	$\bar{x}_{i.}$
Period 1	19.1	50.1	123.0	192.2	64.1
Period 2	23.4	166.1	407.4	596.9	199.0
Period 3	29.5	223.9	398.1	651.5	217.2
Period 4	23.4	58.9	229.1	311.4	103.8
Period 5	16.6	64.6	251.2	332.4	110.8
$x_{.j}$	112.0	563.6	1408.8	$2084.4 = x_{..}$	
$\bar{x}_{.j}$	22.4	112.7	281.8	$139.0 = \bar{x}_{..}$	

The number of moths per observation is the geometric mean of the 3 days in each period.   The analysis of variance table is

Source	Degrees of freedom	Sum of squares	Mean square	F
Total	15	546,154.16		
Mean	1	289,648.22		
Columns (traps)	2	173,333.00	86,666.50	26.65*
Rows (periods)	4	52,065.92	13,016.48	3.40 n.s.
Remainder	8	30,607.02	3,825.88	

* Significant at the 1 percent probability level.
  n.s. = not significant.

This is a model-I analysis of variance, because both the traps and the time periods were fixed by the experimenter.   Therefore, the treatment effects can be tested only if it can be assumed that there is no interaction.   If there is no interaction, both columns and rows mean squares are tested against the remainder mean square.   Assuming no interaction the test finds that the trap used has a significant effect on the number of moths caught.   The period effects are not quite significant at the 5 percent probability level.   If the periods and the traps were both samples from a large number of possible traps and periods, the two mean squares could be tested against the remainder mean square without assuming that interaction is absent.

9-2.5   Interaction

The meaning of interaction can be illustrated graphically.   Suppose that larval survival is measured at two temperature levels and two humidities.   A possible set of data is

	Humidity 1	Humidity 2
Temperature 1	50	70
Temperature 2	30	50

Although the highest larval survival was at temperature 1 and humidity 2, the difference between the two temperature treatments is 20 at both humidity treatments.   In Fig. 9-2 this is represented by a pair of parallel lines, each line representing the change in larval survival with changing temperature for a constant humidity.   The effect of temperature does not depend on humidity, and the effects of both temperature and humidity on larval survival are additive.

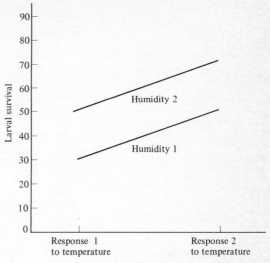

FIGURE 9-2
Larval survival plotted against temperature for constant values of humidity. The two lines representing constant humidities are parallel, indicating that the effects of temperature and humidity on larval survival are additive and that there is no interaction.

Suppose, however, that the data were

	Humidity 1	Humidity 2
Temperature 1	50	100
Temperature 2	30	50

(Fig. 9-3). In this case the effect of temperature depends on the relative humidity. In Fig. 9-3 the lines are not parallel. This is interaction. If temperature 1 and humidity 2 were low temperature and high humidity, respectively, the interpretation of these data would indicate that low temperatures favor larval survival, and that the favorability of low temperature is greater at higher humidities. Therefore, larval survival is not simply the sum of the two treatment effects.

A significant interaction term can result for several reasons. One possible reason is synergism between the two treatments, such that the analysis of variance model would be

$$X_{ij} = \mu + a_i + b_j + (a_i b_j) + e_{ij}$$

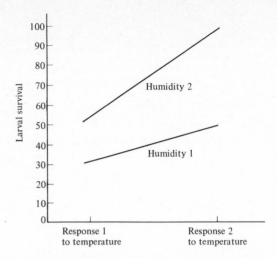

FIGURE 9-3

A graphic demonstration of interaction to be compared with Fig. 9-2. The lines representing constant humidities are not parallel, indicating that there is some sort of interaction between temperature and humidity in determining larval survival.

The term of the analysis of variance model $(a_i b_j)$ is a convention to indicate an additive interaction term and does not mean that the term is equal to the product of a_i and b_j. The response of the dependent variable is due to its overall mean, additive treatment effects a_i and b_j, an error term, and an interaction term representing an additional increment due to the synergism of the two main treatments. However, a common interaction is due not to an additive synergistic effect, but multiplicative effects of the treatments corresponding to a model such as

$$X_{ij} = \mu a_i b_j e_{ij}$$

If this is true, one of the primary assumptions of the analysis of variance model, i.e., the additivity of treatment effects, has been violated. Therefore, if a significant mean square for interaction is found, it is rather pointless to go on and test for the significance of the two treatment effects, because the interaction may not be additive as assumed by the analysis of variance model. If a significant mean square for interaction is found, interaction is present, but we do not know its form.

In the light trap experiment it was necessary to assume that additive interaction effects were absent, because it was a model-I analysis of variance. The linear model of the situation, therefore, is $X_{ij} = \mu + a_i + b_j + e_{ij}$. It is possible, however, that the correct model is $X_{ij} = \mu a_i b_j e_{ij}$, that is, that the treatment effects are multiplicative rather than additive. The remainder sum of squares, assuming additive interactions are absent, represents only the variance of the error terms if the situation is truly additive. However, if the treatment effects are multiplicative, not additive, the remainder sum of squares includes the discrepancies of the observations from the additive model. Therefore, in a two-way analysis of variance without replication the remainder sum of squares should be tested to see if the treatment effects are truly additive and not multiplicative. If the treatment effects are multiplicative, testing the treatment mean squares is meaningless. In a replicated two-way classification the interaction sum of squares is due to both additive interactions and deviations from the additive model. Therefore, in a replicated two-way classification the interaction mean square, if significant, should be tested to see if the interaction is additive before the main effects are tested.

To illustrate Tukey's test for an unreplicated two-way classification consider the light trap experiment above. From the data two series of values, d_i and d_j, are calculated as

$$d_i = \bar{x}_{i.} - \bar{x}_{..} \qquad d_j = \bar{x}_{.j} - \bar{x}_{..}$$

where $\bar{x}_{..}$ is the overall mean, 139.0, the mean of all the numbers of moths. These values of d_i and d_j are

$$\begin{aligned} d_i = \ &-74.9 & d_j = \ &-116.6 \\ &60.0 & &-26.2 \\ &78.2 & &142.8 \\ &-35.2 & \\ &-28.1 & \end{aligned}$$

From these values of d_i and d_j the statistics w_i are calculated as

$$w_i = \sum x_{ij} d_j$$

or for w_1 and w_2

$$\begin{aligned} w_1 &= (19.1)(-116.6) + (50.1)(-26.2) + (123.0)(142.8) \\ &= 14{,}025 \\ w_2 &= (23.4)(-116.6) + (166.1)(-26.2) + (407.4)(142.8) \\ &= 51{,}096 \end{aligned}$$

and so forth, giving $w_1 = 14{,}025$, $w_2 = 51{,}096$, $w_3 = 47{,}543$, $w_4 = 28{,}444$, and $w_5 = 32{,}243$. The quantity N is

$$N = \sum w_i d_i$$

or $N = (14,025)(-74.9) + (51,096)(60.0) + (47,543)(78.2) + (28,444)(-35.2) + (32,243)(-28.1) = 3,825,900$. The quantity D is

$$D = (\sum d_i^2)(\sum d_j^2)$$

or $D = (17,354)(34,674) = 601,700,000$. Given N and D the sum of squares in the remainder due to nonadditivity is equal to N^2/D, or $(3,825,900)^2/601,700,000 = 24,327$. Returning to the analysis of variance table

Source	Degrees of freedom	Sum of squares	Mean square	F
Traps	2	173,333.00		
Periods	4	52,066.92		
Remainder	8	30,607.02		
Nonadditivity	1	24,327.00	24,327.00	27.1*
Remainder	7	6,280.02	897.00	

* Significant at the 1 percent probability level.

The mean square for nonadditivity is tested over the mean square of the second remainder term. The F value is highly significant. There are significant multiplicative interactions between the treatments, the model is not additive, and so the two treatment terms were not properly tested. Therefore, the conclusions reached were meaningless. The proper course to follow in this case is to transform the observations to logarithms. If the multiplicative model is transformed to logarithms, the model becomes additive with additive treatment effects. The analysis is then carried out on the transformed data. Transformation will be discussed in more detail later.

In the replicated earthworm example, the interaction term was significant. Tukey's test can be applied to determine whether or not the interaction was truly additive or represented multiplicative interaction effects. If the interaction was multiplicative, the tests of the treatment effects would not have been appropriate. In the replicated two-way classification the replicates in each treatment combination are added, and the test for nonadditivity carried out as before. In the earthworm example the sum of squares for nonadditivity is 195.69, and the analysis of variance table becomes

Source	Degrees of freedom	Sum of squares	Mean square	F
Interaction	8	711.93		
Nonadditivity	1	195.65	195.65	2.65
Remainder	7	516.28	73.75	

The mean square for nonadditivity is tested against the mean square for the remainder of the interaction term. The resulting F value is not significant. Apparently the treatment and interaction effects are additive, and the additive model is correct. If the F statistic had been significant, the data would have been converted to logarithms and the analysis again conducted to see if the transformation results in a nonsignificant mean square for nonadditivity. If even in transformed form the term for nonadditivity is significant, the model of the situation is neither additive nor multiplicative, but some more complex interaction.

One of the most useful two-way classifications is the *randomized complete blocks* experimental design. Suppose one was interested in the effect of three levels of nitrogen on the biomass of the plants in an old field. Suppose, also, that the field slopes toward a river. The slope of the field, from left to right, say, results in a moisture gradient in the field, also from right to left. If we replicate each treatment five times, a randomized one-way classification analysis of variance experiment will be constructed by dividing the field into 15 parts and randomly allocating the 15 replicates of the three treatments among these 15 parts. It is reasonable to assume, however, that moisture will also affect the biomass of the plants. If the treatment replicates are simply allocated at random in the field, the error mean square will be inflated by the differences in biomass caused by differences in moisture. If it were possible to remove a sum of squares from the analysis of variance due to moisture differences among parts of the field, the error mean square would be reduced in size, our tests for the significance of the differences among treatment means would be more powerful, and the confidence intervals smaller than if the error variance due to moisture were ignored.

The effect of the moisture gradient can be removed by dividing the field from left to right into five parts or *blocks.* Each treatment is replicated once in each block. Suppose that the data turned out to be

	Block 1	Block 2	Block 3	Block 4	Block 5	$x_{i.}$
Nitrogen 1	25.5	22.9	20.7	19.2	16.8	105.1
Nitrogen 2	23.6	21.8	19.8	18.9	16.1	100.2
Nitrogen 3	24.2	21.3	18.7	17.8	15.9	97.9
$x_{.j}$	73.3	66.0	59.2	55.9	48.8	$x_{..} = 303.2$

There are $r = 5$ blocks and $t = 3$ treatments. The analysis of variance table is:

Source	Degrees of freedom	Sum of squares
Total	rt	$\displaystyle\sum_{i=1}^{t}\sum_{j=1}^{r} X_{ij}{}^2$
Mean	1	$X_{..}{}^2/rt$
Blocks	$(r-1)$	$\left(\displaystyle\sum_{j=1}^{r} X_{.j}{}^2/t\right) - \text{SS mean}$
Treatments	$(t-1)$	$\left(\displaystyle\sum_{i=1}^{t} X_{i.}{}^2/r\right) - \text{SS mean}$
Error	$(r-1)(t-1)$	SS total − SS mean − SS blocks − SS treatment

In the biomass experiment:

Source	Degrees of freedom	Sum of squares	Mean square	F
Total	15	6253.36		
Mean	1	6128.68		
Nitrogen levels	2	5.41	2.71	15.91*
Blocks	4	117.91	29.48	173.41*
Error	8	1.36	.17	

* Significant at the 1 percent probability level.

There are highly significant differences in the effects of both treatment and blocks on the biomass of the plants. If we had simply randomized the treatment replications across the field, the error mean square would have been much larger than .17, and conceivably we would not have found a significant difference among the treatment means. More importantly, if we had not blocked the field, the confidence intervals for the differences in means would be much larger than they are in this randomized complete blocks experiment. However, if the gradient effect had been very small or nonexistent, the loss of degrees of freedom due to removing the blocks sum of squares may have actually increased the error mean square, because the use of blocking in this example cost us 4 degrees of freedom which otherwise would have been associated with the error term of the analysis of variance table.

Note that the randomized complete blocks experiment is equivalent to a two-way classification without replication. Therefore, in treating the error sum of squares as error we are implicitly assuming that there is no interaction between block and treatment effects. Water conceivably could affect the efficiency of a nitrogen level in increasing or decreasing biomass; i.e., there may be a significant interaction between the nitrogen treatment and the moisture content of the ground.   In order to test for interaction, each treatment is replicated c times in each block.   Note that this is exactly the same experimental design used to analyze the earthworm problem.   The statements made earlier about testing the mean squares for significance in a two-way classification apply equally well here.

The block design can be used to illustrate the importance of randomization in experiments.   Suppose that in ignorance the field were divided into three blocks and all five replicates of the first treatment placed in the first block, all five replicates of the second treatment placed in the second block, and all five replicates of the third treatment placed in the third block.   If we were to carry out an analysis of variance on these data, the analysis would show a highly significant difference among the means of the three nitrogen levels. However, we already know that the major determinant of plant biomass was the moisture gradient in the field and not the nitrogen levels.   Therefore, we incorrectly conclude that the mean differences observed are due to the action of the nitrogen levels, when in fact the majority of the difference in biomass is due not to differences in nitrogen level but differences in moisture. In fact, there is no way to separate the treatment from block differences, i.e., the block and treatment effects are *confounded*, unless the treatment replicates are allocated at random throughout the field, or the randomized complete blocks design is used.

In some experiments the field may contain opposing gradients, perhaps a moisture gradient from north to south and an iron gradient from east to west. An experimental design termed a *latin square* can be used to remove sums of squares for both gradients from the total sum of squares.   The latin square is discussed in most statistics texts.

9-2.6   Assumptions of the Analysis of Variance

There are four important assumptions in the analysis of variance:

1 The error terms of the model are independently and normally distributed with a constant variance and a mean of zero.

2 The treatment effects, the overall mean, and the interaction are all additive.

3 The variances of the observations for each treatment or treatment combination are equal.

4 The observations are normally distributed.

In many cases the variance of a series of observations increases with the mean of the series.   Therefore, before an analysis of variance is performed the variances of the treatments or treatment combinations should be tested for equality. A convenient test for equal variances in a one-way classification is Bartlett's test for the homogeneity of variances. The test will be applied to the one-way classification example in Table 9-1. The variances of the observations for the three treatments are 6.272, 2.844, and 2.452, leading one to suspect that perhaps the variances are not equal. If the variances are not equal, we will not be justified in applying the technique of the analysis of variance. The computational steps given below are for unequal sample sizes, but are also applicable to equal sample sizes. The number of observations for each treatment is 15, so there is $15 - 1 = 14$ degrees of freedom for each treatment. The sum of degrees of freedom is

$$\sum_{i=1}^{t} (r_i - 1) = (14) + (14) + (14) = 42$$

where r_i is the number of observations of each treatment. A weighted average variance is computed as

$$s^2 = \frac{\sum_i (r_i - 1)s_i^2}{\sum_i (r_i - 1)}$$

$$= \frac{(14)(6.272) + (14)(2.844) + (14)(2.452)}{42}$$

$$= 3.86$$

Take the logarithm of the weighted average variance, that is, $\log_{10} 3.86 = .582063$, and compute

$$\sum_i (n_i - 1)\log_{10} s_i^2 = (14)(.79406) + (14)(.453930) + (14)(.389520)$$

$$= 22.9251$$

The statistic χ^2 is

$$\chi^2 = 2.3026\left[\sum_i (n_i - 1)\log_{10} s^2 - \sum_i (n_i - 1)\log_{10} s_i^2\right]$$

$$= 2.3026(24.4466 - 22.9251)$$

$$= 3.5034$$

The parameter 2.3026 is for conversion of the logarithms to the base e from base 10. The χ^2 statistic should be corrected by the factor c.

$$c = 1 + \frac{1}{3(t - 1)}\left[\sum_i \frac{1}{r_i - 1} - \frac{1}{\sum_i (r_i - 1)}\right]$$

$$= 1 + \frac{3/14 - 1/42}{6}$$

$$= 1.0317$$

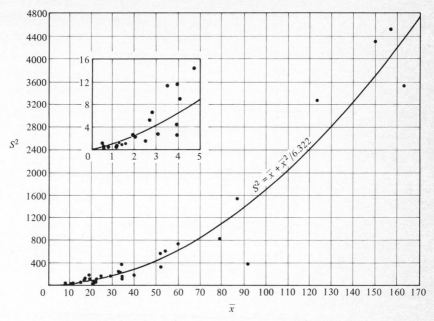

FIGURE 9-4

Relationship between variance and mean for spruce budworm egg masses on standing trees during 1950–1954 in northwestern New Brunswick. Each point is plotted based on a cluster of 100 trees. s^2 = variance and $\bar{x}$ = mean number of egg masses per 10 square feet of branch surface. Inset shows points below $\bar{x} = 5$ on an expanded scale. (*From Morris, 1954.*)

The adjusted χ^2 statistic is $\chi^2/c = 3.3958$. This adjusted statistic is compared to a table of the chi-square distribution with $t - 1$ degrees of freedom. In this example there are 2 degrees of freedom, and the chi-square statistic is not significant at the 95 percent probability level. Apparently there are no significant differences among the variances. The Bartlett test is sensitive to deviations from normality in the data.

Suppose, however, that the variances were significantly different. A glance at the variance seems to indicate that as the mean increases so does the variance. If this is true, the problem can sometimes be solved by transforming the data to logarithms, i.e., taking the logarithms to the base 10 of the observations. If the data in Table 9-1 are transformed to logarithms, the variances are .01597, .01560, and .08097. The variances no longer increase with the mean. If the logarithmic transformation removes the problem of unequal variances, the analysis of variance is carried out on the transformed data. If there are zero entries in the table, the transformation should be $\log(x + 1)$ rather than $\log x$. Figure 9-4 shows a more obvious example of

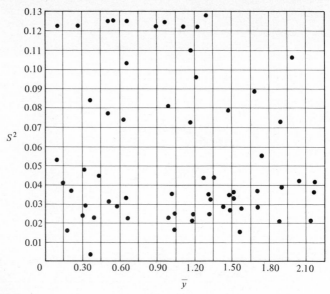

FIGURE 9-5

The same data as in Fig. 9-4 following a logarithmic transformation. (*From Morris, 1954*.)

the necessity of a logarithmic transformation, the relationship between the mean and variance in the number of spruce budworms, *Choristoneura fumiferana*, on standing trees in northwestern New Brunswick. The variance increases with the mean. If the logarithms of the numbers of egg masses are taken and plotted (Fig. 9-5), the dependence of the variance on the mean disappears.

Nonnormality of the data is not too serious a violation of the analysis of variance; i.e., the test is *robust*. However, nonnormality should still be checked for and corrected if necessary. If the data are not normally distributed, a transformation can be used to make them more normally distributed. The use of a transformation, particularly a logarithmic transformation, often simultaneously achieves normality, causes homogeneity of variances, and eliminates nonadditive treatment effects.

One of the more convenient ways of checking for normality is with probability paper. Figure 9-6 shows a piece of probability paper with a cumulative normal probability distribution marked on one axis, and an arithmetic scale on the other. The use of probability paper will be illustrated with the data in Table 5-2. The number of red mites per leaf on 150 apple leaves was determined. The data are recorded as a frequency distribution, and in addition the cumulative frequencies and percent cumulative frequencies

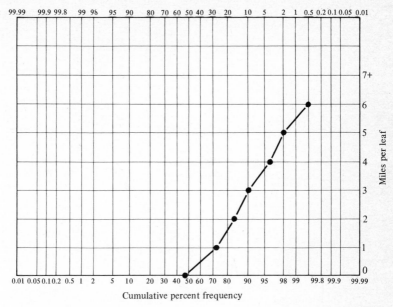

FIGURE 9-6
The mites per leaf plotted on probability paper against cumulative percent frequency. The relationship ignoring the upper part of the frequency distribution is curvilinear, indicating that the number of mites per leaf is not normally distributed. The data are from Table 5-2.

are also calculated, as in Table 9-4. The percent cumulative frequency is plotted against the number of mites per leaf on the probability paper (Fig. 9-6). If the data are normally distributed, a straight line will result. In this case the line is not straight, so the data are not normally distributed.

The correct transformation to use in this case is a square root transformation. Square root transformations are usually applied to counts of individuals with an approximately Poisson distribution. If zero entries are present, the square roots are of $x + \frac{1}{2}$, rather than of simply x. The square roots of the number of mites per leaf plus $\frac{1}{2}$ because of the presence of zero entries are plotted against percent cumulative frequency in Fig. 9-7. Some linearity has been achieved, and if the last point is ignored, the data are approximately normally distributed. The curve at the upper percentage part of the graph is almost always found in transformed data, because of a paucity of observations at this end of the probability distribution, i.e., above seven mites per leaf. In practice this is not very important.

The logarithmic transformation is also useful in alleviating deviations from normality. The transformation is particularly useful with frequency

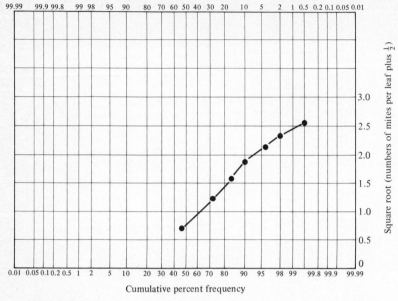

FIGURE 9-7
The same data as in Fig. 9-6, but with a transformation square root $(Y + \frac{1}{2})$. The upper part of the frequency distribution has again been ignored. The data are not completely normal, but some normalization has occurred.

distributions skewed to the right. Probability paper is available with a logarithmic instead of an arithmetic scale.

The arcsine transformation is used to normalize percentages and proportions. The transformation is the arcsine of the square root of the percentage. The arcsine transformation is tabulated in Beyer (1968). An example of the use of the transformation is given in Fig. 9-8 and Table 9-5.

Table 9-4   THE NUMBER OF MITES PER APPLE LEAF AS PRESENTED IN TABLE 5-2, TO ILLUSTRATE THE SQUARE ROOT TRANS-FORMATION TO NORMALIZE DATA

Mites per leaf	Number of leaves	Cumulative frequency	Percent cumulative frequency
0	70	70	.465
1	38	108	.720
2	17	125	.835
3	10	135	.900
4	9	144	.960
5	3	147	.980
6	2	149	.995
7+	1	150	1.000

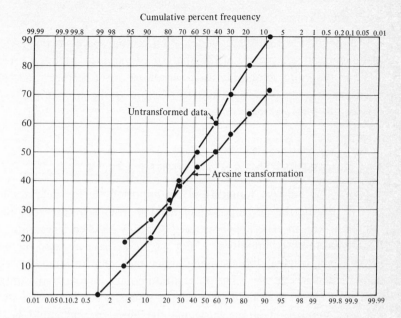

Cumulative percent frequency

FIGURE 9-8

Percent fertility of eggs of the CP strain of D. *melanogaster* raised in 100 vials of 10 eggs each plotted on probability paper before and after transformation of the data using an arcsine transformation.   The data are presented in Table 9-7.

Table 9-5   APPLICATION OF THE ARCSINE TRANSFORMATION TO NORMALIZE DATA CONSISTING OF PERCENTAGES OR PROPORTIONS.   PERCENT FERTILITY OF EGGS OF THE CP STRAIN OF *D. MELANOGASTER* RAISED IN 100 VIALS OF 10 EGGS EACH

Percent fertility	Arcsine $(p)^{1/2}$	Frequency	Cumulative frequency
0	0	1	1
10	18.44	3	4
20	26.56	8	12
30	33.21	10	22
40	39.23	6	28
50	45.00	15	43
60	50.77	14	57
70	56.79	12	69
80	63.44	13	82
90	71.56	9	91
100	90.00	9	100

SOURCE: From Sokal, R. R., and Rohlf, F. J., "Biometry: The Principles and Practice of Statistics in Biological Research," W. H. Freeman and Company, San Francisco, 1969.

9-2.7 A Closing Comment

It is impossible within the confines of an ecology book to do justice to the usefulness and complexity of the analysis of variance. I have covered only a few of the more commonly used variations of the method, and the reader is advised to consult a statistics text before trying to put the analysis of variance to practical use.

9-3 REGRESSION AND CORRELATION

Suppose that relative humidity is an important factor determining larval survival during the development of a larva from eclosion from the egg to pupation. In an experiment sets of 100 newly hatched larvae are subjected to different relative humidities. After the larvae have pupated the number of larvae of the original 100 surviving under each relative humidity regime can be plotted as in Fig. 9-9. The observation points are scattered, but the points tend to occur in a straight line. If the points occur in a perfectly straight line, the number of larvae surviving the larval period could be perfectly predicted by knowing the relative humidity. The points do not lie on a straight line. Even though the points do not form a perfectly straight line, it may still be desirable to express the number of larvae surviving as a linear function of relative humidity. This is the basis of *linear regression*.

9-3.1 Linear Regression

The equation of a straight line is

$$Y = \alpha + \beta X$$

where Y is the dependent variable, larval survival, and X is the independent variable, relative humidity. The constant α is the intercept of the line on the Y axis, and represents the number of larvae surviving in the absence of any humidity in the air at all. The parameter β is the slope of the line. The latin equivalents of α and β, a and b, are used to represent estimates of the two population parameters. Therefore, in theoretical models the parameters are written as α and β, but when estimates of the parameters are given the letters a and b are used.

Because the observation points do not lie completely on the line, a term must be added to the model to account for the deviations of the points from the line. This is the error term ε_i. The error terms of the model are assumed to be independent of each other, and are normally distributed with

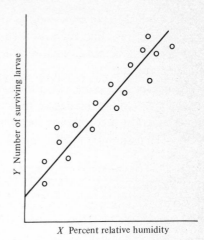

FIGURE 9-9
A hypothetical example of a linear relationship between a dependent variable, number of surviving larvae, and an independent variable, percent relative humidity.

a zero mean and a constant variance. The variance of the error terms, also called the residual variance, is the variance of the dependent variable not accounted for by changes in the independent variable. The greater the scatter of the data points about the regression line, the higher the residual variance. Each individual observation on the dependent variable can be represented as the value of the dependent variable in the absence of the independent variable α, plus the effect of the independent variable βX_i, plus the random effects of other unknown factors ε_i. The linear regression model is

$$Y_i = \alpha + \beta X_i + \varepsilon_i$$

In regression a line is fit to the data points in such a way that the sum of the squared vertical distances of the observation points to the regression line are as small as possible. This is *least-squares* regression. In regression analysis the parameters of the regression model are estimated, and the significance of the independent variable as a predictive factor of the variance of the dependent variable is determined.

In this type of experiment the values of the independent variable are set by the experimenter. The value of the independent variable must be measured exactly, without error. Suppose that the densities of two species of insects are estimated, not counted, in a series of quadrats. One species is made the dependent variable, and the other the independent variable. However, both species are estimated, and, therefore, the independent variable is measured with error; i.e., the estimate is not the true population density, only an estimate. In this situation ordinary linear regression cannot be used. In fact, even if the species serving as the independent variable can be measured exactly, the resulting regression equation, although correct, does not make much biological sense because it may not be true that changes in the first species are due to changes in the density of the second. Both species densities may be caused by a third factor such as temperature or a predator. In such situations the degree to which the two species vary together can be determined by a correlation method discussed later.

The assumptions of a simple linear regression are:

1 The values of the independent variable X are fixed, and are measured without error. The value of X is taken, and the set of observations or experimental results, the resulting value of Y, the dependent variable, is recorded. There is enough latitude in the assumption to use temperature in the field as an independent variable.

2 The mean values of Y, given various values of X, lie on a straight line.

3 For a given value of X, the values of Y are independently and normally distributed.

4 The samples along the regression line have a common variance, which is the variance of the error terms in the regression. This means that the spread of the points around the regression line is not broader at one end of the line than at the other.

The first step in the regression analysis is to estimate the two parameters α and β. First calculate the *corrected sum of squares and cross products*, $\sum x^2$ and $\sum xy$

$$\sum x^2 = \sum_i^n X_i^2 - \frac{(\sum_i^n X_i)^2}{n}$$

$$\sum xy = \sum_i^n X_i Y_i - \frac{(\sum_i^n X_i)(\sum_i^n Y_i)}{n}$$

Then the two estimates b and a are

$$b = \frac{\sum xy}{\sum x^2} \qquad a = \bar{Y} - b\bar{X}$$

where $\bar{X}$ and $\bar{Y}$ are the means of the observations on the independent and dependent variables. Having calculated these two parameters we wish to test if the regression model explains a significant amount of the variance of the dependent variable. This test is an analysis of variance of the total sum of squares of the dependent variable, that is, $\sum Y^2$, into sums of squares due to the mean value of Y, the regression, and the residual variance.   The sums of squares and their degrees of freedom are

Source	Degrees of freedom	Sum of squares
Total	n	$\sum Y^2$
Mean	1	$(\sum Y)^2/n$
Regression	1	$\beta \sum xy$
Residual	$n-2$	SS total $-$ SS mean $-$ SS regression

The regression mean square and the residual mean square are found by dividing the respective sums of squares by their degrees of freedom.   The residual mean square is denoted by $s_{Y.X}^2$, and is an estimate of the variance of the deviations of the observations from the regression line.   The variance $s_{Y.X}^2$ represents the variance in Y, given the variance attributable to X has been removed by the regression.   If the regression sum of squares is divided by a corrected total sum of squares (corrected total SS = SS total $-$ SS mean), the result is the proportion of the corrected total sum of squares accounted for by the regression line.   This proportion or percent is an indication of how effective the independent variable is in explaining the variance of Y.   The null hypothesis in regression is $H_0: \beta = 0$, versus the alternative hypothesis $H_1: \beta \neq 0$.   Clearly if β equals 0, the calculated regression line will be horizontal to the X axis and any value of X will result in the same prediction for Y, explaining none of the variance of Y.   The test is conducted by calculating the ratio regression mean square/residual mean square.   This ratio is treated as an F statistic with 1 and $n-2$ degrees of freedom.   If the value of F is not significant, the null hypothesis is accepted, and it is concluded that the independent variable X does not explain a significant proportion of the variance of Y.

The standard error of the regression coefficient b is

$$SE(b) = \left(\frac{s_{Y.X}^2}{\sum x^2}\right)^{1/2}$$

The hypothesis $H_0: \beta = 0$ can be tested independently of the analysis of variance table by using the t statistic

$$t = \frac{b - 0}{SE(b)}$$

and comparing the statistic with a table of the t distribution with $n-2$ degrees of freedom.   If β is not significantly different than 0 the t statistic will not be significant, and the null hypothesis will be accepted.   If the

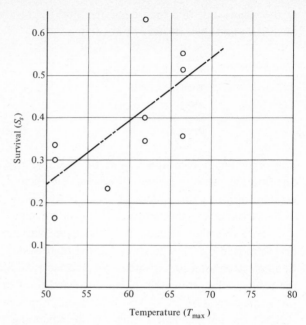

FIGURE 9-10

Relationship between the survival of small larvae (S_s) of the spruce budworm, *C. fumiferana*, and the mean maximum temperature in degrees Fahrenheit during the 4-day period bracketing emergence from hibernation.

t statistic is significant, the parameter β is not equal to 0 and the null hypothesis will be rejected.

In Fig. 9-10 the percent survival of small larvae of the spruce budworm, *Choristoneura fumiferana*, during the 4-day period bracketing emergence from hibernation is plotted against mean maximum temperature. The observations on the dependent variable, larval survival, and the independent variable, maximum temperature, are listed in Table 9-6. The estimated value of β is $b = \frac{5.3}{336.0} = .0164$. Given this estimate of β, an estimate of α is $a = .383 - (.0164)(60) = -.6009$. The estimate of α brings out an important point. Never extrapolate beyond the range of the available data. The parameter α represents the mean value of Y when X is zero. Clearly larval survival cannot be negative. Therefore, the relationship between X and Y beyond the range of X in the table, 52 to 66°F is not linear. The use of the regression equation

$$Y = -.6009 + .0164X$$

is valid only within the range of values of $X = 52$ to 66.

The analysis of variance table is

Source	Degrees of freedom	Sum of squares	Mean square	F
Total	10	1.6535		
Mean	1	1.4669		
Linear regression	1	.0904	.0904	7.5333
Residual	8	.0962	.0120	

The proportion of the corrected total sum of squares accounted for by the regression is $.0904/(1.6535 - 1.4669) = 48$ percent.   The value of F has 1 and 8 degrees of freedom.   The probability of a value of F this large arising by chance if the null hypothesis is true is only 2.5 percent.   The null hypothesis is, therefore, rejected, and it is concluded that the regression accounts for a significant proportion of the variance in larval survival.

In predicting Y for a given value of the independent variable X_i, the standard error of the estimated value of Y should also be calculated in order to indicate the variability of the estimate.   The standard error of the predicted value of Y is

$$\mathrm{SE}(\hat{Y}) = \sqrt{s_{Y.X}^2 \left[\frac{1}{n} + \frac{(X_i - \bar{X})^2}{\sum x^2}\right]}$$

Table 9-6 OBSERVED PERCENT SURVIVAL OF SMALL LARVAE OF THE SPRUCE BUD-WORM, *C. FUMIFERANA*, AT DIFFERENT MEAN MAXIMUM TEMPERATURES

Y, larval survival	X, mean maximum temperature, °F
.16	52
.30	52
.34	52
.24	57
.34	63
.40	63
.63	63
.36	66
.51	66
.55	66
Total = 3.83	Total = 600
Mean = .383	Mean = 60

SOURCE: Data from Morris, R. F. (ed.): The Dynamics of Epidemic Spruce Budworm Populations, *Mem. Entomol. Soc. Can.*, no. 31, 1963.

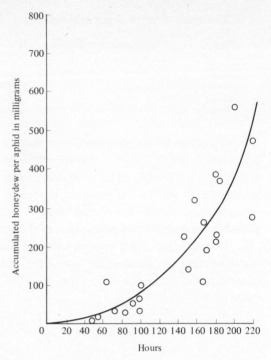

FIGURE 9-11
Milligrams of honeydew accumulated per aphid (*M. pisi*) as a function of age. (*Data courtesy of B. J. Rathcke.*)

The standard error of the estimated value of the dependent variable depends on the value of X. The estimated value of Y, given $X = 60$, is $Y = -.6009 + (.0164)(60) = .3831$. The standard error of the estimate of Y is

$$\text{SE}(\hat{Y}) = \sqrt{.0120\left[\frac{1}{10} + \frac{(60 - 60)^2}{336}\right]}$$
$$= \pm.0346$$

9-3.2   Curvilinear Regression

In simple linear regression the relationship between the dependent and independent variables is linear. In other cases this is often not true. For example, the milligrams of honeydew accumulated by the aphid *Macrosiphum pisi* with time under uncrowded conditions are shown in Fig. 9-11. The relationship is roughly exponential, not linear. When faced with curvilinear relationships between the dependent and independent variables, there are three general alternatives: (*1*) transform the data so that the relationship between the transformed variables is linear, (*2*) try to fit the data to the curve produced by the polynomial expression $Y = \alpha + \beta_1 X + \beta_2 X^2 + \beta_3 X^3 + \cdots + \beta_k X^k$, and (*3*) use a method of least squares known as iterative regression. Of the three methods only the first two will be discussed in this section.

Some commonly occurring curvilinear relationships are shown in Fig. 9-12. Curve type 1 is the straight line. Types 2 and 4 are known as logarithmic curves, and can be made linear by taking logarithmic transformations of the independent, the dependent, or both variables. The regression model appropriate for each curve is listed alongside the figure. The aphid data represent an example of a type-2 curve. The relationship between accumulated honeydew and hours can be made linear, *rectified*, by taking the logarithms to the base 10 of the values of the dependent variable Y. If the logarithms of the Y values are plotted against X, the relationship will be linear (Fig. 9-13). The regression is carried out by regressing log Y on X in the usual way. A type-2 curve is given by the function $Y = \alpha\beta^X$. For the type-3 curve the data are rectified by taking the logarithms of both X and Y. This type of curve is given by the function $Y = \alpha X^\beta$. Last, curves of type 4 are rectified by taking the natural logarithms of the values of X. The function of this curve is $e^Y = \alpha X^\beta$.

The data for the aphid experiment are presented in Table 9-7. The logarithms of the dependent variables were taken, and a linear regression of log Y on X performed. The estimate of log β was .0091, and of log α .8013. The regression equation is log $Y = .8013 + .0091X$. Taking antilogs gives the equivalent function $Y = (6.33)(1.021)^X$ which serves as a model of the amount of accumulated honeydew as a function of time.

The curvilinear relationship is often not one of the logarithmic-type curves. In some cases the second-degree polynomial

$$Y_i = \alpha + \beta_1 X_i + \beta_2 X_i^2 + \varepsilon_i$$

will fit the data satisfactory. The fitting of a second-degree polynomial is a multiple regression with the two independent variables X and X^2. In Fig. 9-12 is an example of a second-degree polynomial curve (type 5). Data can also be fit by higher polynomials. For example, curve type 6 in Fig. 9-12 is the function $Y = \alpha + \beta_1 X + \beta_2 X^2 + \beta_3 X^3$. Theoretically, higher polynomials can give better fits to curvilinear data than the second-degree polynomial. However, if multiple regression methods are used with polynomial terms over X^3, rounding errors can become overwhelming enough to make the analysis meaningless. In certain situations polynomial equations

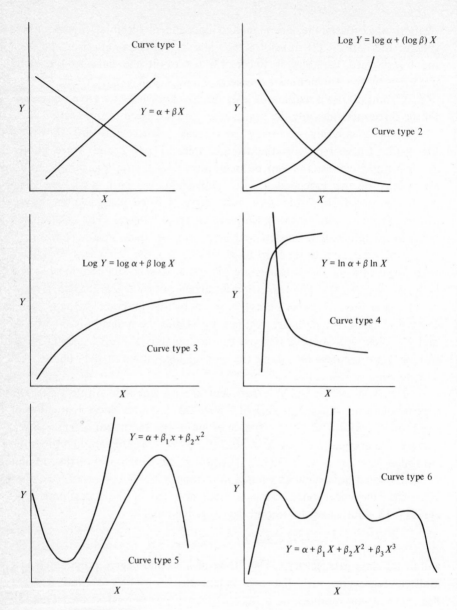

FIGURE 9-12
Six types of commonly encountered curvilinear relationships and the proper
regression model for each.

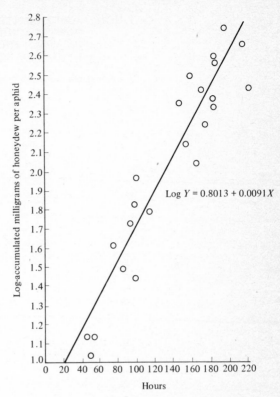

$$\text{Log } Y = 0.8013 + 0.0091X$$

FIGURE 9-13

Milligrams of honeydew accumulated in the aphid *M. pisi* as a function of age. The dependent variable is log *Y*. The line is the fitted regression line.

can be fit by using *orthogonal polynomials*. Orthogonal polynomials are discussed in most statistics texts.

The second-degree polynomial equation is treated as a multiple regression with two independent variables X and X^2. In Table 9-8 the average dry weight of the aphid *M. pisi* for each day up to day 14 is listed. The data are plotted in Fig. 9-14. The relationship is decidedly curvilinear. To estimate the parameters of the polynomial regression equation, first calculate the matrix of corrected sum of squares and cross products $\sum y^2$, $\sum x_1^2$, $\sum x_2^2$, $\sum yx_1$, $\sum yx_2$, $\sum x_1x_2$. For example, $\sum x_1x_2$ is

$$\sum x_1x_2 = \sum X_1X_2 - \frac{(\sum X_1)(\sum X_2)}{n}$$

where X_1 is equal to X, and X_2 is X^2.

Table 9-7   ACCUMULATED MILLIGRAMS OF HONEY-DEW PER APHID FOR DIFFERENT LENGTHS OF TIME. THE APHID IS *M. PISI* LIVING UNDER UNCROWDED CONDITIONS

Y, accumulated honeydew in milligrams	$\log Y$	X, hours
225	2.3522	181
373	2.5717	182
315	2.4983	156
92	1.9638	100
31	1.4914	82
224	2.3502	180
63	1.7993	114
388	2.5988	180
177	2.2480	173
224	2.3502	142
561	2.7490	192
14	1.1461	42
112	1.6021	75
14	1.8325	96
40	2.1461	152
68	1.4472	98
140	1.0414	49
28	1.7243	93
11	2.6776	214
53	2.4378	218
476	2.4219	169

| 274 | | |
| 258 | | |

SOURCE: Data courtesy of B. J. Rathcke.

Table 9-8   ACCUMULATED DRY WEIGHT IN MILLIGRAMS PER APHID AFTER VARYING LENGTHS OF TIME. THE APHID IS *M. PISI* LIVING UNDER CROWDED CONDITIONS

X, days	X^2	Y, average dry weight in milligrams	$\log Y$
0	0	23.8	1.3766
3	9	86.5	1.9370
4	16	136.7	2.1356
5	25	203.0	2.3075
7	49	355.8	2.5512
8	64	397.0	2.5988
9	81	384.0	2.5843
10	100	450.0	2.6532
12	144	552.0	2.7419
14	196	368.0	2.5658

SOURCE: Data courtesy of B. J. Rathcke.

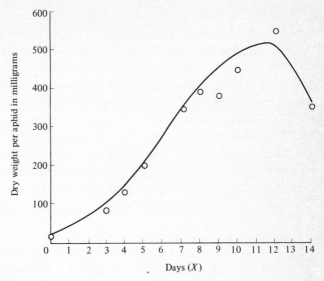

FIGURE 9-14
Dry weight per aphid in milligrams as a function of time for the aphid *M. pisi*. The curve was fitted by eye. (*Data courtesy of B. J. Rathcke.*)

Estimates of the two regression coefficients β_1 and β_2 are

$$b_1 = \frac{(\sum x_2^2)(\sum x_1 y) - (\sum x_1 x_2)(\sum x_2 y)}{(\sum x_1^2)(\sum x_2^2) - (\sum x_1 x_2)^2}$$

$$b_2 = \frac{(\sum x_1^2)(\sum x_2 y) - (\sum x_1 x_2)(\sum x_1 y)}{(\sum x_1^2)(\sum x_2^2) - (\sum x_1 x_2)^2}$$

The estimate of α is

$$a = \bar{Y} - b_1 \bar{X} - b_2 \bar{X^2}$$

The significance of the two regression coefficients can be tested by the analysis of variance table

Source	Degrees of freedom	Sum of squares	Mean square (MS)	F
Total	n	$\sum Y^2$		
Mean	1	$(\sum Y)^2/n$		
Corrected total	$n-1$	SS total $-$ SS mean		
Regression	2	$b_1 \sum x_1 y + b_2 \sum x_2 y$	SS/2	MS Regression/ MS residual
Linear (b_1)	1	$(\sum x_1 y)^2/\sum x_1^2$	SS/1	MS Linear/ MS residual
Quadratic $(b_2 \| b_1)$	1	SS regression $-$ SS linear	SS/1	MS Quadratic/ MS residual
Residual	$n-3$	SS total $-$ SS mean $-$ SS regression	SS/$(n-3)$	

In the table linear (b_1) is the sum of squares of the regression sum of squares accounted for by linear regression, i.e., a straight line. The entry quadratic $(b_2|b_1)$ indicates the sum of squares accounted for by the quadratic term X^2 after the sum of squares due to linear regression have been removed. A significant value of quadratic $(b_2|b_1)$ indicates that the addition of the X^2 term improves the fit of the regression equation to the data; i.e., there is a decided nonlinear component in the data.

The measured dry weight in milligrams per aphid is plotted in Fig. 9-14 and listed in Table 9-8. The curvilinear relationship is so complex that a logarithmic transformation of the dependent variable was carried out first to simplify the curve somewhat. The transformed data are shown in Fig. 9-15. The matrix of corrected sums of squares and cross products is

$$
\begin{array}{c}
\begin{array}{ccc} y & x_1 & x_2 \end{array} \\
\begin{array}{c} y \\ x_1 \\ x_2 \end{array}
\begin{bmatrix}
1.6149 & 14.3007 & 169.0000 \\
14.3007 & 166.0000 & 2347.0000 \\
169.0000 & 2347.0000 & 26386.0000
\end{bmatrix}
\end{array}
$$

The estimates of β_1 and β_2 are

$$
b_1 = \frac{(36386.0000)(14.3007) - (2347.0000)(169.0000)}{(166.0000)(26386.0000) - (2347.0000)^2}
$$

$$
= .2326
$$

and $b_2 = -.0104$. The estimate of α is

$$
a = 2.3453 - (.2326)(7.2) - (-.0104)(68.4) = 1.3819
$$

The polynomial regression equation is $\log Y = 1.3819 + .2326X - .0104X^2$. The analysis of variance table is

Source	Degrees of freedom	Sum of squares	Mean square	F
Total	10	56.6141		
Mean	1	54.9992		
Regression	2	1.5687	.7844	118.84*
Linear (b_1)	1	1.2320	1.2320	185.67*
Quadratic $(b_2|b_1)$	1	.3487	.3487	52.83*
Residual	7	.0462	.0066	

* Significant at the 1 percent probability level.

There is a significant regression of days on aphid dry weight, and the addition of the quadratic term is an improvement over the simple linear regression. The fitted regression line is shown in Fig. 9-15.

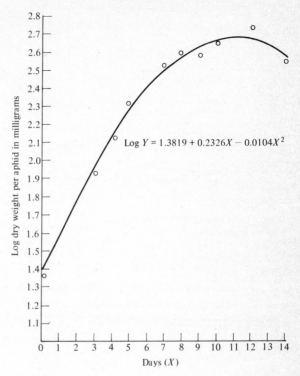

FIGURE 9-15
A fitted polynomial curve to the data in Fig. 9-14, but with a logarithmic transformation of the dependent variable.

Some of the most commonly occurring curves in ecological work have an upper or lower asymptote, as in Fig. 9-16.   An asymptotical curve might result if the number of eggs laid per female were plotted as a function of increasing time.   Clearly there is a maximum number of eggs the female can lay no matter how much time passes.   An equation commonly used to fit data with an upper or lower asymptote is

$$Y = \alpha + \beta(\rho^X)$$

In this equation there is a third parameter ρ in addition to the two parameters α and β.   In *asymptotic regression* three variables are created from the data: $X_0 = 1$, $X_1 = r_1 X$, and $X_2 = X r_1^{X-1}$.   The measurements on the independent variable are X, and r_1 is the initial estimate of ρ.   In Table 9-9 a numerical example from Snedecor and Cochran (1967) is given.   The three new variables are listed with the observed values of X and Y.   The values

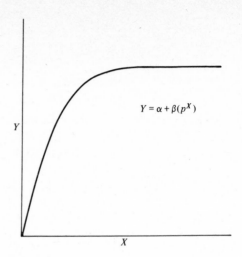

$$Y = \alpha + \beta(p^X)$$

FIGURE 9-16
Example of an asymptotic curve with the
function generating it.

of X, the independent variable, should change by unity, e.g., days, or be coded
to do so.   Designating the values of the dependent variable as Y_0, Y_1, ..., Y_{n-1},
and initial estimates of ρ for sample sizes of 4 to 7 are

$$n = 4: \quad r_1 = \frac{4Y_3 + Y_2 - 5Y_1}{4Y_2 + Y_1 - 5Y_0}$$

$$n = 5: \quad r_1 = \frac{4Y_4 + 3Y_3 - Y_2 - 6Y_1}{4Y_3 + 3Y_2 - Y_1 - 6Y_0}$$

$$n = 6: \quad r_1 = \frac{4Y_5 + 4Y_4 + 2Y_3 - 3Y_2 - 7Y_1}{4Y_4 + 4Y_3 + 2Y_2 - 3Y_1 - 7Y_0}$$

$$n = 7: \quad r_1 = \frac{Y_6 + Y_5 + Y_4 - Y_2 - 2Y_1}{Y_5 + Y_4 + Y_3 - Y_1 - 2Y_0}$$

**Table 9-9   A NUMERICAL EXAMPLE OF
FITTING A REGRESSION
MODEL TO AN ASYMPTOTIC
CURVE**

X	Y	X_0	X_1	X_2
0	57.5	1	1.00000	0
1	45.7	1	.55000	1.00000
2	38.7	1	.30250	1.10000
3	35.3	1	.16638	.90750
4	33.1	1	.09151	.66550
5	32.2	1	.05033	.45753
Total	242.5	6	2.16072	4.13053

In this example $n = 6$, and the first approximation to ρ, r_1, is

$$r_1 = \frac{(4)(32.2) + (4)(33.1) + (2)(35.3) - (3)(38.7) - (7)(45.7)}{(4)(33.1) + (4)(35.3) + (2)(38.7) - (3)(45.7) - (7)(57.5)}$$

$$= .552$$

In terms of the three artificial variables the regression equation is $Y = aX_0 + b_1X_1 + cX_2$, where a and b are the sample estimates of α and β, and $c = b(r_2 - r_1)$. The second approximation to ρ, r_2, is $r_2 = r_1 + (c/b)$. To estimate a, b, and c a matrix of the sums of squares and cross products of the three artificial X variables is calculated

$$\mathbf{M} = \begin{bmatrix} x_0{}^2 & x_0 x_1 & x_0 x_2 \\ x_1 x_0 & x_1{}^2 & x_1 x_2 \\ x_2 x_0 & x_2 x_1 & x_2{}^2 \end{bmatrix}$$

The values of a, b, and c are found by postmultiplication of the inverse of $\mathbf{M}$ by the column vector

$$\begin{bmatrix} x_0 y \\ x_1 y \\ x_2 y \end{bmatrix}$$

In the example

$$\mathbf{M} = \begin{bmatrix} 6.00000 & 2.16072 & 4.13053 \\ 2.16072 & 1.43260 & 1.11767 \\ 4.13053 & 1.11767 & 3.68578 \end{bmatrix}$$

$$\mathbf{M}^{-1} = \begin{bmatrix} 1.62101 & -1.34608 & -1.40843 \\ -1.34608 & 2.03212 & .89229 \\ -1.40843 & .89229 & 1.57912 \end{bmatrix}$$

and

$$\mathbf{M}^{-1} \begin{bmatrix} 242.50000 \\ 104.86457 \\ 157.06527 \end{bmatrix} = \begin{bmatrix} 30.723 \\ 26.821 \\ .05024 \end{bmatrix} = \begin{bmatrix} a \\ b \\ c \end{bmatrix}$$

The second approximation to ρ, r_2, if r_1 is .55, is

$$r_2 = .55 + \frac{.05024}{26.821}$$

$$= .55187$$

If desired, the estimate of ρ can be improved by further iteration. This asymptotic regression is a special form of the general process of iterative regression mentioned earlier. Taking r_2 as a satisfactory estimate of ρ the asymptotic regression equation is

$$Y = 30.723 + 26.821(.55187^X)$$

The parameter ρ is always between 1 and 0. If the value of b is positive, as in the example, the curve approaches the lower asymptote given by a, in this case 30.723. If the value of b is negative, the curve approaches an upper asymptote. If the regression equation were $Y = 30.723 - 26.821(.55187^X)$ the curve would increase with time to the upper asymptote 30.723.

9-3.3 Correlation

A regression equation implies a functional relationship between the dependent and independent variables. However, suppose that the numbers of two species of organisms are counted in a series of quadrats and designated species X and species Y without making a distinction between dependent and independent variables. If both species counts are normally distributed, and a plot of X on Y is roughly linear, Y can be regressed on X, or alternatively, X can be regressed on Y. In some cases the changes in one species may be due to changes in the second, but more often interactions of the two species are reciprocal, or are caused by one or more other environmental factors. The two regression equations, although mathematically correct and perhaps of empirical value, make little biological sense. It is more logical, therefore, to determine the degree to which the two species vary in the same degree rather than to express each species as a function of the other.

If the two regression equations are created, the parameters of the two equations differ. However, the proportions of the corrected total sum of squares of the "dependent" variables accounted for by the regression of X on Y and Y on X are the same. Therefore, in terms of corrected sums of squares

$$\frac{(\sum xy)^2/\sum x^2}{\sum y^2} = \frac{(\sum xy)^2/\sum y^2}{\sum x^2} = \frac{\text{regression SS}}{\text{corrected total SS}}$$

This quantity is denoted by r^2 and is the coefficient of determination. The square root of r^2 is calculated as

$$r = \frac{\sum xy}{(\sum x^2 \sum y^2)^{1/2}}$$

and is called the correlation coefficient between the two variables X and Y;

it is a measure of the degree to which two variables vary together. The correlation coefficient can be calculated for any set of observations on two variables, but in order to test the significance of the correlation coefficient both variables must be normally distributed. If the two variables are not normally distributed, the value of r is simply a measure of equality of variation. The value of r can range from $+1$ for complete association to -1 for complete negative association.

The correlation coefficient is much the same thing as a *covariance*. In terms of corrected sums of squares and cross products the variance of a variable Y is $\sum y^2/(n-1)$. The variance of X is $\sum x^2/(n-1)$. The covariance of X and Y is $\sum xy/(n-1)$, and like the correlation coefficient is a measure of the equality of the fluctuations in the two variables. If s_X and s_Y are the standard deviations of X and Y, and s_{XY} is the covariance between the two variables

$$r = \frac{s_{XY}}{(s_X s_Y)}$$

The size of a covariance can range between plus and minus infinity, and is dependent not only on the changes in the two variables but also on the units of measurement. For example, if petiole lengths of two flower species were measured in millimeters, the covariance between the two variables would be much larger than if the lengths were measured in inches. The correlation coefficient does not depend on the units of measure.

The correlation coefficient will be illustrated with some data from Grieg-Smith (1957). In Table 9-10 the numbers of two species of plants in nine adjacent 10-square-foot plots are listed. Remember in correlation no distinction is made between dependent and independent variables. Assuming both species numbers are normally distributed, the estimated correlation coefficient between the numbers of these two plant species is

$$r = \frac{-1614}{(1152)(2870)^{1/2}}$$

$$= -.8886$$

There is a distinct negative association between these two species of plants, one species tending to be common if the other species is rare, and vice versa. No functional relationship is implied by the coefficient, however. The negative association is observed but not explained. The correlation coefficient is very variable if sample size is small. Because of this variability it is usually difficult to detect a significant but small correlation unless sample size is quite large.

The significance of the correlation coefficient is tested if both variables are normally distributed. The first obvious hypothesis is H_0: $\rho = 0$, versus the alternative hypothesis H_1: $\rho \neq 0$. The parameter ρ is the true population value of the correlation coefficient of which r is an estimate. The hypothesis of no significant difference of r from 0 is accepted if the t statistic

$$t = |r| \sqrt{\frac{n-2}{1-r^2}}$$

is not significant with $n - 2$ degrees of freedom. In the example of the two plant species $t = 5.12558$. The probability of a value of t this large arising by chance if the null hypothesis is true is much less than 1 percent. Therefore, the correlation between the two species of plants is significantly different from zero. If we wished to test for a correlation coefficient only significantly larger or only significantly smaller than zero, i.e., H_0: $\rho > 0$ or H_0: $\rho < 0$, all of α, the probability of accepting the wrong hypothesis, is put in one tail of the t distribution. In the t table if a 5 percent probability level is desired for a one-tailed test, look under 10 percent, and so forth. Tables for checking the significance of a correlation coefficient directly can be found in Beyer (1968) and other compendiums of statistical tables.

Sometimes the correlation coefficient between two variables is calculated separately in two independent, random samples. Can the two correlation

Table 9-10 NUMBERS OF TWO SPECIES OF PLANTS EXPRESSED AS COVER COUNTS IN NINE 10-SQUARE-FOOT PLOTS

X, Festuca ovina	Y, Cirsium acaule
99	10
95	4
83	22
82	13
68	35
64	26
62	21
49	36
46	37

SOURCE: Data from Grieg-Smith, P., The Use of Random and Contiguous Quadrats in the Study of the Structure of Plant Communities, *Ann. Bot. Lond. N.S.*, vol. 16, 1957.

coefficients be considered equal?   The null hypothesis is $H_0: \rho_1 = \rho_2$, versus the alternative hypothesis $H_1: \rho_1 \neq \rho_2$.   To test the null hypothesis the statistic d is calculated as

$$d = \frac{z_1 - z_2}{\left(\dfrac{1}{n_1 - 3} + \dfrac{1}{n_2 - 3}\right)^{1/2}}$$

where n_1 and n_2 are the sample sizes of the first and second samples, and z_1 and z_2 are transformations of the two sample estimates of ρ_1 and ρ_2, i.e., r_1 and r_2, given by the equation

$$z = \frac{1}{2} \ln\left(\frac{1 + r}{1 - r}\right)$$

The hypothesis of equal correlations is accepted if $|d| \leq z_{\alpha/2}$, where $z_{\alpha/2}$ is found from a table of the standardized normal distribution function, e.g., in Beyer (1968).   This z is not the same as the transformation z above, but is a standardized normal variate.   Because the test is two-tailed as formulated, one-half of the confidence level α is placed in each of the two tails.   When α is .05, the test is $|d| \leq z_{.025}$.   The alternative hypothesis is accepted if $|d| > z_{.025}$.   Tables of the z transformation are included in Beyer (1968).

Using the z transformation of the correlation coefficient a confidence interval may be placed around an estimated ρ as

$$P\left\{\tanh\left[z - \frac{z_{\alpha/2}}{(n - 3)^{1/2}}\right] \leq \rho \leq \tanh\left[z + \frac{z_{\alpha/2}}{(n - 3)^{1/2}}\right]\right\} = 1 - \alpha$$

9-3.4   Multiple Regression

It is a rare situation indeed if only one independent variable significantly influences a dependent variable such as larval survival.   The linear, one-variable regression equation is $Y_i = \alpha + \beta X_i + \varepsilon_i$.   The dependent variable Y can also be expressed as a linear, additive function of more than one independent variable.   If there are four independent variables, perhaps temperature, humidity, rainfall, and wind velocity, the multiple regression equation might be

$$Y_i = \alpha + \beta_1 X_{1i} + \beta_2 X_{2i} + \beta_3 X_{3i} + \beta_4 X_{4i} + \varepsilon_i$$

In this equation X_1, X_2, X_3, and X_4 are the four independent variables, and β_1, β_2, β_3, and β_4 are known as partial regression coefficients.   The assumptions made in multiple regression are generally the same as in linear regression with one independent variable.   The dependent variable is assumed to be normally distributed, the independent variables are measured without error, and the relationship between the independent variables and the dependent variable must be linear and additive, or can be made so by transformation. Also, the deviations from the regression are normally and independently distributed with zero mean and variance σ_e^2. The number of independent variables must be small relative to the sample size. A regression analysis with 80 independent variables and 100 observations will lead to meaningless results.

The regression model postulates a linear, additive relationship between the independent and dependent variables. Completely linear, additive relationships are rare in practice. It is not surprising that many multiple regression models fail miserably as predictors of the dependent variable. The first step in any multiple regression analysis is to determine the existence and extent of interactions among the independent variables by the analysis of variance. One should also check for curvilinear relationships between each of the independent variables and the dependent variable. After these interactions and curvilinear relationships have been discovered, the variables are transformed, if possible, to result in a linear, additive regression model. After the correct regression model has been found the analysis can be carried out. Sometimes this is not possible, and multiple regression cannot be used. If one is content simply to supply the computer with raw data and a multiple regression program, the results will not be satisfactory. The regression model also assumes that the independent variables are independent of each other; i.e., one variable is not a linear function of another. In practice this requirement is not too rigid, and does not seriously affect the predictive value of the regression equation.

Although multiple regression analysis can be carried out with desk calculators for two and three independent variables, regressions with more than three independent variables are computer material. Most computing centers have multiple regression programs of one form or another. The calculations given below are all in matrix notation.

The initial step in the multiple regression analysis is the calculation of the covariance matrix between all the variables. This is done by first calculating the matrix of corrected sums of squares and cross products. With three independent variables, X_1, X_2, and X_3, and one dependent variable Y, the matrix A is

$$\mathbf{A} = \begin{bmatrix} \sum y^2 & \sum yx_1 & \sum yx_2 & \sum yx_3 \\ \sum x_1 y & \sum x_1^2 & \sum x_1 x_2 & \sum x_1 x_3 \\ \sum x_2 y & \sum x_2 x_1 & \sum x_2^2 & \sum x_2 x_3 \\ \sum x_3 y & \sum x_3 x_1 & \sum x_3 x_2 & \sum x_3^2 \end{bmatrix}$$

The covariance matrix of variances and covariances of the variables is

$$S = \frac{A}{n-1}$$

where n is the number of observations. With three independent variables and one dependent variable

$$S = \begin{bmatrix} \text{var } Y & \text{cov}(YX_1) & \text{cov}(YX_2) & \text{cov}(YX_3) \\ \text{cov}(X_1 Y) & \text{var}(X_1) & \text{cov}(X_1 X_2) & \text{cov}(X_1 X_3) \\ \text{cov}(X_2 Y) & \text{cov}(X_2 X_1) & \text{var}(X_2) & \text{cov}(X_2 X_3) \\ \text{cov}(X_3 Y) & \text{cov}(X_3 X_1) & \text{cov}(X_3 X_2) & \text{var}(X_3) \end{bmatrix}$$

The matrix has been partitioned into four parts which will be represented as

$$S = \begin{bmatrix} s_1^2 & s_{12}' \\ s_{12} & S_{22} \end{bmatrix}$$

where s_1^2 is the variance of the dependent variable, s_{12} and s_{12}' are the covariances between the dependent and independent variables, and S_{22} represents the covariances and variances between the independent variables. Given these four partitions of the covariance matrix, the vector of estimates of the partial regression coefficients is

$$B = S_{22}^{-1} s_{12}$$

The parameter α is estimated as

$$a = \bar{Y} - b_1 \bar{X}_1 - b_2 \bar{X}_2 - b_3 \bar{X}_3$$

The partial regression coefficients β_i are written formally as $\beta_{1i.23...i-1}$. For example, β_3 is $\beta_{14.23}$. The notation $\beta_{14.23}$, the partial regression coefficient between the dependent variable and the third independent variable, represents the coefficient of linear regression between the third independent variable and the dependent variable with the first and second independent variables held constant. This is not the same as the simple linear regression between the two when the other two independent variables are ignored. The parameter $\beta_{14.23}$ represents the variance of the dependent variable Y explainable by X_3 after the variance accounted for by X_1 and X_2 has been removed. The general procedure of multiple regression is to first remove the variance of the dependent variable explainable by X_1. Of the remaining variance, the residual variance, in Y, X_2 will explain some more. After this variance is removed part of the residual variance may be subtracted by the regression of Y on the third independent variable. The significance of the partial regression coefficient of each independent variable depends on the order of the independent variables in the regression equation. If the independent variables are reordered, the significance of the partial regression coefficients will change.   For example, independent variable 7 may explain a significant amount of the variance in Y in a simple linear regression, but if placed in the seventh position in a multiple regression, it may not explain a significant amount of the residual variance left after the variance attributable to the first six independent variables has been removed.   A practical procedure in multiple regression is to arrange the independent variables in an order such that X_1 is the variable accounting for the largest proportion of the corrected total sum of squares of Y, X_2 is the variable removing the maximum amount of residual variance, and so forth.   A computer program called *stepwise multiple regression* carries out this procedure automatically.

Once the parameters of the multiple regression equation have been estimated, a logical question is, "Do the independent variables account for a significant proportion of the variance of the dependent variable Y?"   The proportion of the corrected total sum of squares of the dependent variable $\sum y^2$ accounted for by the independent variables is the square multiple correlation coefficient, $R^2_{1.234...q}$, where q is the total number of independent variables plus the dependent variable.   The dependent variable is to the left of the dot and the independent variables to the right.

$$R^2_{1.234...q} = \frac{\mathbf{s}_{12}'\mathbf{S}_{22}^{-1}\mathbf{s}_{12}}{s_1^{\ 2}}$$

With three independent variables and one dependent variable the squared multiple correlation coefficient is $R^2_{1.234}$.   To test the significance of $R^2_{1.234}$ calculate the F statistic

$$F = \left(\frac{n-q}{q-1}\right)\left(\frac{R^2}{1-R^2}\right)$$

which has an F distribution with $q-1$ and $n-q$ degrees of freedom.   Two equivalent null hypotheses are tested: (*1*) H_0: $\sigma_{12} = 0$, and (*2*) H_0: $\beta = 0$, where σ_{12} and β are the true population vectors estimated by $\mathbf{s}_{12}$ and $\mathbf{B}$. If the calculated F statistic is greater than the tabulated F value with $q-1$ numerator degrees of freedom and $n-q$ denominator degrees of freedom, the null hypotheses will be rejected, and it will be concluded that some or all of the independent variables account for a significant proportion of the variance of Y.

A test for the significance of each independent variable amounts to testing each of the partial regression coefficients to see if it differs significantly from zero.   In notation the null hypothesis for the significance of the third independent variable is H_0: $\beta_{14.23} = 0$, versus the alternative hypothesis

$H_1: \beta_{14.23} \neq 0$. Because the significance of an independent variable depends on its position in the regression equation, the significance of the partial regression coefficients are tested successively. Suppose that there are four independent variables. The regression between X_1 and Y is estimated, and the proportion of the corrected total sum of squares $\sum y^2$ is designated $R_1{}^2$, equivalent in terms of multiple regression to $R_{1.2}^2$. The second independent variable is then added, and $R_{1.23}^2$ calculated. Designate $R_{1.23}^2$ as $R_2{}^2$. The inclusion of the second independent variable may or may not have explained a significant amount of the residual variance of the dependent variable. $R_2{}^2$ will always be larger than $R_1{}^2$, but it may not be significantly larger. To see if $R_2{}^2$ is significantly larger than $R_1{}^2$ calculate the F statistic

$$F = \frac{(n-3)(R_2{}^2 - R_1{}^2)}{(1 - R_2{}^2)}$$

The F statistic has 1 and $n-3$ degrees of freedom. If the F value is significant, it is concluded that the addition of the second independent variable has reduced the residual variance of the dependent variable significantly. If the second independent variable is significant, we can then add a third independent variable. The squared multiple correlation coefficient is $R_{1.234}^2 = R_3{}^2$. As before, calculate the F statistic

$$F = \frac{(n-4)(R_3{}^2 - R_2{}^2)}{(1 - R_3{}^2)}$$

with 1 and $n-4$ degrees of freedom. If the F statistic is significant, the third independent variable has explained a significant amount of the residual variance of the dependent variable left after the variance explained by X_1 and X_2 has been removed.

In general, if k independent variables have already been tested for significance and included in the regression equation, the addition of m additional variables may be tested to see if they significantly reduce the residual variance by the F statistic

$$F = \frac{(n-k-m-1)(R_{k+m}^2 - R_k{}^2)}{m(1 - R_{k+m}^2)}$$

with m and $n-k-m-1$ degrees of freedom. If only one additional independent variable is added at each step

$$F = \frac{(n-k-2)(R_{k+1}^2 - R_k{}^2)}{(1 - R_{k+1}^2)}$$

with 1 and $n-k-2$ degrees of freedom.

If there are four independent variables, and the successive F tests show that in the order they have been included in the regression equation independent variables 2 and 4 are not significant, variables 2 and 4 may be dropped from the regression equation. However, if an independent variable is dropped, the partial regression coefficients of the remaining independent variables must be reestimated.

To illustrate the use of multiple regression suppose that the biomass of a plant species produced per square meter in 18 plots is the dependent variable, and that the amounts of three nutrients in the soil are three independent variables. Suppose that the matrix of corrected sums of squares and cross products is

$$\mathbf{A} = \begin{array}{c} Y \\ X_1 \\ X_2 \\ X_3 \end{array} \begin{array}{cccc} Y & X_1 & X_2 & X_3 \\ \left[\begin{array}{cccc} 12389.66 & 3231.48 & 2216.44 & 7593.00 \\ 3231.48 & 1752.96 & 1085.61 & 1200.00 \\ 2216.44 & 1085.61 & 3155.78 & 3364.00 \\ 7593.00 & 1200.00 & 3364.00 & 35572.00 \end{array} \right] \end{array}$$

By dividing $\mathbf{A}$ by 17, $(n-1)$, the covariance matrix is

$$\mathbf{S} = \left[\begin{array}{cccc} 728.80 & 190.09 & 130.38 & 446.65 \\ 190.09 & 103.12 & 63.86 & 70.59 \\ 130.38 & 63.86 & 185.63 & 197.88 \\ 446.65 & 70.59 & 197.88 & 2092.47 \end{array} \right]$$

The vector $\mathbf{B}$ is

$$\mathbf{B} = \left[\begin{array}{ccc} 103.12 & 63.86 & 70.59 \\ 63.86 & 185.63 & 197.88 \\ 70.59 & 197.88 & 2092.47 \end{array} \right]^{-1} \left[\begin{array}{c} 190.09 \\ 130.38 \\ 446.65 \end{array} \right] = \left[\begin{array}{c} 1.785 \\ -.083 \\ .161 \end{array} \right]$$

The estimate of α is $a = 81.28 - (1.785)(11.94) - (-.083)(42.11) - (.161)(123.00) = 43.66$ where 11.94, 42.11, and 123.00 are the means of the observations of the three independent variables. The multiple regression equation is

$$Y = 43.66 + 1.785\,X_1 - .083\,X_2 + .161\,X_3$$

The squared multiple correlation coefficient $R^2_{1.234}$ is

$$R^2_{1.234} = \frac{[190.09 \quad 130.38 \quad 446.65]\mathbf{B}}{728.80}$$

$$= .55$$

indicating that 55 percent of the variance in plant biomass is accounted for by the three independent variables.   The F test for the significance of the squared multiple correlation coefficient is

$$F = \left(\frac{18 - 3}{2}\right)\left(\frac{.55}{1 - .55}\right)$$

$$= 9.2$$

With 2 and 15 degrees of freedom this F value is significant at about the .5 percent probability level.   In testing the significance of each of the partial regression coefficients successively in the given order, only b_1 proves to be greater than zero.   Independent variables X_2 and X_3 do not explain a significant amount of the residual variance left after the variance attributable to X_1 is subtracted.   However, the third independent variable approaches significance.

One of the chief difficulties in applying multiple regression is the presence of interactions among the independent variables.   An additive interaction between two independent variables is

$$Y = \alpha + \beta_1 X_1 + \beta_2 X_2 + \beta_3 X_1 X_2$$

or the sum of the actions of the two variables plus an extra effect due to their interaction.   When an additive interaction is found or suspected, the term $X_1 X_2$ is treated as a third independent variable in the multiple regression. However, as the number of independent variables increases, the number of possible additive interactions increases tremendously.   With three independent variables

$$Y = \alpha + \beta_1 X_1 + \beta_2 X_2 + \beta_3 X_3 + \beta_4 X_1 X_2 + \beta_5 X_1 X_3 + \beta_6 X_2 X_3$$
$$+ \beta_7 X_1 X_2 X_3$$

resulting in a multiple regression with seven rather than only three independent variables.   However, the above method is an effective way of improving a regression model adulterated by additive interaction effects.

An interaction may also be multiplicative.   A possible model is

$$Y = \alpha X_1{}^{\beta_1} X_2{}^{\beta_2}$$

The model is made linear and additive by the transformation

$$\log Y = \log \alpha + \beta_1 \log X_1 + \beta_2 \log X_2$$

To test if this transformation has rectified the data, plot $\log Y$ on $\log X_1$ for constant values of $\log X_2$.   If the rectification has been successful, a straight line will result for every constant value of X_2. If a series of straight lines do not result, the transformation has not been successful.

Curvilinear but additive relationships between a dependent and two independent variables can be modeled as

$$Y = \alpha + \beta_1 X_1 + \beta_2 X_1{}^2 + \beta_3 X_2 + \beta_4 X_2{}^2$$

Multiple regression is a powerful tool, but to be successful the underlying relationships between the variables must be known.

9-3.5   Partial Correlation

When there is a group of variables and no distinction is made between independent and dependent variables, partial correlations are used to study the effect of holding some of the variables constant on the correlations between the remaining variables.   If there are four variables, then analogous to the partial regression coefficients $r_{14.23}$ is the correlation between variables 1 and 4, given that variables 2 and 3 are held constant.   The s variables are divided into p variables of interest, and q variables to be held constant.   Then calculate the covariance matrix between all the variables and partition it.

$$S = \begin{bmatrix} S_{11} & S_{12} \\ S_{12}{}' & S_{22} \end{bmatrix}$$

The matrix S_{11} contains the variances and covariances among the p variables, S_{22} the variances and covariances among the q variables, and S_{12} the covariances between the p and q variables.   Then calculate the matrix of partial covariances $S_{1.2}$ as

$$S_{1.2} = S_{11} - S_{12}S_{22}{}^{-1}S_{12}{}'$$

The matrix $S_{1.2}$ represents the variances and covariances among the p variables, given that the variance attributable to the q variables has been removed.   Suppose that the covariance matrix of four species densities of birds is

$$
S =
\begin{array}{c}
 \\
1 \\
2 \\
3 \\
4
\end{array}
\begin{array}{cc}
\overset{\displaystyle p}{} & \overset{\displaystyle q}{} \\
1 \quad 2 & 3 \quad 4 \\
\begin{bmatrix}
20 & 12 & 7 & 9 \\
12 & 18 & 3 & 6 \\
7 & 3 & 17 & 8 \\
9 & 6 & 8 & 25
\end{bmatrix}
\end{array}
$$

We wish to know the variances and covariances among the first two species of birds after subtracting the variance attributable to species 3 and 4.

$$S_{1.2} = \begin{vmatrix} 20 & 12 \\ 12 & 18 \end{vmatrix} - \begin{bmatrix} 7 & 9 \\ 3 & 6 \end{bmatrix} \begin{bmatrix} 17 & 8 \\ 8 & 25 \end{bmatrix}^{-1} \begin{bmatrix} 7 & 3 \\ 9 & 6 \end{bmatrix}$$

$$= \begin{bmatrix} 15.5864 & 9.5331 \\ 9.5331 & 16.4799 \end{bmatrix}$$

The covariance between species 1 and 2 has been reduced from 12 to 9.5331 by holding species 3 and 4 constant.   The partial correlations are calculated as

$$R_{1.2} = D^{-1/2} S_{1.2} D^{-1/2}$$

where the matrix $D^{-1/2}$ is a diagonal matrix whose elements are the square roots of the reciprocals of the diagonal elements of $S_{1.2}$.   In the example

$$R_{1.2} = \begin{bmatrix} .253 & 0 \\ 0 & .246 \end{bmatrix} \begin{bmatrix} 15.5864 & 9.5331 \\ 9.5331 & 16.4799 \end{bmatrix} \begin{bmatrix} .253 & 0 \\ 0 & .246 \end{bmatrix}$$

$$= \begin{bmatrix} 1.00 & .61 \\ .61 & 1.00 \end{bmatrix}$$

If variables 3 and 4 are held constant, the partial correlation between variables 1 and 2, $r_{12.34} = .61$.

Partial correlations and partial covariances are particularly useful in the following type of situation.   Suppose that the biomass of several species of plants in a series of random quadrats is measured.   Several important environmental variables such as nitrogen content of the soil, densities of herbivores, and moisture are also recorded.   The observed correlations among the biomasses of the plant species may be due to: (*1*) interactions among the plants, (*2*) the responses of the plants to the environmental variables, and (*3*) chance fluctuations and other unknown variables.   We would like to know how much of the correlation between any two plant species is due to interactions among the species, and how much is due to the other variables of the environment that we measured.   By partialing out the environmental variables we remove the correlation in the species densities due to these variables, leaving only the correlation due to plant interactions and other unmeasured environmental variables.

9-4   DEDUCTIVE MODELING

Regression models are inductive.   An experiment is run or observations are made; the data are plotted, and a regression equation is fit to the data depending on the functional relationship suggested by the data.   Models may also be derived deductively.   In a deductive model we make logical hypotheses about the process being studied, formulate the hypotheses as a mathematical model, and then try to fit the model to the data.   Sometimes a model may be derived by a combination of deductive and inductive methods.   A deductive model is fit to the data and, based on the observed residuals of the model to the data, the model is modified.

The models of population change in Parts One and Two of this text were for the most part derived deductively.   The derivation of deductive models usually involves rather advanced calculus.   However, the general principle involved can be illustrated by a few simple examples.   Suppose that lead from cars is being added to the roadside at a constant rate a.   The instantaneous change in the amount of lead in the soil with time can be represented by the differential equation

$$\frac{dP}{dt} = a$$

The differential dP/dt is an instantaneous rate, and is the slope of the line tangent to the curve representing the amount of lead present at time t.   In this case dP/dt is a constant a, and so the slope of the curve is constant at any point t.   Because the slope is constant, the functional response between amount of lead and time is a straight line.   Applying the methods of calculus, the differential equation is solved for an explicit value of P at time t, P_t.   However, there is an infinite number of lines with slope a, and, therefore, the initial value of P at time $t = 0$ must be stipulated to arrive at a unique solution of the differential equation.   The solution is $P_t = P_0 + at$, the equation of a straight line.   In exponential population growth the rate of change in the density of the population is assumed to be proportional to the density of the population.   The proportionality constant is r, and the differential equation is $dN/dt = rN$.   Again, to derive a unique solution of the differential equation, the initial value of N must be stipulated.   Solving the equation by calculus we find $N_t = N_0 e^{rt}$.   In logistic growth the rate of increase of the population is not constant, but depends on the density of individuals.   We assume that the decrease in r is directly proportional to the density of the population; i.e., rate of increase $= r - aN$, rather than a constant r.   Therefore, the differential equation for limited population growth is $dN/dt = N(r - aN)$, rather than $dN/dt = Nr$.   If a variable such as larval survival is a function of several variables, the formulation of deductive models can become complex.

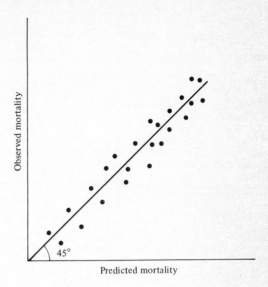

FIGURE 9-17
A plot of observed and predicted values of larval mortality to illustrate the clustering of the data points around a line drawn at a 45-degree angle to the two axes. There are no significant directional deviations, indicating that the model includes most of the significant factors affecting the dependent variable.

Watt (1968) discusses the formulation of deductive models in some detail, and the reader may refer to his book for a presentation of the methods of deductive modeling.

9-4.1   Testing the Fit of a Model to Data

After a model has been formulated, either inductively or deductively, and the parameters of the model estimated, the model should be tested against a new set of data.   Suppose larval mortality is considered a function of temperature. For a series of experiments at different temperatures there will be a series of observed mortality values and a series of predicted mortalities.   The data points, the observed mortality versus the predicted mortalities, for each experiment or observation are plotted, and a line is drawn through the graph at a 45-degree angle to the X or Y axis.   If the points are randomly scattered around the line, as in Fig. 9-17, there is no reason to reject the fit of the model to the data.   However, if there are patches of aberrant points, as in

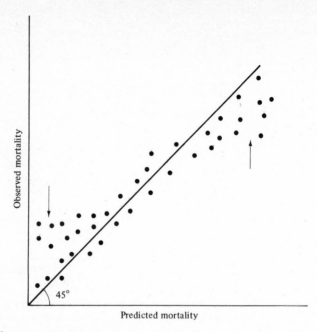

FIGURE 9-18
A plot similar to Fig. 9-17, but this time there are significant deviations of the
data points from the line, indicating that some significant factor affecting the
dependent variable has been omitted.

Fig. 9-18, it may be that some significant factor has been omitted or the
correct functional relationship has been missed.   In this case the model is
either corrected, or if a factor is missing, the model is reformulated to
include it.

10

THE MODELING OF A NATURAL POPULATION

The purpose of this chapter is to try to tie together some of the information presented in Chap. 9 by reviewing one of the truly classic research programs in ecology, the modeling of the spruce budworm population in northwestern New Brunswick. The discussion covers the research of Morris et al. (1963) between the years 1945 and 1959 in their study area 1.

Although normally a fairly rare insect, the spruce budworm, *Choristoneura fumiferana*, occurs in periodic severe outbreaks of great economic importance. One of these outbreaks took place during the years 1945–1959, when the research was being done. Therefore, the model of the spruce budworm developed is applicable only during the outbreak, epidemic, phase of the moth.

Section 10-1 outlines the development of the model, and analyzes both the successes of the model and some of its shortcomings. Section 10-2 takes up an alternate approach to population modeling called key factor analysis. Section 10-3 considers the analysis of a series of nonindependent observations by autoregression and time series analysis.

10-1   THE DEVELOPMENT OF THE SPRUCE BUDWORM MODEL

The first step in the development of a model of numerical change in the spruce budworm population was the formation of a series of life tables for the insect. This mode of modeling is sometimes called the life table method. Three developmental stages were sampled regularly on the same trees in a set of permanent plots: (1) eggs, (2) third- to fourth-instar larvae, and (3) pupae. These quantitative counts, along with an assessment of the environmental factors causing mortality between successive stages, provided the basic data used to develop the life tables and the model. The life table was broken down into: (1) eggs, (2) small larvae—instar I, (3) large larvae—instars III and IV, (4) pupae, (5) moths, and (6) females times 2. The females-times-2 component was added to reflect the effect of an unequal sex ratio on the percent survival during a generation. Eighty life tables were obtained. In Table 10-1 the average of these 80 tables is shown. The life table begins with 200 eggs as a normal fecundity value. The N_x column lists the number of individuals surviving through the succession of stages from 200 eggs to 2.71 moths. Because of a slight inequality of the sex ratio the effective number of adults is twice the number of females, 2.52. The $M_x F$ column lists the mortality factors operating on each stage, breaking the total numbers lost at each stage into the number lost to each source of mortality. The S_x column lists the percent of each stage surviving to reach the next. The last row of the life table gives the total survival in each generation. Of 200 eggs only .5 percent of them survive to become egg-laying adults, resulting in a complement of 100 eggs to begin a new generation. Most of the life tables were formulated during periods of decreasing density, which accounts for the decrease from 200 to 100 eggs in a generation.

The form of the basic model was the trend of the population size as a function of the survival of each of the age groups, the proportion of females, and the fecundity of the females. If 50 percent of the eggs and 50 percent of the small larvae survive, the percent of the eggs surviving to become large larvae will be $(.5)(.5) = .25$, because the probability of two independent stochastic events both happening is the product of their probabilities. In the same way the percent of the eggs surviving from laid egg to adult is

$$\Pr(\text{egg to egg-laying adult}) = S_E S_S S_L S_P S_A$$

where S_E is the probability of survival of eggs to eclosion, S_S is the probability of survival of small larvae, S_L is the probability of survival of large larvae, S_P is the probability of survival of pupae, and S_A is the probability of survival of adults up to and including the time of oviposition.   The population trend from one generation to the next is the number of eggs in the $(n + 1)$st generation divided by the number of eggs in the nth generation.

$$I = \frac{N_E(n + 1)}{N_E(n)} \tag{10-1}$$

The number of eggs laid by the female adults to become $N_E(n + 1)$ will depend on the proportion of females, P, the maximum number of eggs each female can lay, F (200 in the study), and the proportion of the maximum fecundity achieved by the females of a given generation, P_F.   The actual fecundity of the females depends on pupal size, and is influenced by feeding conditions during the larval stage of the female moths.   Actual fecundity is, therefore, $P_F F$. In place of Eq. (10-1) we can write

$$I = S_E S_S S_L S_P S_A P P_F F \tag{10-2}$$

Table 10-1   THE MEAN LIFE TABLE FOR THE SPRUCE BUDWORM POPULATION IN AREA 1 DURING THE OUTBREAK

x	N_x	$M_x F$	M_x	$100M/N$	S_x
Eggs	200	Parasites	18.0	9	
		Predators	12.0	6	
		Others	8.0	4	
		Total	38.0	19	.81
Instar I	162	Fall and spring dispersal	132.8	82	.18
Instar III	29.2	Parasites	11.7	40	
		Disease	6.7	23	
		Other	6.7	23	
		Total	25.1	86	.14
Pupae	4.10	Parasites	.53	13	
		Predators	.16	4	
		Other	.70	17	
		Total	1.39	34	.66
Moths	2.71	Sex (46.5% females)	.19	7	.93
Females × 2	2.52	Reduction in fecundity	.50	20	.80
Generation			198.7	99.49	.005

x = age interval; N_x = number alive at beginning of x; $M_x F$ = factor responsible for M_x; M_x = number dying during x; $100M/N = M_x$ as percent of N_x; S_x = survival rate within x.

SOURCE: After Morris, R. F. (ed.), The Dynamics of Epidemic Spruce Budworm Populations, Mem. Entomol. Soc. Can., no. 31, 1963.

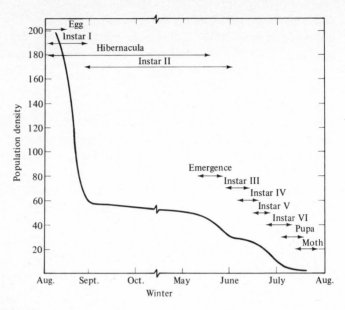

FIGURE 10-1

The survivorship curve for the spruce budworm population in area 1. Population densities are based on mean survival values for each age interval. (*From Morris, 1963.*)

It was not possible to separate adult dispersal from adult mortality, so S_A includes losses to mortality and losses and gains due to dispersal. It is theoretically possible to have S_A larger than 1 if large numbers of adults immigrate into the area, but few emigrate. If dispersal is known, the additional term $N_D/N_E(n)$ can be added to Eq. (10-2). In this study the last term was not used because dispersal could not be separated from S_A, even though it was important.

The overall model is, therefore, Eq. (10-2), and each of its terms (for example, S_E) is a component to be modeled as discussed in Chap. 9. The index I is the same as the survival from generation to generation if the F term is dropped, so that from Eq. (10-2) the probability of survival over the generation becomes

$$.005 = .812 \times .182 \times .141 \times .656 \times .510 \times .932 \times .801 \qquad (10\text{-}3)$$
$$\quad\; S_G \quad\;\; S_E \quad\;\; S_S \quad\;\; S_L \quad\;\; S_P \quad\;\; S_A \quad\; 2P \quad\;\; P_F$$

using the statistics in Table 10-1.

To study the relative importance of each of the terms to generation

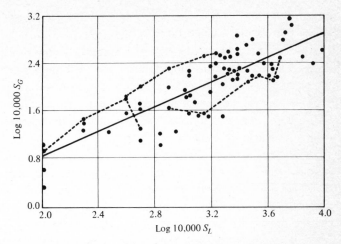

FIGURE 10-2

The relationship between generation survival (S_G) and the survival of large larvae (S_L).   The broken lines above the regression line connect the points contributed by plot G5; those below the line connect those contributed by plot K1.   (*From Morris, 1963.*)

survival a regression analysis may be carried out by transforming Eq. (10-3) into a linear equation by using logarithms.

$$\log 10{,}000S_G = \log 10{,}000S_E + \log 10{,}000S_S + \log 10{,}000S_L$$

$$+ \log 10{,}000S_P + \log 10{,}0002P + \log 10{,}000P_F \quad (10\text{-}4)$$

The S_A term was deleted because adult mortality could not be separated from dispersal.   Each survival term was multiplied by 10,000 to avoid negative logarithms.   A simple linear regression and correlation coefficient was calculated between S_G and each of the other S terms, using the 80 life tables as data.   For example, the regression of log $10{,}000S_L$ against log $10{,}000S_G$ is shown in Fig. 10-2.   The dotted lines connect data points arising from two of the plots.   When the individual data points were labeled according to plot and year, the variate values for a given plot tended to fall consistently above or below the regression line.   By measuring the deviations from the regression line, and correcting each point according to the mean deviation from the regression for the appropriate plot, plot differences were eliminated from the regression.   Table 10-2 lists the results of the analysis. If the variance due to plots in the regression of S_G and S_L is corrected for, the regression explains 86 percent of the variance in S_G.   The most critical period in the life history of the spruce budworm is the large-larvae stage, mortality at the large-larvae stage determining to a large extent the future of the population. If mortality is high (the survival of large larvae was on the average only 14 percent), great variability in the mortality rate exerts a large effect on the future population size. Although mortality at the small-larvae stage was almost as large (survival was only 18 percent), the mortality rate was very consistent from year to year. As a result the proportion of the variance in generation survival explained by the regression of log $10,000S_S$ on log S_G was only 28 percent.

The negative correlation of S_E on S_G is a coincidence. In three of the years in which S_S was high, egg parasitism was also high. If parasitism is corrected for (see below), the percentage of variance in S_G explained by S_E is almost nil. Mortality factors acting on a high survival rate like S_E must change drastically in order to have a significant effect on S_G. When all variables were considered, the square partial correlation coefficients were: $S_E = .000$, $S_S = .224$, $S_L = .702$, $S_P = .028$, $2P = .002$, and $P_F = .044$, leading to the same conclusions reached before.

With the general model decided upon and the data gathered, the next step is the modeling of each of the components. In this particular study

Table 10-2 REGRESSION STATISTICS WHEN GENERA-TION SURVIVAL S_G IS TREATED AS THE DEPENDENT VARIABLE AND AGE-INTERVAL SURVIVAL AS THE INDEPEN-DENT VARIABLE. SURVIVAL IN EACH CASE WAS EXPRESSED AS log $10,000S$, AND SIMPLE, NOT MULTIPLE, REGRESSION METHODS WERE USED (r = CORRELATION COEFFICIENT; b = SLOPE; a = INTERCEPT)

Survival	Adjustment	r^2	r	a	b
S_E	None	.114	−.34	13.30	−2.89
	Parasitism	.001	.03	1.20	.31
S_S	None	.280	.53	−1.02	.92
S_L	None	.680	.82	−1.23	1.03
	Plots	.859	.93	−1.47	1.10
S_P	None	.122	.35	−5.76	2.04
$2P♀$	None	.010	−.10	7.50	−1.38
P_F	None	.063	.25	−2.38	1.13

$r = .23$ significant at the 5 percent probability level.
$r = .30$ significant at the 1 percent probability level.

SOURCE: After Morris, R. F. (ed.), The Dynamics of Epidemic Spruce Budworm Populations, Mem. Entomol. Soc. Can., no. 31, 1963.

it was decided to use primarily regression models, although the large-larvae survival component was formulated by using a combination of deductive and inductive methods. The derivation of the large-larvae component of the model will be used as an example.

10-1.1 Large-Larvae Survival S_L

In modeling S_L Watt (in Morris, 1963) reasoned that S_L could be thought of as a function of the multiplicative effect of the survivals from each source of mortality. If M_i is the mortality caused by the ith factor (say parasitism) the survival of large larvae from this factor will be

$$S_L = \frac{N_L - M_i}{N_L}$$

where N_L is the initial number of large larvae. If several mortality factors are operating, larval survival will be the product of the survivals from each mortality factor, because the probability of two events both happening is the product of their probabilities. The total larvae survival model is

$$S_L = \left(\frac{N_L - M_1}{N_L}\right)\left(\frac{N_L - M_2}{N_L}\right) \cdots \left(\frac{N_L - M_n}{N_L}\right)$$

where there are n mortality factors.

Mortality due to predacious insects, birds, and spiders was examined first. If any of the three groups of predators is to be a significant mortality factor to the spruce budworm population during an outbreak, one or all of the groups must show a significant increase in density with increasing budworm population. The relationship between the population densities of each of these three groups and spruce budworm density is shown in Fig. 10-3. Even though there appears to be some increase in the densities of all three groups with increasing budworm density, the increase is small compared to the increase in the budworm population. This was also true of parasitism. A more desirable way of measuring parasitism is the number of adult parasites available to attack larvae, rather than total percent larval parasitism. Unfortunately, measuring parasite densities in the field was difficult and was confounded by the parasites' attacking of alternate hosts. No significant correlation was found in this study between larval survival and parasitism or predators. Quite likely this would not be true during the endemic population periods of the spruce budworm.

Apparently larval survival actually increases, not decreases, at higher larval densities until other factors such as starvation come into play.

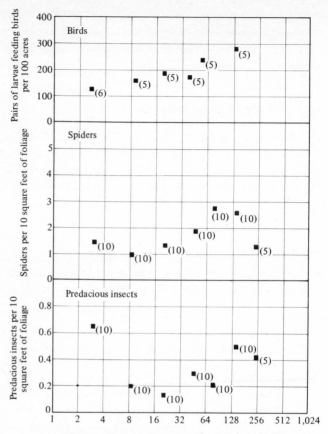

FIGURE 10-3
The numerical response of birds, spiders, and predacious insects to changes in the population density of spruce budworm larvae. The numbers in brackets just below each plotted point indicate the number of sets of data pooled to obtain the plotted values. (*From Morris, 1963.*)

Figure 10-4 demonstrates this to be true, even though the scatter of the data points is great. Watt investigated this point further by "correcting" larval survival for the effects of parasitism. This correction was carried out by determining the mortality caused by parasitism M_{par}, and calculating a corrected survival as

$$S_L' = \frac{S_L}{(N_L - M_{par})/N_L}$$

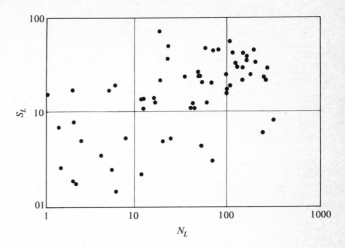

FIGURE 10-4
The survival of large larvae (S_L) as a function of larval density (N_L). Each point represents the mean datum from one plot in 1 year. (*From Morris, 1963.*)

where S_L' is the corrected value of S_L for each data point after the effect of parasitism has been removed. These corrected values of S_L' were plotted against N_L as shown in Fig. 10-5. It was decided that the data could best be fit by the equation

$$S_L' = a + N_L e^{-(b+cN_L)}$$

where a, b, and c are constants. This function fitted to the data points by iterative regression is

$$S_L' = .096 + N_L e^{-(4.866+.009N_L)} \qquad (10\text{-}5)$$

Equation (10-5) is the model of $(N_L - M_{\text{density}})/N_L$, representing the effect of larval density on survival.

For each data point the mortality due to increasing density can be calculated from Eq. (10-5), and S_L corrected for the effects of both parasitism and density, in the same manner as S_L' was arrived at, i.e.,

$$S_L'' = \frac{S_L}{[(N_L - M_{\text{par}})/N_L][(N_L - M_{\text{density}})/N_L]}$$

The effect of "phenological development" was then considered, and defined as the difference in days between the time of 25 percent growth of balsam fir shoots, the preferred host of the budworm, and the end of the fifth larval instar. This index was intended to express the availability of suitable

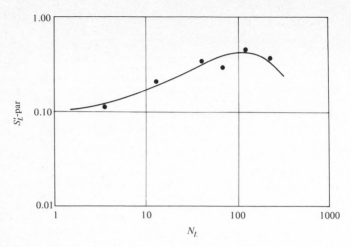

FIGURE 10-5

The survival of large larvae "corrected" for parasitism ($S_{L\text{-par}}$) as a function of larval density (N_L). The solid line is calculated from $S'_L = .096 + N_L e^{-(4.8666 + .009N_L)}$. The left circle represents a mean of 11 observed data sets; the remaining five circles represent means of 10 data sets each. (*From Morris, 1963.*)

food for the larvae. The regression equation expressing the survival of large larvae as a function of phenological development Z was

$$S_L'' = 1.5797 - .03959Z$$

Larval survival may again be corrected, this time for parasitism, density, and phenological development. This new index of larval survival, S_L''', was plotted against the ratio "mean temperature last 14 days of large larval period/mean relative humidity last 14 days of large larval period," symbolized as T/H. The regression equation expressing S_L''' as a function of T/H was

$$S_L''' = -3.851 + 6.299 \frac{T}{H} - 1.970 \frac{T^2}{H^2}$$

where the appropriate substitutions are $X_1 = T/H$ and $X_2 = T^2/H^2$. Because cold-wet or hot-dry conditions are unfavorable to budworm larvae, the effect of humidity and temperature is an interaction, not a straight additive effect.

Correcting larval survival for parasitism, density, phenological development, and temperature-humidity, S_L'''' was plotted against F_i, the isolation or distance of a stand of trees from adjoining stands. The equation used to fit the data was

$$S_L'''' = .6688 + .009F_i$$

The analysis was terminated at this point.

The entire model is

$$S_L = (S_{par})(.096 + N_L e^{-(4.87 + .009N_L)})(1.5797 - .0396Z)$$
$$\times \left(-3.85 + 6.30 \frac{T}{H} - 1.97 \frac{T^2}{H^2} \right)(.67 + .009F_i)$$

or

$$S_L = S_{par} S_N S_Z S_{T/H} S_{F_i}$$

To test the predictive value of the model the following substitutions were made: (1) $X_1 = S_{par}$, (2) $X_2 = S_N$, (3) $X_3 = S_Z$, (4) $X_4 = S_{T/H}$, and (5) $X_5 = S_{F_i}$. The model of large-larvae survival becomes

$$S_L = X_1 X_2 X_3 X_4 X_5$$

After taking the logarithms of S_L, X_1, X_2, X_3, X_4, and X_5 a multiple linear regression was run (Table 10-3). Only S_N, S_Z, and S_{F_i} were significant in the order given. The S_{par} and $S_{T/H}$ components of the model of large-larvae survival could have been dropped.

Table 10-3 ANALYSIS OF VARIANCE TABLE FOR THE MULTIPLE REGRESSION ANALYSIS OF S_L AS A FUNCTION OF LARGE LARVAL DENSITY, PHENOLOGICAL DEVELOPMENT, ISOLATION, PARASITES, AND TEMPERATURE-HUMIDITY INDEX

Source of variation	Degrees of freedom	Sum of squares	F
Density	1	22.453	30.91***
Phenological development	1	3.958	8.64**
Isolation	1	11.974	6.86*
Parasites	1	5.076	.32
Temperature-humidity index	1	2.483	.07
All independent variables	5	31.354	

*** Significant at the .1 percent probability level.
** Significant at the 1 percent probability level.
* Significant at the 5 percent probability level.

SOURCE: After Morris, R. F. (ed.), The Dynamics of Epidemic Spruce Budworm Populations, *Mem. Entomol. Soc. Can.*, no. 31, 1963.

10-1.2   A Summary of the Model and Its Properties

Each component of Eq. (10-3) was similarly modeled. The entire model is simply the product of each of the submodels of the components. The overall success of the model depends upon the reality of each of the components. For example, the S_L component of the model can be used to illustrate the problem of interaction. In formulating his model of S_L Watt postulated, in effect, multiplicative interaction between independent variables. Mott (in Morris, 1963) reviewed the problem of interaction in the S_L component, first dropping the two subcomponents S_{par} and $S_{T/H}$. The deletion of these two terms reduced the number of variables, and made it possible to set up a regression equation of the form

$$S_L' = \alpha + \beta_1 f(N_L) + \beta_2 F_i + \beta_3 Z + \beta_4 f(N_L)F_i + \beta_5 f(N_L)Z$$
$$+ \beta_6 f(N_L)ZF_i + \beta_7 ZF_i$$

where

$$f(N_L) = .096 + N_L e^{-(4.87 + .009N_L)}$$

carrying out the multiple regression results in Table 10-4. Each of the single terms in the regression is significant, but of the interactions only the interaction between F_i and Z is significant.

Mott concludes that interaction between the variables and its form should have been more carefully considered in all the analyses. However, a great many sets of data are necessary to test for the significance of interaction terms if there are more than two or three variables, because of the rapidly increasing number of interaction terms. This rapid increase in the number of interaction terms results in a vast reduction in the degrees of freedom left to the error term, and reduces the power of the test.

The entire model accounts for 41 percent of the variance of the survival in a generation S_G. Mott then asks whether the remaining 59 percent is due entirely to sampling error, or whether some significant independent variables had not been discovered. Would it be possible to improve the model by searching for other undiscovered mortality factors given the same sampling program?

This question can be answered in part by calculating the standard error of each survival rate. The standard error of each survival rate can be computed from the standard error of the two population means from which it is computed, for example, $S_L = N_L/N_P$. These standard errors differ from year to year and from plot to plot, although generally rates calculated from low population levels will be less precise than those from high population levels.   The standard error of the survival rate S for eggs, small larvae, and large larvae can be computed as

$$\sigma_S = (S) \left[\left(\frac{\sigma_{N_1}}{N_1} \right)^2 + \left(\frac{\sigma_{N_2}}{N_2} \right)^2 \right]^{1/2} \tag{10-6}$$

where, for example, N_1 is the number of large larvae, and N_2 is the number of pupae if the standard error of S_L is being estimated.   Equation (10-6) overestimates the standard error, because of covariance between N_1 and N_2, making the test that follows more rigorous.   After the standard errors of the survival rates are computed, each deviation from the fitted model can be expressed in standard error units as $(S_{obs} - S_{est})/\sigma_S$.   If expressed this way, the deviations will not be normally distributed, with about 68 percent of them between -1 and $+1$, about 95 percent between -2 and $+2$, and about 99 percent between -3 and $+3$.   The deviations in standard error units fitted to survival of small larvae S_S, and large larvae S_L, have been summarized as frequency distributions in Fig. 10-6, along with the expected frequency distributions.   There are clearly too many deviations on the left.   Therefore,

Table 10-4   ANALYSIS OF VARIANCE OF LARGE-LARVAE SURVIVAL CORRECTED FOR PARASITISM (S_L') AS A FUNCTION OF EXPRESSIONS OF N_L, F_i, AND Z, AND INTERACTIONS BETWEEN THESE FACTORS

Source of variation	Degrees of freedom	Sum of squares	F
All variables and interactions	7	1.2182	6.99***
Effect $f(N_L)$	1	.7109	28.6***
Added effect F_i	1	.1835	7.4**
Added effect Z	1	.1474	5.9*
Added effect $[F_i \times f(N_L)]$	1	.0187	.8
Added effect $[Z \times f(N_L)]$	1	.0078	.3
Added effect $[F_i \times Z \times f(N_L)]$	1	.0256	1.0
Added effect $(F_i \times Z)$	1	.1243	5.0*
Error	53	1.3205	

$R^2 = .4800$

*** Significant at the .1 percent probability level.
** Significant at the 1 percent probability level.
* Significant at the 5 percent probability level.

SOURCE: From Morris, R. F. (ed.), The Dynamics of Epidemic Spruce Budworm Populations, *Mem. Entomol. Soc. Can.*, no. 31, 1963.

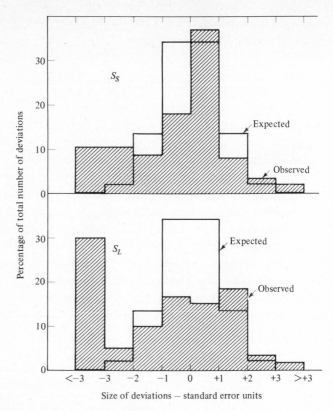

FIGURE 10-6
Frequency distributions of deviations between observed and calculated survival rates for small larvae (S_S) and large larvae (S_L) when the deviations are expressed as standard error units. (*From Morris, 1963.*)

the residual error unexplained by the model is not due entirely to sampling error, but also to one, or perhaps more, undiscovered independent variables.

The lack of independence of observations, an assumption of regression analysis, is often one of its shortcomings in building models. For example, if a series of measurements were taken in a sample plot at intervals of time, the population estimates would be correlated as shown in Sec. 10-3, violating the assumption of independence. Another violation of this assumption arises in calculating small-larvae survival as N_S/N_L and then proceeding to choose N_S^{-1} as one of the independent variables included in the model of small-larvae survival. Even though the legitimacy of this move is questionable, the use of N_S as an independent variable makes biological and practical sense.

The assumption of normality of the error terms is happily satisfied, but the distribution of survival rates for small and large larvae is not normal. The situation could have been rectified with a $\log(x + 1)$ transformation, but the inclusion of the constant 1 strongly biased the data. Therefore, the survival rates were not transformed.

Finally it must be stressed that the model is applicable only within the range of data values used to produce it. The model cannot be used for endemic populations of the spruce budworm, only for population change during outbreak periods. Many of the independent variables used in the component models are difficult to interpret. For example, population density was one of the independent variables used in the models of the components, but density represents a multitude of possible factors such as competition, predation, disease, dispersal, and so on. It is difficult to measure these factors directly. Sometimes only the effects of these factors can be measured, e.g., the decrease in survival with increasing density.

10-2   KEY FACTOR ANALYSIS

Morris (1963) has approached the modeling of the spruce budworm population in a simpler manner. He postulates that, although important population changes may be affected by many factors, these changes may be essentially predicted by a very few. In developing his *key factor analysis* he made use of the fact that the population density at some time t is almost always dependent on the past history of the population, particularly the previous generation. More specifically, a population rising in density in one generation is likely to continue increasing in the next. The same is true of a decreasing population. Therefore, the population density at time t is usually significantly correlated with the density of the population at time $t - 1$. It is possible to predict with some accuracy the population density of the $(t + 1)$st generation by knowing only the population density of the tth generation. This is done intuitively all the time. If in 1970 there were 200 bass in a pond and the density had been rising by 10 fish a year since 1965, the logical estimate of the population in 1971 would be 210 fish. However, because of chance elements the population might crash or perhaps even double. Neither, however, is as likely as approximately 210 fish.

The ideal stage in the life history of the spruce budworm to use as a measure of density was the large-larvae stage N_L, because this is the stage at which the future of the population is largely determined. Measuring the density of only large larvae eliminates the problems of estimating numbers

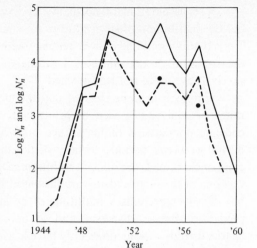

FIGURE 10-7
The logarithms of the numbers of large larvae for plot G2-4 from 1944 to 1960. The solid line is the number of larvae, and the dotted line is log N'', the number of large larvae corrected for predation and parasitism. The two dots for 1954 and 1957 represent the expected population in the absence of moth invasion the previous fall. (*From Morris, 1963.*)

of eggs, pupae, etc., with all the attendant errors. The number of large larvae per 10 square feet was measured each year from 1945 to 1960. The sampling period included measurements of density at low population levels.

The logarithms of the numbers of large larvae for one plot from 1945 to 1960 are shown in Fig. 10-7. If N_n, the number of large larvae in the nth generation, is plotted against N_{n+1}, a curvilinear relationship results which can be made linear by taking the logarithms of both N_n and N_{n+1}. A regression equation was fit to these data

$$\log N_{n+1} = \log \alpha + \beta_1 \log N_n$$

The regression equation for these data was

$$\log N_{n+1} = .98 + .76 \log N_n \tag{10-7}$$

The percent of the variance in $\log N_{n+1}$ explained by $\log N_n$ is 57 percent.

Morris suspected that parasitism was a key factor in determining the future density of large larvae. If there were estimates of the number of large larvae killed by parasites, the proportion surviving parasitism would be $S_{par} = 1 -$ proportion of larvae killed by parasites in generation n. The future density of large larvae, N_{n+1}, would be a function of the interaction between N_n and S_{par}, or

$$\log N_{n+1} = \log \alpha + \beta_1 \log(N_n S_{par})$$

In this particular case the regression equation is

$$\log N_{n+1} = 1.11 + .75 \log(N_n S_{par})$$

with an r^2 of 57 percent.   The inclusion of parasitism in the model has not improved its predictive value.   Apparently parasitism was not an important factor in the survival of large larvae, a fact concluded earlier.   If parasitism had been significant, it might have been desirable to test for the significance of predation and parasitism

$$\log N_{n+1} = \log \alpha + \beta_1 \log(N_n S_{par} S_{pred}) \qquad (10\text{-}8)$$

If weather is a key factor, there will be some wide deviations of individual data points from the regression line given by Eq. (10-8).   These deviations may be plotted against an index of weather, such as mean maximum daily temperature T_{max} during the period of larval development in generation n. Because there are both positive and negative deviations, they can be expressed in terms of temperature as $T_{max} - \bar{T}_{max}$, where $\bar{T}_{max}$ is the temperature at zero deviation.   In the example when T_{max} is regressed against the deviations from the regression in Eq. (10-8), the value of T_{max} at which the dependent variable, the deviations, is zero is 66.5.   Therefore, $\bar{T}_{max}$ is 66.5.   The regression equation can be reformulated as

$$\log N_{n+1} = \log \alpha + \beta_1 \log(N_n S_{par} S_{pred}) + \beta_2(T_{max} - \bar{T}_{max}) \qquad (10\text{-}9)$$

The explained variance in $\log N_{n+1}$ is 68 percent.   Because S_{par} and S_{pred} are not significant, they could have been left out of the equation.

With only a knowledge of mean maximum temperature and the density of the large-larvae population in the previous generation it was possible to explain 68 percent of the variation of $\log N_{n+1}$.   If S_{par} and S_{pred} are deleted from the regression

$$\log N_{n+1} = .98 + .76 \log N_n + .18(T_{max} - \bar{T}_{max}) \qquad (10\text{-}10)$$

Figure 10-8 shows the plot of N_{n+1} estimated versus N_{n+1} observed, from which the observed squared correlation is $r^2 = .68$.   By eliminating certain plots and observations as being nonrepresentative, Morris managed to increase r^2 to 79 percent, quite a respectable figure.

Morris ran a simulation study using Eq. (10-10).   Beginning with an assumed large-larvae population of .01 larvae per 10 square feet in 1925, and the mean maximum temperature data for the years 1925–1959, Fig. 10-9 resulted.   The changes in population density from year to year are remarkably

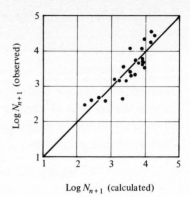

FIGURE 10-8
A plot of the expected and observed population densities of large larvae. The expected densities were calculated from Eq. (10-10). (*From Morris, 1963.*)

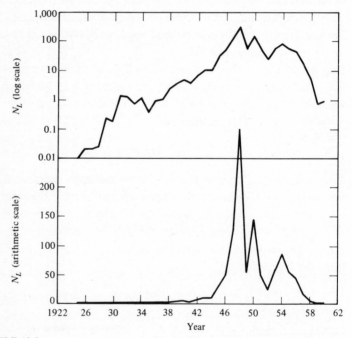

FIGURE 10-9
Larval population trends (N_L) between 1925 and 1960 as simulated by the key factor population model using the temperature data plotted in Fig. 10-9 and Eq. (10-10). (*From Morris, 1963.*)

close to what actually happened.   The occurrence of the outbreak in the late forties was correctly predicted, although the simulation predicted the outbreak to start a few years earlier than it actually did.   Considering the simplicity of the method the results are encouraging.

10-3   TIME SERIES ANALYSIS

Population density at time t is often strongly correlated with the density at time $t - 1$.   The time period may be years, months, days, generations, and so forth.   There may also be correlations between t and $t - 2$, or t and $t - 3$, and so on, although the correlation between t and some previous time usually decreases as time passes.   Correlations between the population density at time t and times $t - 1, t - 2, t - 3$, etc., are called *serial* or *autocorrelations*.

The questions time series analysis can help to answer are:

1   Are the changes in numbers from time to time random fluctuations, or does the past history of the population determine at least in part its future?

2   Are there regular oscillations of the population?

To analyze the first question the serial correlation coefficients between the population density at time $t = 0$ and at regular intervals of time in the future $t = 1, 2, 3, \ldots, n$ can be calculated from the equation

$$r_t = \left[\frac{n}{n-t} \right] \left[\frac{\sum_{i=1}^{n-t} X_i X_{i+t} - (n-t)^{-1} \sum_{i=1}^{n-t} X_i \sum_{i=t+1}^{n} X_i}{\sum_{i=1}^{n} X_i^2 - n^{-1} (\sum_{i=1}^{n} X_i)^2} \right]$$

where n is the number of observations from $i = 1$ to n, X_i is each of the observations, and r_t is the serial correlation.   Obviously $r_0 = 1$.

Moran (1952) has calculated the serial correlation coefficients up to $t = 12$ for a red grouse population in a group of Scottish estates from 1866 to 1938 ($n = 52$).   The serial correlation, for example, between the population density at t and $t + 8$ is .057.   The other serial correlations are listed in Table 10-5. Because the observations are not independent, it is not possible to test each of the serial correlations for significance as ordinary correlation coefficients. An approximate test of the significance of r_1 can be performed by comparing r_1 with its expected value based on the hypothesis that the observations are independent of each other.   The expected value of the correlation coefficient if the true serial correlation is zero is

$$E(r_1) = \frac{-1}{n-1}$$

with a variance

$$\text{var}(r_1) = \frac{(n-2)^2}{(n-3)^3}$$

For the grouse data $E(r_1) = -.0196$, and $\text{var}(r_1) = .02125$. The estimated value of r_1 and its standard error based on a hypothesis of independence of terms in the series is $r_1 = -.01964 \pm .1458$. The observed value of r_1 is much larger, falling outside the range of the expected value of r_1. Therefore, the observed correlation between t and $t+1$ is highly significant. The remaining serial correlations may be tested as shown below.

The serial correlations can be plotted as in Fig. 10-10, a *correlogram*. If a population does not oscillate in a regular periodic manner, the correlogram rapidly damps down to zero as t increases. Stochastic fluctuations will prevent complete damping of a correlogram, and a distinction must be made between chance fluctuations around the baseline in the correlogram and larger periodic fluctuations of the correlogram, as in Fig. 10-12. Although the fluctuations of the correlogram are not as consistent as in Fig. 10-12, the red grouse correlogram does not damp down, having a second peak at about $t = 10$.

Table 10-5 THE CALCULATED SERIAL CORRELATION COEFFICIENTS FOR THE RED GROUSE POPULATION IN THE ATHOLL DISTRICT OF SCOTLAND FOR THE YEARS 1866–1938

t	r_t
1	.643
2	.215
3	−.003
4	−.062
5	−.001
6	.040
7	.026
8	.057
9	.235
10	.299
11	.222
12	−.020

SOURCE: From Moran, P. A. P.: The Statistical Analysis of Game-Bird Records, *J. Anim. Ecol.*, vol. 21, 1952.

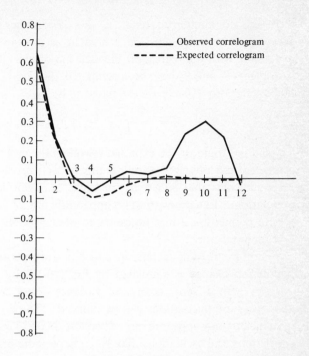

FIGURE 10-10
The observed correlogram of the red grouse population *L. scoticus* in the Atholl district of Scotland, and the estimated correlogram from Eq. 10-11. (*Data from Mackenzie, 1952.*)

This peak is indicative of a tendency, although not a large one, of the red grouse population to undergo a cyclic type of population change. In contrast, the correlogram for the ptarmigan, *Lagopus mutus*, in the same area does damp down. Regular cycles of buildup añd decline do not occur in the ptarmigan.

If the successive densities of a species such as the red grouse or the ptarmigan are serially correlated, the density of the population can be predicted from its past history. As a simple model of population change *autoregression* equations can be used to predict the population density at time t as a function of its density at $t - 1$, $t - 2$, and earlier dates. Commonly only the $t - 1$ and $t - 2$ terms are used, giving an autoregression equation

$$(X_t - \bar{X}) = \beta_1(X_{t-1} - \bar{X}) + \beta_2(X_{t-2} - \bar{X}) \qquad (10\text{-}11)$$

In áutoregression the data values are usually corrected for the mean, so that $E(X) = 0$. In any case the expectation of X must always be constant; i.e., there are no long-term trends in the data, either up, down, or regularly periodic. The autoregression is treated as a simple two-variable multiple regression, with the regression through the origin. Once the two regression parameters β_1 and β_2 have been estimated, the serial correlations generated by the function Eq. (10-11) are estimated to be

$$\rho_1 = \frac{b_1}{1 - b_2} \qquad \rho_t = b_1\rho_{t-1} + b_2\rho_{t-2}$$

This is the origin of the expected correlogram in Fig. 10-10. Although the estimated and observed correlograms fit well up to r_8, the lack of fit after $t = 8$ indicates that some sort of cyclic behavior exists in the red grouse population. Equation (10-11) is apparently not an adequate model of the red grouse population if it is necessary to predict population density more than 8 years in advance.

Autoregressions cannot produce a truly periodic series. However, if population change is simulated by Eq. (10-11) by adding random normal deviates with a zero mean and variance equal to the residual variance, quasi-periodic fluctuations can be induced. These oscillations are not truly periodic but may appear to be. Therefore, in observed series with pronounced oscillations, the oscillations may be truly periodic or quasi-periodic, i.e., they may be due to the serial correlations of the data and random error terms, rather than to truly periodic tendencies of the population. The primary use of the correlogram is in detecting truly periodic fluctuations of the population as distinct from quasi-periodic ones. If the population fluctuates in a quasi-periodic manner, Eq. (10-11) can be used to simulate population change by including random error terms at each step in the simulation.

Moran (1953) used time series analysis to analyze the famous 10-year cyclic fluctuations in the Canadian lynx. Elton and Nicholson (1942) list the number of lynx trapped by the Hudson Bay Company from 1821 to 1934. Moran considers these records to be good proportional indicators of the true population density of the lynx during this period. These records are graphed in Fig. 10-11. A strong, regular cycle appears to occur every 10 years. When the data are transformed to logarithms, the series fluctuates symmetrically around the mean ($\bar{X} = 2.9036$ expressed as logarithms). The logarithms to the base 10 of the original data will be indicated as x_t. Because of the symmetrical variation around the mean in the transformed data, i.e., the variance of the observations about the mean may be normally distributed, a regression equation of the form

$$x_t - \bar{x} = \beta_1(x_{t-1} - \bar{x}) + \beta_2(x_{t-2} - \bar{x}) + \beta_3(x_{t-3} - \bar{x}) + \cdots + \beta_k(x_{t-k} - \bar{x}) + \varepsilon_t$$

$$(10\text{-}12)$$

was used in an attempt to model the changes in the lynx data.

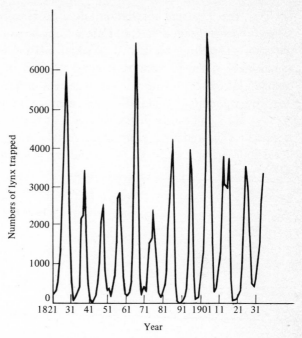

FIGURE 10-11
Number of Canadian lynx trapped by the Hudson Bay Company in the years
from 1824 to 1934 to show the 10-year cycle.   (*Data from Elton and Nicholson, 1942.*)

Using the transformed data the serial correlations were estimated
(Table 10-6), and plotted as a correlogram (Fig. 10-12).   This correlogram
has a distinct periodic nature and does not damp down.   Evidently the lynx
population is truly periodic, fluctuating with a period of 10 years.   Therefore, the period of the fluctuations in Fig. 10-11 is 10 years.

How many terms should be included in an autoregression?   This
question can be answered by including as many terms as wanted and then
calculating the *first-order* partial correlation coefficients (see Chap. 9).   In
Table 10-7 the estimated partial correlation coefficients are listed for the
lynx data for the first six terms in Eq. (10-12).   Also listed are the partial
correlation coefficients significant at the 5 percent probability level.   Usually
a partial correlation coefficient is tested by entering the table of the
distribution of the correlation coefficient with $n - k - 2$ degrees of freedom,
where k variables are held constant.   Because of serial correlation, however,
the significance of the serial partial correlation coefficient is not the same
as an ordinary partial correlation coefficient.   Quenouille (1949) has shown,

however, that a partial correlation based on serially correlated data can be tested for a significant difference from zero by providing three extra degrees of freedom. Therefore, to test if a serial partial correlation coefficient is significantly different from zero enter a table of the distribution of the correlation coefficient with $n - k + 1$ degrees of freedom.

By examining Table 10-7 you can see that r_1, $r_{13.2}$, $r_{15.234}$, and $r_{16.2345}$ are significantly different than 0 at the 5 percent probability level. Because $r_{15.234}$ and $r_{16.2345}$ are small relative to r_1 and $r_{13.2}$, Moran (1953) decided to include only $x_{t-1} - \bar{x}$ and $x_{t-2} - \bar{x}$ as terms in Eq. (10-12).

Table 10-6 CALCULATED SERIAL CORRELATION COEFFICIENTS FOR THE NUMBERS OF CANADIAN LYNX TRAPPED BY THE HUDSON BAY COMPANY FROM 1821 TO 1934

t	r_t
1	.79516
2	.34788
3	−.13596
4	−.51332
5	−.65116
6	−.51696
7	−.16895
8	.25296
9	.58422
10	.66419
11	.42315
12	−.01529
13	−.43674
14	−.69531
15	−.70538
16	−.47523
17	−.08560
18	.30214
19	.54795
20	.54308
21	.26726
22	−.14559
23	−.50129
24	−.64342

SOURCE: From Moran, P. A. P., The Statistical Analysis of the Canadian Lynx Cycle, *Aust. J. Zool.*, vol. 1, 1953.

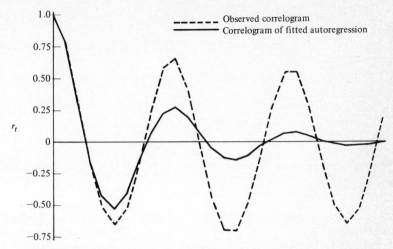

FIGURE 10-12
The observed correlogram for the Canadian lynx, and the estimated correlogram
based on Eq. 10-12. Note how the observed correlogram does not damp down.
(*After Moran, 1953.*)

Table 10-7 VALUES OF THE PARTIAL CORRELATION COEFFICIENTS FOR THE MULTIPLE REGRESSION EQ. (10.19) AS APPLIED TO THE NUMBERS OF CANADIAN LYNX TRAPPED IN A YEAR BY THE HUDSON BAY COMPANY

Suffix of r	r	r significant at the 5 percent level
1	.79516	.1820
13.2	−.77341	.1828
14.23	−.07810	.1836
15.234	−.25037	.1844
16.2345	.19955	.1852
17.23456	.06076	.1860

SOURCE: From Moran, P. A. P., The Statistical
Analysis of the Canadian Lynx Cycle, *Aust. J. Zool.*,
vol. 1, 1953.

The regression equation is

$$x_t - \bar{x} = \beta_1(x_{t-1} - \bar{x}) + \beta_2(x_{t-2} - \bar{x}) + \varepsilon_t$$

The two parameters β_1 and β_2 in a *second-order* autoregression are easily estimated as

$$b_1 = \frac{r_1(1 - r_2)}{1 - r_1^2} \qquad b_2 = \frac{r_2 - r_1^2}{1 - r_1^2}$$

In the lynx example $b_1 = 1.4101$, and $b_2 = -.7734$. The second-order auto-regression is

$$x_t - 2.9036 = 1.4101(x_{t-1} - 2.9036) - .7734(x_{t-2} - 2.9036) \quad (10\text{-}13)$$

The residual variance is estimated as

$$\text{var}(e_t) = \text{var}(x_t)\,\frac{(1 + b_2)[(1 - b_2)^2 - b_1^2]}{1 - b_2}$$

The expected correlogram as calculated from b_1 and b_2 damps down, so Eq. (10-13) is not an adequate long-term predictor of the population changes in the Canadian lynx.   The fit might be improved if the $t - 4$ and $t - 5$ terms were included, but because the population fluctuations are truly periodic no autoregression can truly mimic the behavior of the lynx population.   Periodic time series are best fit by *Fourier series*, and the series are best analyzed by *spectral analysis*.   Both are subjects too complex to include in this text.

Even though Eq. (10-13) may not be a satisfactory model of long-term changes in the lynx population, it may be quite satisfactory as an empirical model of population change from year to year.   In Fig. 10-13 predicted and observed values of the untransformed densities X_t for the years 1927–1934 are plotted.   The population in 1930 is predicted from Eq. (10-13) by knowing the population in 1929 and 1928, and so forth.   The agreement between expected and observed is quite good, indicating that Eq. (10-13) predicts the population density of the Canadian lynx quite well from a knowledge of the population density for the preceding 2 years.

Predictions more than 1 year in advance become increasingly bad.   These predictions are arrived at by predicting the 1927 population from a knowledge of the populations in 1926 and 1925, and then predicting the 1928 population from the known population in 1926 and the predicted density in 1927, and so forth.   With time the deviations of the observed and estimated values of X_t become increasingly large.

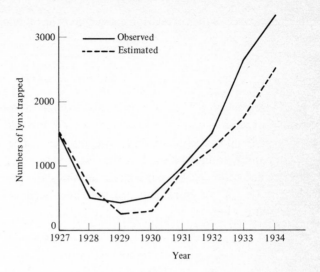

FIGURE 10-13
The estimated and observed values of the untransformed densities of the lynx.

10-3.1   Correlation between Two Serially Correlated Series

In many field situations it may be necessary to calculate the correlation coefficient between two sets of data, both of which are serially correlated. An example might be the counts of two species of insects in the same habitat over a number of years. The method of calculation of the correlation coefficient between two time series is the same as in independent series of observations, but because the distribution of the correlation coefficient between two serially correlated series is not the same as the ordinary correlation coefficient, problems arise in testing the significance of the correlation coefficient.

If both series can be adequately fit by autoregression equations, the standard error of the correlation coefficient between two large serially correlated series is approximately

$$SE(r) \approx (n-1)^{-1}\left(1 + 2\sum_{t=1}^{\infty} \rho_t \bar{\rho}_t\right)$$

where ρ_t is the tth expected serial correlation of the first species, and $\bar{\rho}_t$ is the tth expected serial correlation in the second series. For purposes of testing the significance of the correlation coefficient r the number of degrees of freedom used for the test is

$$n^* = (n - 1)\left(1 + 2 \sum_{t=1}^{\infty} \rho_t \bar{\rho}_t\right)^{-1}$$

An autoregression is fit to each series, and from the autoregression the expected serial correlations computed. Obviously, the test will be valid only if both series can be adequately fit by the autoregressions. The observed serial correlations r_t and $\bar{r}_t$ cannot be used in the test. Moran calculated the correlation coefficients between the population densities of four bird species: (1) the red grouse, (2) the ptarmigan, (3) the caper, and (4) the blackgame. The correlations were

	Grouse	Ptarmigan	Caper	Blackgame
Grouse	1.000	.504**	.504**	.347*
Ptarmigan	...	1.000	.417*	.052
Caper	...	...	1.000	.189
Blackgame	...	...	...	1.000

** Significant at the 1 percent probability level.
* Significant at the 5 percent probability level.

In nonserially correlated series the significance of these correlation coefficients is based on $n - 2$ degrees of freedom ($n = 53$). However, for serially correlated series the degrees of freedom to use is n^*. For the four species of birds the degrees of freedom were

	Grouse	Ptarmigan	Caper	Blackgame
Grouse	...	32	27	35
Ptarmigan	...	...	34	41
Caper	...	...	...	37

10-3.2 Checking for Trends in Data

If generation survival has been measured over a series of years, it might be necessary to check the data for long-term trends toward greater or lesser survival. If there are no long-term trends, the time series is said to be *stationary*. Sokal (1966) studied the percent survival to the pupal stage in 32 generations of a strain of *Drosophila melanogaster* selected for pupation site. The data (Fig. 10-14) can be represented as

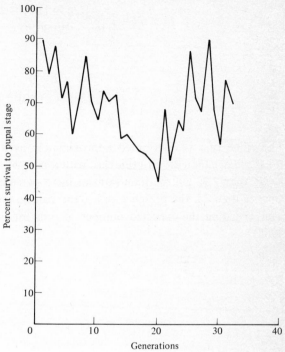

FIGURE 10-14
Percent survival to the pupal stage in a strain of *D. melanogaster* selected for peripheral pupation sites plotted against the number of generations of selection. (*After Sokal and Rholf, 1970.*)

Generation: increase or decrease

1	2	3	4	5	6	7	8	9	10	11	12	13	14	15
−	+	−	+	−	+	+	−	−	+	−	+	−	+	

16	17	18	19	20	21	22	23	24	25	26	27	28	29	30	31	32
−	−	−	−	−	+	−	+	−	+	−	−	+	−	−	+	−

Groups of generations with like signs are underlined; each is called a run. The number of generations n was 32 and the observed number of runs was 23. A t test can be used to compare the observed number of runs and the number of runs expected if there are no significant trends in the data. The approximate value of t is

$$t = \frac{r - [(2n - 1)/3]}{[(16n - 29)/90]^{1/2}}$$

or

$$t = \frac{23 - (63/3)}{[(16)(32) - 29/90]^{1/2}}$$
$$= .86$$

The value of t is distributed approximately as a standard normal variate if n is greater than 25.   In this case with a normal approximation a value of t larger than 1.96 will be significant at the 5 percent probability level.   If n is smaller than 25, the presence of a trend can be found directly by consulting a table listing the expected number of runs for n less than 25 (e.g., Beyer, 1968).

11

SAMPLING AND THE ESTIMATION OF POPULATION PARAMETERS

Whenever certain characteristics of a population such as density, fecundity, migration, or survival are estimated by samples from the population, as they almost always are, errors of sampling arise. The true population mean is estimated by a sample mean. If the population is repeatedly sampled, the estimated mean will vary. The difference between the estimated mean and the true mean is the sample error introduced into a measurement of some population parameter. In addition, human error in measuring and heterogeneity in the environment increase the average deviation of an estimated mean from the true population parameter. The purpose of the first section of this chapter is to discuss some of the more general methods of designing a sampling program to reduce errors of estimation as much as possible. Sampling theory is a complicated subject. For a more thorough discussion of sampling the reader should consult Cochran (1963) or one of several other excellent books on the subject. Morris (1954) has written an account of the development of the sampling program for the spruce budworm project, illustrating the use of many of the techniques discussed and pointing out many of the problems that can, and did, arise in a large-scale sampling program for a natural population.

The remaining sections are concerned with various methods of estimating population density, death rate and birthrate, immigration, and emigration, using mark-recapture and related methods.

11-1   DEVELOPMENT OF A SAMPLING PROGRAM

11-1.1   Random Sampling

In a completely homogeneous area where every point in the area is exactly like every other (such as an average American city), and where there is no reason to believe heterogeneity exists between organisms at one place and those at another, random sampling can be used.   In estimating the number of seeds produced by some species of plant, random sampling can be carried out by tagging or somehow marking every plant in the population.   A sample of size n is taken by using a table of random numbers.   Suppose 100 plants have been marked and a sample of 10 is to be taken.   From a table of random numbers two-digit numbers are picked, reading up or down columns or across rows, for example, 58, 24, 24, 15, 55, 39, 04, 95, 41, and 01. Because by chance 24 is repeated, a replacement is drawn, for example, 49. The 10 plants chosen from the possible 100 plants are those numbered 58, 24, 15, 39, 04, 95, 41, 01, and 49.   Mean seed production is estimated from these 10 plants.

If the homogeneous area is next to another area where seed production is less, perhaps because of less water, it will be incorrect to estimate the seed production of the total two areas by sampling only in the first area.   This is obvious.   If this is done, a *bias* is introduced into the data.   If the mean number of seeds per plant is estimated several times from several successive samples, the sum of the deviations of these estimated means from the true population mean should sum to zero.   If the sum of the deviations is zero, the estimate of the mean is *unbiased*.

The estimated standard error of the sample mean if simple random sampling is used is

$$\text{SE}(\bar{y}) = \frac{s}{n^{1/2}}\left(1 - \frac{n}{N}\right)^{1/2}$$

where $\text{SE}(\bar{y})$ is the estimated standard error, s is the standard deviation of the data, n is the sample size, and N is the population size.   Note that if n equals N, the standard error is zero because the true population mean is known, barring errors in measurement.   For attribute data such as alive or dead, the standard error is

$$\text{SE}(\hat{p}) = \left[\frac{N - n}{(n - 1)N} pq \right]^{1/2}$$

where p is the percentage of the first attribute, and q is the percentage of the second.

The size of the standard error depends almost entirely on the sample size, and little on the fraction of the population sampled.   Therefore, a sample size of 100 results in nearly the same sample standard error as a sample taken from a population of size 1,000,000 or 1,000.   If out of a population of 1,000 plants 10 are taken to measure seed production, the standard error of the estimated mean seed production given a standard deviation of 1.78 will be

$$\text{SE}(\bar{y}) = \left(\frac{1.78}{10^{1/2}} \right)\left(1 - \frac{10}{1,000} \right)^{1/2}$$
$$= \pm .56$$

11-1.2   Sample Size in Random Sampling

How large a sample should be taken in simple random sampling?   The first step in answering this question is to set up a desirable confidence interval about the sample mean; i.e., choose an interval $\pm L$ about the true population mean, and state that the probability of the sample mean falling outside of the range of $\pm L$ is to be 5 percent, 1 percent, or some other level of assurance.   The sample size necessary to assure that the sample mean will be within $\pm L$ of the true population mean 95 percent of the time is $n = 4s^2/L^2$, where n is the sample size, and s^2 is the sample variance estimated either by some sort of preliminary sampling or from sampling in a similar situation.

If we estimate the number of seeds set per plant by using a simple random sampling program, and decide it is necessary that the estimated mean fall within ± 5 seeds of the true population mean with 5 percent chance that it does not, the necessary sample size if the variance has been estimated as 100 in a similar experiment in the adjacent field will be $n = (4)(100)/5^2 = 16$ plants.

If the sample size exceeds 10 percent of the population size, a corrected sample size can be calculated as

$$n' = \frac{n}{1 + (n/N)}$$

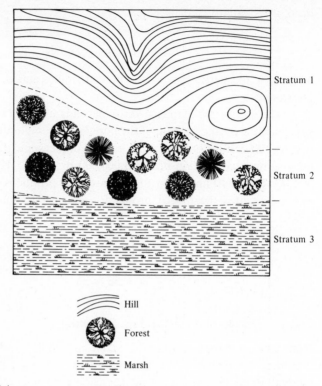

Hill

Forest

Marsh

FIGURE 11-1
A diagrammatic representation of the division of a heterogeneous habitat into homogeneous strata.

11-1.3 Stratified Sampling

Suppose sampling is being carried out in an area such as the one shown in Fig. 11-1. If some population parameter such as fecundity is being estimated, the mean fecundity may differ from place to place because of the heterogeneity of the environment. If an animal is adapted to wet conditions, it is reasonable to assume a priori that its fecundity in the marshy area of the habitat will be greater than in the hills. If the mean fecundity is estimated by simple random sampling, the standard error of the estimated mean fecundity will depend not only on the average deviation of the estimate from the mean, but on the differences between means. The optimum way to sample in this area is with *stratified sampling*. By using stratified sampling the effect of differences in strata is removed from the estimate of the standard error of the mean.   In stratified sampling a heterogeneous environment is divided into *strata*, or homogeneous parts.   In Fig. 11-1 the area has been divided by the dashed lines into three strata: one containing the hills, a second the forest, and a third the marsh.   The first step in any sampling program is to check for significant heterogeneity in the sample area.

In checking for heterogeneity it is useful to divide the sampling area into units of some sort, and use an analysis of variance (see Chap. 9) to check for significant mean differences among the units.   If the area in Fig. 11-1 is suspected of being heterogeneous, it can be divided into the three units shown, and a one-way analysis of variance run to check for differences in the mean fecundity among the three areas or strata.   The strata are the treatments of the analysis of variance.   Morris (1954), in sampling for number of spruce budworm larvae per 10 square feet of branch surface, decided to check for significant differences in the number of larvae among different trees, among the vertical levels of the trees, and among the four sides of a tree (termed quadrats in his article).   After transformation of the data to $y = \log(x + 1)$ to stabilize the variances, a three-way analysis of variance was calculated, resulting in Table 11-1.   This is a model-I analysis of variance with no replication, so all mean squares are tested against the residual three-factor interaction mean square as a test of significance.   Both trees and levels are significant, indicating significant differences in larval density among different trees and among the levels of a tree.   There are no significant differences among the sides of a tree.   In sampling each tree stratified sampling can be used, each vertical level being a stratum.   Similarly

Table 11-1   ANALYSIS OF VARIANCE OF THE DIFFERENCES IN LARVAL DENSITIES BETWEEN TREES, LEVELS WITHIN A TREE, AND OF THE SIDES OF A TREE (QUADRATS) OF THE SPRUCE BUDWORM, *C. FUMIFERANA*, IN AREA 1 IN NORTHWESTERN NEW BRUNSWICK

Source of variance	Degrees of freedom	Sum of squares	Mean square	F
Trees	3	1.3864	.4621	4.36*
Levels	3	1.1099	.3700	3.50*
Trees × Levels	9	.4387	.0487	
Quadrats	3	.0672	.0224	
Quadrats × Trees	9	.7618	.0846	
Quadrats × Levels	9	.5449	.0605	
Quadrats × Trees × Levels	27	2.8496	.1055	

* Significant at the 5 percent probability level.

SOURCE: From Morris, R. F., A Sequential Sampling Technique for Spruce Budworm Egg Surveys, *Can. J. Zool.*, vol. 32, 1954.

in sampling trees each tree is also a stratum.   Sampling in situations in which both trees and levels are stratified is rather complicated, and only the simpler case of one-way stratification will be discussed.

The population is divided into a number of strata, and a sample is taken independently in each part.   The estimate of the population mean is

$$\bar{y}_{st} = \frac{\sum N_h \bar{y}_h}{N}$$

where N_h is the number of individuals in the hth stratum, $\bar{y}_h$ is the estimated mean in the hth stratum, and N is the total number of individuals in all strata.   To estimate the population mean the number of individuals N_h in each stratum must be known.   If the same proportion of the population is sampled in each part, proportional allocation, and n_h is the number of individuals sampled in the hth stratum, the estimate of the mean is

$$\bar{y}_{st} = \frac{\sum n_h \bar{y}_h}{n} = \bar{y}$$

or the mean of all the individuals sampled in all the strata.

Indicate the proportion of individuals in each stratum as $W_h = N_h/N$. The standard error of the estimated mean is

$$SE(\bar{y}_{st}) = \left[\sum W_h^2 \frac{s_h^2}{n_h} \left(1 - \frac{n_h}{N_h} \right) \right]^{1/2}$$

where s_h^2 is the sample variance in stratum h.   If the sample in each stratum is less than 10 percent of the population, this equation reduces to approximately

$$SE(\bar{y}_{st}) = \left(\sum \frac{W_h^2 s_h^2}{n_h} \right)^{1/2}$$

i.e., the term $1 - (n_h/N_h)$ becomes negligible.   If allocation is proportional, the equation for the standard error reduces to

$$SE(\bar{y}_{st}) = \left(\frac{1 - f}{n} \sum W_h s_h^2 \right)^{1/2}$$

where $f = n/N$.

Proportional allocation is not the only and not necessarily the best method of stratified sampling.   In fact, the optimum allocation of sample sizes involves taking n_h, the sample size in each stratum, proportional to $N_h s_h/c_h$, where s_h is the standard deviation of the sampling units in the hth stratum, and c_h is the cost of sampling per unit or individual in the hth stratum. Allocating sample sizes in this way results in the smallest possible standard error of the mean for a given total cost of taking the sample.   As a result of this rule a larger sample is taken from a stratum that is unusually variable, and a smaller sample from a stratum in which sampling is more expensive on the average.   If the cost of sampling a unit is the same in all the strata, the percent of the total sample taken in the hth stratum is $N_h s_h / \sum N_h s_h$.

An example of the calculation of the optimum allocation of sample sizes among strata is shown in Table 11-2.   The strata were created on the basis of caribou density.   Each stratum was divided into a series of 4×4 mile units, and the number of caribou per unit estimated.   The efficiency of stratification and optimum allocation of sample sizes in reducing the standard error of the estimated mean can be found by calculating the standard error of the mean with optimum allocation, which is

$$SE(\bar{y}_{st}) = \left(\sum \frac{W_h^2 s_h^2}{n_h} - \sum \frac{W_h s_h^2}{N} \right)^{1/2}$$

In the caribou experiment the standard error of the mean number of caribou per 4×4 mile quadrat was 8.79.   This standard error can be compared with the standard error had a simple random sample been taken

$$SE(\bar{y}_{ran}) = \left\{ \frac{N-n}{Nn} \left[\sum W_h s_h^2 - \sum \frac{W_h s_h^2}{nN} + \sum \frac{W_h^2 s_h^2}{n_h} \right. \right.$$
$$\left. \left. + \sum W_h \bar{y}_h^2 - \left(\sum W_h \bar{y}_h \right)^2 \right] \right\}^{1/2}$$

Table 11-2   THE APPLICATION OF OPTIMUM ALLOCATION IN A STRATIFIED SAMPLING PROGRAM OF CARIBOU INTO STRATA, THE STRATA BEING BASED ON DIFFERENCES IN DENSITY OF CARIBOU. EACH SAMPLING UNIT IS A 4×4 MILE QUADRAT

Stratum	N_h	s_h	$N_h s_h / \sum N_h s_h$	Optimum sample size
A	400	3,000	.428	96
B	30	2,000	.022	5
C	61	9,000	.195	44
D	18	2,000	.013	3
E	70	12,000	.299	67
F	120	1,000	.043	10
Total	699	29,000	1.000	225

SOURCE: From Siniff, D. B., and Skoog, R. B.: Aerial Sampling of Caribou Using Stratified Random Sampling, *J. Wildl. Manage.*, vol. 28, 1964.

In the caribou example the standard error of the estimated mean given random sampling was 29.41. Stratification of the sampling and the optimum allocation of the units among the strata has reduced the standard error of the estimate by more than a factor of 3.

11-1.4 Systematic Sampling

Often in a situation in which random sampling is justified it is simply not possible to tag or otherwise identify all the individuals in the population. Two alternatives are open: systematic sampling, or two-stage sampling. Systematic sampling is applicable most easily when the organisms are somehow arranged or lined up. To draw a 10 percent sample of the total population, the first individual is chosen by picking a number at random from 1 to 10 (say 3), and then taking individuals at 10-unit intervals after that, for example, 3, 13, 23, 33, and so on. Not only is systematic sampling easier than simple random sampling, but often it is more accurate because it spreads the sample evenly over the entire population. There is a danger, however, of systematic trends seriously biasing the sample. It might be that every tenth leaf on a tree is on the tip of a branch, so that the leaves more toward the center of the tree are underrepresented and those toward the outside overrepresented. In addition, there are no general reliable methods of estimating the standard error of the sample mean in systematic sampling. Cochran (1963) gives estimates of the standard error of the mean for specific cases.

11-1.5 Two-Stage Sampling

Another way of circumventing the problem of marking every individual in the population is to divide the area into smaller units, such as quadrats, and to take a random sample of the quadrats. This is the first stage of sampling, and the quadrats are referred to as *primary units*. In each of the chosen quadrats a random sample of individuals is taken; the quadrat is *subsampled*. This is the second stage.

To estimate the mean seed production of a grass species in a field in which several thousand individuals occur, the field is first divided into a series of quadrats, quadrats are chosen at random, and each of the randomly chosen quadrats is subsampled. If there is approximately the same number of individuals in all the quadrats, the standard error of the estimated mean number of seeds per individual will be

$$SE(\bar{y}_{ss}) = \frac{1}{n_1^{1/2}} \left[\frac{\sum (\bar{y}_i - \bar{y})}{n_1 - 1} \right]^{1/2}$$

where n_1 is the number of quadrats selected, $\bar{y}_i$ is the mean number of seeds per individual in the ith quadrat, and $\bar{y}$ is the mean number of seeds in all the quadrats chosen. This estimate of the standard error is adequate if the number of quadrats chosen does not exceed 10 percent of the total number of available quadrats.

In the one-way analysis of variance there was a series of treatments, and for each treatment a series of replicates. This design can be thought of in terms of quadrats and individuals, the treatments being the quadrats and the individuals being the replicates in each treatment. The total variance of all the observations is equal to differences among the quadrats σ_1^2, and differences among the individuals within a quadrat σ_2^2. By computing the mean squares for differences among quadrats and differences within quadrats the estimated variance components s_1^2 and s_2^2 are

$$s_1^2 = \frac{\text{Mean square among} - \text{mean square within}}{n_2}$$

where n_2 is the number of individuals sampled in each quadrat, and s_2^2 is the mean square within. Suppose that 10 plants per quadrat in 100 quadrats had been sampled. A possible analysis of variance table is

Source	Degrees of freedom	Mean square
Among plots	9	3.7604
Within plots	90	1.7860

Then $s_1^2 = (3.7604 - 1.7860)/10 = .01974$, and $s_2^2 = 1.7860$. With s_1^2 and s_2^2 known the standard error of the estimated mean is

$$\text{SE}(\bar{y}_{ss}) = \left(\frac{s_1^2}{n_1} + \frac{s_2^2}{n_1 n_2}\right)^{1/2} \tag{11-1}$$

In the hypothetical example $\text{SE}(\bar{y}_{ss}) = [.01974/100 + 1.7860/(10)(100)]^{1/2} = .0445$. This is a very small standard error, and perhaps a larger standard error could be tolerated in order to reduce the number of samples taken. If only 10 quadrats and 10 individuals per quadrat were sampled, what would be the standard error of the estimate?

$$\text{SE}(\bar{y}_{ss}) = \left[\frac{.01974}{10} + \frac{1.7860}{(10)(10)}\right]^{1/2}$$
$$= .1408$$

In fact, Eq. (11-1) can be used to calculate the possible combinations of n_1 and n_2, given some tolerable standard error of the estimate. Suppose .2 is chosen as an acceptable standard error. The possible combinations of n_1 and n_2 can be found from the iterative solution of

$$\left(\frac{.01974}{n_1} + \frac{1.7860}{n_1 n_2}\right)^{1/2} = .2$$

in our hypothetical example. The use of Eq. (11-1) requires, of course, initial estimates of $s_1{}^2$ and $s_2{}^2$ arrived at by prior sampling.

Because the iterative solution of Eq. (11-1) gives a large number of possible combinations of n_1 and n_2, the problem still remains of which combination is best, since they all result in the same standard error. The answer depends on the relative cost of adding an extra quadrat (increasing n_1) versus the cost of adding additional individuals in each quadrat (increasing n_2). It might be easier, and in terms of man-hours cheaper, to sample an additional individual in each quadrat rather than go to the trouble of choosing an additional quadrat, tagging all the individuals, and taking a random sample of size n_2 from it.

In many subsampling studies the cost of the sample can be approximated by

$$\text{Cost} = c_1 n_1 + c_2 n_1 n_2$$

where c_1 is the average cost per primary unit of those elements of the cost depending solely on the number of primary units and not on the amount of subsampling in each primary unit. On the other hand, c_2 is the average cost per subunit of those parts of the cost that are directly proportional to the total number of subunits. If these two costs are known, even approximately, it is possible to estimate the best combination of n_1 and n_2 minimizing the variance of the estimate of the mean and the total cost. If the total amount of money or hours that can be spent is C, and the desired variance of the mean is V, the trick is to minimize the product VC. The product VC is equal to

$$VC = \left(\frac{s_1{}^2}{n_1} + \frac{s_2{}^2}{n_1 n_2}\right)(c_1 n_1 + c_2 n_1 n_2) \qquad (11\text{-}2)$$

Equation (11-2) can be expanded to

$$VC = (s_1{}^2 c_1 + s_2{}^2 c_2) + n_2 s_1{}^2 c_2 + \frac{s_2{}^2 c_1}{n_2} \qquad (11\text{-}3)$$

The expression minimizing Eq. (11-3) is

$$n_2 = \left(\frac{c_1 s_2{}^2}{c_2 s_1{}^2}\right)^{1/2} \qquad (11\text{-}4)$$

The number of individuals to be sampled in each quadrat can be estimated from Eq. (11-4), with some prior knowledge of c_1 and c_2. If in the hypothetical example $c_1 = 2.00$ and $c_2 = 1.00$, the number of individuals to be sampled in each quadrat will be $n_2 = [(2.00)(1.7860)/(1.00)(.0197)]^{1/2} = 13.45$ plants per quadrat. The number of quadrats to be sampled is found by substituting n_2 into Eq. (11-3) and setting V and C at some desired levels. It is possible that, if V and C are set unreasonably low, there will be no realistic value of n_1 which will satisfy this equation. If the cost is lowered, the variance of the sample mean must necessarily increase for some set combination of n_1 and n_2, and vice versa.

11-1.6   Cluster Sampling

Sometimes plants or animals occur in clusters. It is possible to choose a number of clusters at random and measure the particular variables being studied on each of the units or individuals in the cluster. This type of sampling is *cluster sampling*. For the statistical aspects of cluster sampling see Cochran (1963).

11-2   MARK-RECAPTURE METHODS

The purpose of mark-recapture studies is to estimate various population parameters such as density, birthrates and death rates, emigration, and immigration. Because these parameters are estimated from samples, the variance and bias of the estimate must be determined, as well as the estimates themselves. The basic principle behind mark-recapture studies is best illustrated by the Lincoln or Peterson index, one of the first mark-recapture estimates. In its simplest form the mark-recapture method consists of taking a random sample from the population, marking the individuals, and releasing them. After a period of time a second sample is taken, and the number of marked and unmarked individuals are counted. If there are no gains or losses to the population during the time interval, and both samples consist of 100 individuals, the proportion of marked individuals in the second sample is an estimate of the percentage of the total population that the initial sample made up. If the number of marked individuals in the second sample is 10, it will be concluded that the original sample of 100 made up 10 percent of the population. Therefore, the total population size is 1,000 individuals. This is true given the following conditions:

1   The marked individuals become randomly mingled with the remainder of the population.

2 There are no losses or gains to the population due to deaths, births, immigration, or emigration.

3 The marked individuals are not affected by the marking.

4 The two samples are both taken randomly, and all individuals are equally available for capture.

5 Sampling is carried out at discrete intervals of time. The period of time taken in sampling must be small relative to the time interval between the two samples.

6 Being captured once does not affect the probability of an individual being subsequently captured (e.g., some small mammals become trap-addicted).

Most of the mark-recapture methods are subject to all these conditions except condition 2.

There is a multitude of mark-recapture methods available, enough to serve as a subject of a separate book. The discussion in this chapter will be limited to a few of the most generally used methods, or those most applicable to field situations. For further discussion the reader may consult Cormack (1969).

The discussion begins with the Lincoln index and its modifications, and proceeds to methods utilizing several sampling and marking periods rather than just two. The multiple-recapture method formulated by Jolly (1965) is discussed in some detail, because it allows for stochastic fluctuation in birthrates, death rates, and immigration rates. Of all the methods available, Jolly's is perhaps the most realistic and useful. A method of estimating population size based on the removal of individuals rather than marking is also discussed.

11-2.1 The Lincoln or Peterson Index

The simplest of the mark-recapture methods, the Lincoln index, consists of taking a sample of size a, marking the individuals, and releasing them. After the individuals are randomly mixed with the remainder of the population, a second sample of size n is taken, r of which are found to be marked. Given that both samples were large, and that a large number of individuals was recaptured, the estimated population size x is

$$\hat{x} = \frac{an}{r}$$

The variance of the estimate is approximately

$$\mathrm{var}(\hat{x}) = \frac{a^2 n(n - r)}{r^3}$$

If the estimate of population density is to be effective, the product of the two samples *an* should be greater than four times the true population density.

The Lincoln index is deceptively simple. In addition to the very restrictive assumptions listed above, the Lincoln estimate is biased, and the exact variance of the estimate is exceedingly difficult to compute (see Robson and Riegert, 1964). The estimate and its variance as presented are only approximations to the true population density, and unless the two sample sizes and the number of recaptures are large relative to the population size, the approximation may not be very good. If the number of recaptures is small, the bias of the estimated density will be relatively large, on the order of $1/E(r)$. For example, if $x = 1000$ and $n = a = 100$, the expected value of r, the number of recaptures, will be 10. The bias of the estimate is, therefore, about 10 percent. Bailey (1951) proposes the modified estimate

$$\hat{x}' = \frac{a(n + 1)}{r + 1}$$

This estimate is also biased for small samples, but less so than the unmodified estimate. The variance of the above estimate x' is approximately

$$\text{var}(\hat{x}') = \frac{a^2(n + 1)(n - r)}{(r + 1)^2(r + 2)}$$

This variance estimate is for large samples, and has a bias on the order of $r^2 e^{-r}$.

Both estimates of the population density regard the total size of both samples as fixed. This form of sampling is termed *direct sampling*. In *inverse sampling* the second sampling is continued until some predetermined number of marked individuals m has been captured. The estimate of the total population size with inverse sampling is

$$\hat{x}_{\text{inv}} = \frac{n(a + 1)}{m} - 1$$

and the exact variance of the estimate is

$$\text{var}(\hat{x}_{\text{inv}}) = \frac{(a - m + 1)(\hat{x} + 1)(\hat{x} - a)}{m(a + 2)}$$

All these variations of the Lincoln estimate are subject to the six assumptions listed at the beginning of Sec. 11-2.

11-2.2   Triple-Catch Method

Bailey (1951) has developed a mark-recapture method estimating population size, the birthrate, and the death rate, based on a series of three samples. The birthrate and the death rate are assumed to be constants. In Bailey's

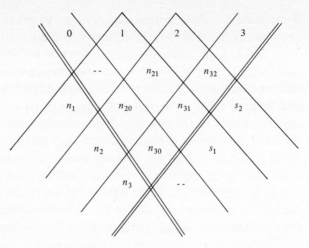

FIGURE 11-2
Trellis diagram for listing data in Bailey's triple-catch method. The notation is
defined in the text.

method a sample is taken, and the individuals are marked and released.   A
second sample is taken, the number of marked individuals noted, and all the
individuals marked and released.   Finally, a third sample is taken, and the
number of marked and unmarked individuals recorded.   For notation n_j is
the total catch on the jth day, of which n_{ji} were first marked on the ith day
while n_{j0} are unmarked.   Further let s_j freshly marked animals be released
on the jth day.   These data can be set out in a trellis diagram for convenience (Fig. 11-2).

From Eq. (1-1) we can set $\lambda = e^{\beta t}$ and $\mu = e^{-\gamma t}$, where λ and μ
are the birthrate and death rate per time interval.   Also t_1 and t_2 are the
time intervals between the first and second and the second and third samples.
If the number of recaptures is relatively large, estimates of x, λ, and μ
will be

$$\hat{x} = \frac{s_2\, n_2\, n_{31}}{n_{21} n_{32}} \qquad \hat{\lambda} = \frac{n_{21} n_3}{n_2\, n_{31}} \qquad \hat{\mu} = \frac{s_2\, n_{31}}{s_1 n_{32}}$$

where x is the density of the population at the beginning of the sampling
program.   The instantaneous birth and death rates are determined from

$$e^{-\gamma t_1} = \hat{\mu}$$

and

$$e^{\beta t_2} = \hat{\lambda}$$

The variance of the estimate of the total population size x is

$$\text{var}(\hat{x}) = \hat{x}^2 \left(\frac{1}{n_{21}} + \frac{1}{n_{32}} + \frac{1}{n_{31}} + \frac{1}{n_2} \right)$$

and the variances of the estimates of the birthrates and death rates are

$$\text{var}(\hat{\beta}) = \frac{1/n_{21} + 1/n_{31} - 1/n_2 - 1/n_3}{t_2^{\,2}}$$

$$\text{var}(\hat{v}) = \frac{1/n_{32} + 1/n_{31}}{t_1^{\,2}}$$

It is convenient in all these calculations to make the time interval between samples one, such as 1 day or 1 week. Bailey (1951) also gives the variances of the estimates of λ and μ.

Bailey also proposes adjusted estimates of x, λ, and μ if the number of recaptures is small. These estimates are

$$\hat{x} = \frac{s_2(n_2 + 1)n_{31}}{(n_{21} + 1)(n_{32} + 1)}$$

$$\hat{\lambda} = \frac{n_{21}(n_3 + 1)}{n_2(n_{31} + 1)}$$

$$\hat{\mu} = \frac{s_2 \, n_{31}}{s_1(n_{32} + 1)}$$

For these adjusted estimates the approximate variances of the estimates are

$$\text{var}(\hat{x}) = - \frac{s_2^{\,2}(n_2 + 1)(n_2 + 2)n_{31}(n_{31} - 1)}{(n_{21} + 1)(n_{21} + 2)(n_{32} + 1)(n_{32} + 2)} + \hat{x}^2$$

$$\text{var}(\hat{\lambda}) = \hat{\lambda}^2 - \frac{n_{21}(n_{21} - 1)(n_3 + 1)(n_3 + 2)}{n_2(n_2 - 1)(n_{31} + 1)(n_{31} + 2)}$$

$$\text{var}(\hat{\mu}) = \hat{\mu}^2 - \frac{s_2^{\,2}n_{31}(n_{31} - 1)}{s_1^{\,2}(n_{32} + 1)(n_{32} + 2)}$$

Bailey's triple-catch model assumes that the population is closed; i.e., no immigration or emigration occurs. If the population is not closed, the birthrate and death rate will include individuals lost or gained by dispersal, as well as lost or gained as a result of births and deaths in the population.

11-2.3 Jolly's Stochastic Multiple-Recapture Method

Several mark-recapture models take into account stochastic fluctuations in the population parameters. The stochastic method discussed will be one developed by Jolly (1965). Several samples are taken, the number of marked and unmarked individuals counted, and all the individuals marked and returned to the population. It is not necessary for all the individuals to be returned to the population, although it is convenient. To simplify the discussion the notation used is listed below:

l = number of samples.

N_i = total number of individuals in the population when the ith sample is taken.

n_i = number of individuals in the ith sample.

M_i = total number of marked animals in the population at time i.

m_i = number of marked animals in the ith sample.

s_i = number of individuals released from the ith sample after marking; not all individuals need be released.

ϕ_i = probability that an individual alive at the moment of release of the ith sample will survive until the time of capture of the $(i + 1)$st sample, including emigration and death.

B_i = number of new animals joining the population in the interval between the ith and $(i + 1)$th samples and alive at time $i + 1$; B_0 is defined as equal to N_1.

n_{ij} = number of individuals in the ith sample last captured in the jth sample $(1 \leq j \leq i - 1)$.

n_{i0} = number of unmarked animals in the ith sample.

$\alpha_i = m_i/n_i$.

$a_{ij} = \sum_{k=1}^{j} n_{ik}$ = number of individuals in the ith sample last caught in the jth sample or before.

$Z_i = \sum_{k=1}^{l} a_{k, i-1}$ = marked individuals before time i which are not caught in the ith sample but are caught subsequently.

$R_i = \sum_{k=i+1}^{l} n_{ki}$ = number of s_i individuals released from the ith sample that are caught subsequently.

$N_{i(j)}$ = expected number in the population at time i which first joined the population between times j and $j + 1$, that is, those which are members of B_j $(1 \leq j \leq i - 1)$; by definition $N_{i(i-1)} = B_{i-1}$.

The three series of population parameters to be estimated are N_i, ϕ_i, and B_i, each of which can change in value over the sampling period. In addition the model allows for stochastic fluctuation in the values of N_i and ϕ_i, in the sense that the values of N_i and ϕ_i at any time are affected by stochastic fluctuations apart from variance due entirely to errors in estimating these two parameters. The population size at time period i is estimated from

$$\hat{N}_i = \frac{M_i}{\alpha_i} \qquad (i = 1, 2, 3, \ldots, l - 1)$$

where M_i is

$$M_i = \frac{s_i Z_i}{R_i} + m_i \qquad (i = 2, 3, \ldots, l - 1)$$

and

$$\alpha_i = \frac{m_i}{n_i} \qquad (i = 2, 3, \ldots, l)$$

The estimate of ϕ_i is

$$\hat{\phi}_i = \frac{M_{i+1}}{M_i - m_i + s_i} \qquad (i = 2, 3, \ldots, l - 2)$$

and the estimate of B_i is

$$\hat{B}_i = \hat{N}_{i+1} - \hat{\phi}_i(\hat{N}_i - n_i + s_i) \qquad (i = 2, 3, \ldots, l - 2)$$

The variances of these three estimates are

$$\text{var}(\hat{N}_i) = \hat{N}_i(\hat{N}_i - n_i)\left[\left(\frac{M_i - m_i + s_i}{M_i}\right)\left(\frac{1}{R_i} - \frac{1}{s_i}\right) + \frac{1 - \alpha_i}{m_i}\right] + \hat{N}_i - \sum_{j=0}^{i-1} \frac{N_{i(j)}^2}{B_j}$$

$$\text{var}(\hat{\phi}_i) = \hat{\phi}_i^2\left[\left(\frac{(M_{i+1} - m_{i+1})(M_{i+1} - m_{i+1} + s_{i+1})}{M_{i+1}^2}\right)\left(\frac{1}{R_{i+1}} - \frac{1}{s_{i+1}}\right)\right.$$
$$\left. + \left(\frac{M_i - m_i}{M_i - m_i + s_i}\right)\left(\frac{1}{R_i} - \frac{1}{s_i}\right)\right] + \frac{1 - \hat{\phi}_i}{M_{i+1}}$$

$$\text{var}(\hat{B}_i) = \left[\frac{\hat{B}_i^2(M_{i+1} - m_{i+1})(M_{i+1} - m_{i+1} + s_{i+1})}{M_{i+1}^2}\right]\left[\frac{1}{R_{i+1}} - \frac{1}{s_{i+1}}\right]$$
$$+ \left[\frac{M_i - m_i}{M_i - m_i + s_i}\right]\left[\frac{\hat{\phi}_i s_i(1 - \alpha_i)}{i}\right]^2\left[\frac{1}{R_i} - \frac{1}{s_i}\right]$$
$$+ \frac{(\hat{N}_i - n_i)(\hat{N}_{i+1} - \hat{B}_i)(1 - \alpha_i)(1 - \hat{\phi}_i)}{M_i - m_i + s_i}$$
$$+ [\hat{N}_{i+1}(\hat{N}_{i+1} - n_{i+1})]\left[\frac{1 - \alpha_i + 1}{m_{i+1}}\right] + \hat{\phi}_i^2\hat{N}_i(\hat{N}_i - n_i)\left[\frac{1 - \alpha_i}{m_i}\right]$$

The variance estimates are complex but straightforward, the only computational difficulty being the summation term in the variance estimate of $\hat{N}_i$, that is, values of $N_{i(j)}$. For two successive samples

$$N_{i(j=i-1)} = \hat{B}_j \qquad (11\text{-}5)$$

but for two noncontiguous samples

$$N_{k+1(j)} = \frac{\hat{N}_{k+1} - \hat{B}_k}{\hat{N}_k} N_{k(j)} \qquad k > j \qquad (11\text{-}6)$$

where k represents those values of i removed by more than one interval of time from j. Equation (11-6) is a recurrence relation; for example, $N_{5(3)}$ is found as

$$N_{5(3)} = \frac{N_5 - B_4}{N_5} N_{4(3)}$$

where $N_{4(3)}$ is found from Eq. (11-5). The term $N_{6(3)}$ can then be found by using $N_{5(3)}$ as $N_{k(j)}$.

The estimated variances of $\hat{N}_i$ and $\hat{\phi}_i$ are due both to stochastic fluctuations and to errors in the estimation of the parameters. The variances in the population sizes due to errors of parameter estimation are

$$\text{var}(\hat{N}_i \mid N_i) = \hat{N}_i(\hat{N}_i - n_i)\left[\left(\frac{M_i - m_i + s_i}{M_i}\right)\left(\frac{1}{R_i} - \frac{1}{s_i}\right) + \frac{1 - \alpha_i}{m_i}\right]$$

and similarly the variances due to errors of estimation for $\hat{\phi}_i$ are

$$\text{var}(\hat{\phi}_i \mid \phi_i) = \text{var}(\hat{\phi}_i) - \left[\frac{\hat{\phi}_i^2(1 - \hat{\phi}_i)}{M_{i+1}}\right]$$

The standard errors of the estimates are the square roots of these quantities. The remaining variance after the variance due to errors of estimation is subtracted is the variance of the stochastic fluctuations of N_i and ϕ_i. The variance estimate of $\hat{B}_i$ is due entirely to errors of estimation.

Table 11-3 APPLICATION OF JOLLY'S STOCHASTIC MARK-RECAPTURE METHOD TO A POPULATION OF THE BUG B. ANGULATUS; VALUES OF a_{ij}

n_i	s_i	1	2	3	4	5	6	7	8	9	10	11	12	13
54	54													
146	143	10												
169	164	3	34											
209	202	5	18	33										
220	214	2	8	13	30									
209	207	2	4	8	20	43								
250	243	1	6	5	10	34	56							
176	175	0	4	0	3	14	19	46						
172	169	0	2	4	2	11	12	28	51					
127	126	0	0	1	2	3	5	17	22	34				
123	120	1	2	3	1	0	4	8	12	16	30			
120	120	0	1	3	1	1	2	7	4	11	16	26		
142		0	1	0	2	3	3	2	10	9	12	18	35	
$R_i =$			80	70	71	109	101	108	99	70	58	44	35	

SOURCE: From Jolly, G. M., Explicit Estimates from Capture-Recapture Data with Both Death and Dilution—Stochastic Model, *Biometrika*, vol. 52, 1965.

The calculations can best be demonstrated by an example from Jolly (1965). The adult female population of the capsid *Blepharidopterus angulatus* was sampled on 13 successive occasions at 3- to 4-day intervals in an apple orchard. The observed values of n_i, s_i, and n_{ij} are shown in Table 11-3. The table of n_{ij} values is read across rows. For example, the number of individuals in the tenth sample last captured in the sixth is five. The sum of the columns gives the values of R_i. Table 11-4 lists the value of a_{ij}. These values are computed by summing successively larger groupings of columns from Table 11-3. For example, the third column of Table 11-4 was arrived at as

$$
\begin{array}{ccccccc}
5 & & 18 & & 33 & & 56 \\
2 & & 8 & & 13 & & 23 \\
2 & & 4 & & 8 & & 14 \\
1 & & 6 & & 5 & & 12 \\
0 & & 4 & & 0 & & 4 \\
0 & + & 2 & + & 4 & = & 6 \\
0 & & 0 & & 1 & & 1 \\
1 & & 2 & & 3 & & 6 \\
0 & & 1 & & 3 & & 4 \\
0 & & 1 & & 0 & & 1 \\
\end{array}
$$

Table 11-4 APPLICATION OF JOLLY'S STOCHASTIC MARK-RECAPTURE METHOD TO A POPULATION OF THE BUG *B. ANGULATUS*; VALUES OF a_{ij}

1	2	3	4	5	6	7	8	9	10	11	12	13
10												
3	37											
5	23	56										
2	10	23	54									
2	6	14	34	77								
1	7	12	22	56	112							
0	4	4	7	21	40	86						
0	2	6	8	19	31	59	110					
0	0	1	3	6	11	28	50	84				
1	3	6	7	7	11	19	31	47	77			
0	1	4	5	6	8	15	19	30	46	72		
0	1	1	3	6	9	11	21	30	42	60	95	

$Z_{i+1} =$												
14	57	71	89	121	110	132	121	107	88	60		

SOURCE: From Jolly, G. M., Explicit Estimates from Capture-Recapture Data with Both Death and Dilution—Stochastic Model, *Biometrika*, vol. 52, 1965.

The values of Z_{i+1} are found by summing all the elements of each column except the first.   The elements in the box in column 6 of Table 11-4 are $Z_7 = 110$.

The value of M_7 is

$$M_7 = \frac{(243)(110)}{108} + 112$$

$$= 359.50$$

where the top element in each column of Table 11-4 is the value of m_i; that is, $m_6 = 112$.   From these statistics the parameters can be estimated. Table 11-5 lists these estimates with their standard errors for this particular example.   The standard errors of the estimates decrease as the number of recaptures increases.   The estimates are approximately unbiased, provided the number of recaptures during each sampling is not too small.

11-2.4   A Test for Equal Catchability

One of the most important assumptions implicit in all the mark-recapture methods is that all individuals in the population have the same probability of being caught.   Leslie (in Orians, 1958) supplies a test of this assumption.

Suppose that a large number of individuals is marked, and at regular intervals afterward the population is sampled and the number of marked individuals recorded.   A last sample N is taken, and the number of marked individuals recorded.   From these data it can be found how many of the marked individuals were recovered in each of the previous samples, n_i.   It will also be known how many times each marked animal in sample N has been recaptured. The distribution of the number of recaptures will be denoted as $f(x)$.   Given these two sets of statistics a chi-square test can be calculated between the expected and observed distributions $f(x)$.

An example will illustrate the computations.   Orians (1958) ringed 1,440 adult manx shearwaters in 1946.   In 1952, 32 marked individuals were recovered.   The number of these 32 individuals n_i recaptured in each of the 5 years between 1946 and 1952, together with the observed frequency distribution of x, the number of times each of the 32 birds was recaptured are listed in Table 11-6.

The mean of the distribution $f(x)$ is

$$\bar{x} = \frac{\sum xf(x)}{N}$$

$$= .9688$$

Table 11-5 **APPLICATION OF JOLLY'S STOCHASTIC MARK-RECAPTURE METHOD TO A POPULATION OF THE BUG $B.$ $ANGULATUS$; ESTIMATED POPULATION PARAMETERS**

i	$\hat{\alpha}_i$	$\hat{M}_i$	$\hat{N}_i$	$\hat{\phi}_i$	$\hat{B}_i$	$\mathrm{var}(\hat{N}_i)^{1/2}$	$\mathrm{var}(\hat{\phi}_i)^{1/2}$	$\mathrm{var}(\hat{B}_i)^{1/2}$	$\mathrm{var}(\hat{N}_i|N_i)^{1/2}$	$\mathrm{var}(\hat{\phi}_i|\phi_i)^{1/2}$
1	...	0	...	.649	...	...	.114	...	...	.093
2	.0685	35.02	511.2	1.015	263.2	151.2	.110	179.2	150.8	.110
3	.2189	170.54	779.1	.867	291.8	129.3	.107	137.7	128.9	.105
4	.2679	258.00	963.0	.564	406.4	140.9	.064	120.2	140.3	.059
5	.2409	227.73	945.3	.836	96.9	125.5	.075	111.4	124.3	.073
6	.3684	324.99	882.2	.790	107.0	96.1	.070	74.8	94.4	.068
7	.4480	359.50	802.5	.651	135.7	74.8	.056	55.6	72.4	.052
8	.4886	319.33	653.6	.985	−13.8	61.7	.093	52.5	58.9	.093
9	.6395	402.13	628.8	.686	49.0	61.9	.080	34.2	59.1	.077
10	.6614	316.45	478.5	.884	84.1	51.8	.120	40.2	48.9	.118
11	.6260	317.00	506.4	.771	74.5	65.8	.128	41.1	63.7	.126
12	.6000	277.71	462.8	...	...	70.2	...	...	68.4	...
13	.6690									

SOURCE: From Jolly, G. M., Explicit Estimate from Capture-Recapture Data with Both Death and Dilution—Stochastic Model, *Biometrika*, vol. 52, 1965.

and the sum of the squared deviations from the mean is

$$\sum (x - \bar{x})^2 = \sum x^2 f(x) - \frac{[\sum xf(x)]^2}{N}$$

$$= 69 - 30.03$$

$$= 38.97$$

The expected variance of the distribution $f(x)$ is

$$\sigma^2 = \bar{x} - \sum \frac{n_i^2}{N^2}$$

$$= .9688 - .1943$$

$$= .7745$$

The test statistic χ^2 is

$$\chi^2 = \sum \frac{(x - \bar{x})^2}{\sigma^2}$$

$$= \frac{38.97}{.7745}$$

$$= 50.32$$

Table 11-6 THE OBSERVED FRE-
QUENCY DISTRIBUTION OF RE-
CAPTURES OF 32 ADULT MANX
SHEARWATERS MARKED IN 1946
AND LAST RECAPTURED IN 1952

Year	n_i	x	$f(x)$
1947	7	0	15
1948	7	1	7
1949	6	2	7
1950	4	3	2
1951	7	4	1
		5	0
Total			32

SOURCE: After Orians, G. H., A Capture-
Recapture Analysis of a Shearwater Population,
J. Anim. Ecol., vol. 27, 1958.

The statistic χ^2 is treated as a chi-square with $N - 1$ degrees of freedom, 31 in the example. In this case the probability of a chi-square value of 50.32 or greater is between .025 and .010. It is concluded that the shearwaters were not collected at random; i.e., they were not equally catchable.

If the degrees of freedom is greater than 30, the chi-square distribution in the table is usually listed only for multiples of 10 degrees of freedom. If the number of degrees of freedom is large, the quantity $(2\chi^2)^{1/2} - (df \times 2 - 1)^{1/2}$ is normally distributed with a zero mean and a variance of 1. In the example $[(2)(50.32)]^{1/2} - [(2)(31) - 1]^{1/2} = +2.22$. The probability of as large a deviation as $+2.22$ is found from a table of the standardized normal distribution, and in this case is between .025 and .020. For a further discussion of equal catchability, see Cormack (1966).

11-2.5 Survey-Removal Method

If a population is exposed to a differential kill of the sexes, the population density can be estimated from the sex ratios before and after the kill, as well as from the size of the kill. The method as discussed follows Chapman (1954, 1955) and Chapman and Murphy (1965). Although not truly a mark-recapture method, the parallel is obvious. This method is less satisfactory than typical mark-recapture studies, but suitable data are often easily available. An example might be a deer population in which the population is sampled before and after the hunting season.

A first sample is taken, and in this sample there are n_{x1} males and n_{y1} females. After the first sample is taken, C_x males and C_y females are removed. Following the kill a second sample of n_{x2} males and n_{y2} females is taken. The following two assumptions are now made: (1) there is no natural mortality between the first and second samples, and (2) all individuals in the population are equally likely to be "caught" in the two samples. If these two conditions are fulfilled, the total number of males N_{x1} and the total number of females N_{y1} present at the time the first sample was taken are $N_{x1} = \alpha n_{x1}$ and $N_{y1} = \beta n_{y1}$, where

$$\hat{\alpha} = \frac{C_x n_{y1} - C_y n_{x1}}{n_{x1} n_{y2} - n_{x2} n_{y1}} \qquad \hat{\beta} = \frac{C_x n_{y1} - C_y n_{x1}}{n_{x1} n_{y2} - n_{x2} n_{y1}}$$

There are no standard errors for these estimates.

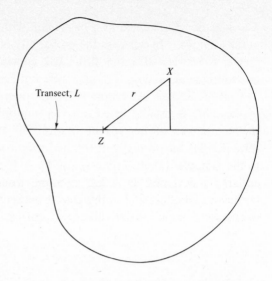

FIGURE 11-3
A diagrammatic representation of the terminology used in the linear transect method
of estimating population density.

11-3   ESTIMATING DENSITY FROM
A LINEAR TRANSECT

The discussion of this technique is based on an article by Gates (1969).   In
the linear transect method an observer walks through an area along a linear
transect of length L (Fig. 11-3).   The transect is randomly placed across the
area, but pragmatically should be placed away from the margins of the area to
avoid edge effects.   The individuals of the population are assumed to be
randomly dispersed.   This is a severe restriction.   The observer walks along
the transect, and when an individual is spotted, a deer or pheasant
perhaps, the distance from the observer's position Z to the animal's position
X is measured.   This distance is r.   The procedure does not require that all
individuals close to the transect be seen, but the probability of spotting an
individual is greater near the transect than far away.   There is a constant λ
representing the visibility of the animals.   The constant is characteristic for a
species and type of cover or terrain.   It is assumed that the probability of
seeing an individual at some given distance from the transect is equal on both
sides of the transect, e.g., the transect does not pass along the margin of a
forest.

The total size of the population in an area of size A is

$$N = \frac{n(2n-1)A}{2L\sum r}$$

where n is the number of individuals seen. If $n = 100$, $L = 30$ meters, $\sum r = 120$ meters, and $A = 1,000$ square meters, then $N = 100(200 - 1)1000/(2)(30)(120) = 2,761$ individuals. The variance of the estimated density is approximately

$$\text{var}(N) = \left[\frac{n}{(2L/A\lambda)}\right]\left[\frac{3n-2}{2(n-1)(2L/A\lambda)} - 1\right]$$

where λ is estimated as $(2n-1)/\sum r$. In this example

$$\text{var}(N) = \left[\frac{100}{(2)(30)/(1.66)(1000)}\right]\left[\frac{300-2}{2(100-1)(2)(30)/(1000)(1.66)} - 1\right]$$

$$= 112436.72$$

The estimated population density with its standard error is $N = 2761 \pm 335$.

11-4 DENSITY OF A RANDOMLY DISPERSED PLANT POPULATION

In the nearest-neighbor method of Chap. 5 the randomness of a population was determined by taking a plant at random and measuring the distance between the randomly chosen plant and its nearest neighbor. Denote the ith distance in n samples as r_i. The density of the population in a unit area if the population is randomly dispersed, and only if the population is randomly dispersed, may be estimated as

$$N = \frac{n}{\pi \sum r_i^2}$$

This estimate of N is biased. An unbiased estimate is

$$N' = \frac{n-1}{\pi \sum r_i^2}$$

The variance of the estimate is

$$\text{var}(N') = \frac{N^2}{n-2} \qquad n > 2$$

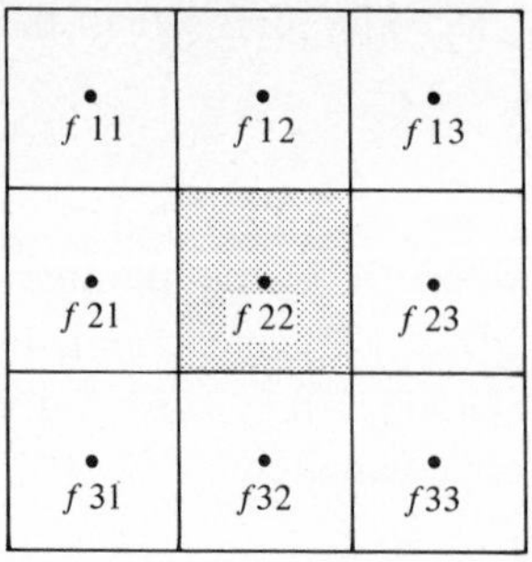

$f\,11$	$f\,12$	$f\,13$
$f\,21$	$f\,22$	$f\,23$
$f\,31$	$f\,32$	$f\,33$

FIGURE 11-4
A hypothetical study area divided into a series of nine squares, one central square surrounded by eight adjoining squares. (*From Dempster, 1957.*)

11-5   ESTIMATION OF DISPERSAL

The purpose of this section is to discuss methods of estimating dispersal in field situations. The first topic will be distinguishing between mortality and dispersal in a spatially dispersed population. The second topic is the estimation of the interchange of individuals between two areas.

11-5.1   Separating Mortality and Dispersal

Divide an area into a number of equal-sized squares (Fig. 11-4). In this area individuals are both dying and dispersing. If the number of individuals in each of these squares is counted or estimated at some later time, the number of individuals in each of the squares will have changed. If the movement of individuals is at random, the overall effect will be a shift in density from areas of high density to areas of low density. However, the change in numbers in each square is also due in part to mortality. The purpose of this estimation procedure is to separate changes in density due to mortality from those caused by dispersal. This method was developed for Dempster (1957) by J. G. Skellam.

If there is no mortality and dispersal is completely at random, Skellam (1951) shows that the change in the density with time in a given square, where f is the number of individuals, is

$$\frac{df}{dt} = \alpha \left(\frac{d^2f}{dx^2} + \frac{d^2f}{dy^2} \right)$$

The notation d^2f/dx^2 is the derivative of df/dx, that is, $d(df/dx)/dx$. If there is mortality μ, the effect of dispersal is overestimated. The effect of mortality on the change in numbers with time must be subtracted; i.e.,

$$\frac{df}{dt} = \alpha \left(\frac{d^2f}{dx^2} + \frac{d^2f}{dy^2} \right) - \mu f \qquad (11\text{-}7)$$

	A	B	C
1	44,230 / 14,600	45,650 / 16,145	6,478 / 5,840
2	47,370 / 22,580	50,500 / 26,475	16,650 / 6,623
3	50,720 / 30,015	103,900 / 52,680	89,160 / 57,860
4	52,060 / 28,560	85,200 / 20,675	70,950 / 32,330
5	48,060 / 36,760	35,065 / 17,570	32,260 / 33,100
6	87,200 / 74,990	124,100 / 27,070	44,040 / 33,630

FIGURE 11-5
The estimated numbers of locusts entering the first (upper numbers) and second (lower numbers) instars in each eighteenth part of the study area. (*From Dempster, 1957.*)

For a central square surrounded by squares of equal area $(d^2f/dx_2 + d^2f/dy^2)$ is set equal to $\frac{1}{3}(2f_{11} - f_{12} + 2f_{13} - f_{21} - 4f_{22} - f_{23} + 2f_{31} - f_{32} + 2f_{33})$. The calculations will be illustrated by an example.

Dempster (1957) estimated the number of first-instar nymphs of the Moroccan locust, *Dociostaurus maroccanus*, in a grid of 18 squares. The total area was 100×200 yards. After the nymphs had passed into the second instar, the numbers of individuals in each square were again estimated. These estimated numbers are shown in Fig. 11-5. The top number in each square is the estimated number at the first sample, and the bottom number represents the estimated population size for the second sample.

For square B2 the change in numbers from the first to the second sample is $df/dt = 26,475 - 50,500 = -24,025$. In addition

$$\frac{d^2f}{dx^2} + \frac{d^2f}{dy^2} = \frac{1}{3}[(2)(44,230) - 45,650 + (2)(6,478) - 47,370 - (4)(50,500)$$
$$- 16,650 + (2)(50,720) - 103,900 + (2)(89,160)]$$
$$= -11,465$$

Substituting both quantities into Eq. (11-7)

$$-24,025 = -11,465\alpha - 50,500\mu$$

Similar equations are formulated for the other central squares *B3*, *B4*, and *B5*. For these four squares the four equations are

$$B2: \qquad -11,465\alpha - 50,500\mu = -24,025$$
$$B3: \qquad -105,707\alpha - 103,900\mu = -51,220$$
$$B4: \qquad -54,125\alpha - 85,200\mu = -64,525$$
$$B5: \qquad +26,207\alpha - 35,065\mu = -17,495$$

The values of α and μ can be found by the simultaneous solution of these four equations. A simpler way to find α and μ is to form the *normal equations* of α and μ. This is done, e.g., for α, by taking the coefficient of α in each of the equations and multiplying through the equation. For square *B2*

$$B2: \qquad (-11,465)^2\alpha - (-11,465)(50,500)\mu = (-24,025)(-11,465)$$

The four equations are added, giving the normal equation for α

$$14,921,738,548\alpha + 15,254,442,345\mu = 8,723,683,325$$

The normal equation of μ is found in the same way, giving

$$15,254,441,345\alpha + 21,834,054,225\mu = 12,646,012,675$$

This procedure reduces four simultaneous equations to two which are easily solved, giving $\hat{\mu} = .58$ and $\hat{\alpha} = -.0029$.

Theoretically α cannot be less than zero, but because of errors of estimation negative numbers can result. At any rate, it is clear that dispersal of first-instar nymphs was almost, if not completely, absent. Fifty-eight percent of the population was lost to mortality during the development of the nymphs from first to second instar. It might be wise to reiterate that dispersal is assumed to be completely at random.

11-5.2 Dispersal between Two Areas

Iwao (1963) has developed a method measuring the reciprocal migration of individuals between two areas. This method is a reformulation of a technique originally devised by Richards and Waloff (1954).

It is first necessary to carry out a mark-recapture study in each area, consisting of a series of three samples. Bailey's triple-catch method is appropriate. The necessary statistics are:

$_xa_1$ = number of marked individuals released in area x on day 1.

$_ya_1$ = number of marked individuals released in area y on day 1.

$_ya_2$ = number of marked individuals released in area y on day 2.

$_{yy}r_{21}$ = recaptures in area y on day 2 marked in area y on day 1.

$_{xy}r_{21}$ = recaptures in area y on day 2 marked in area x on day 1.

and so forth, as determined by the subscripts.   Given these statistics the emigration rate from area x to area y is

$$_{xy}e = \frac{\left[\dfrac{(_{yy}r_{31})(_ya_2) + (_{yy}r_{21})(_{yy}r_{32})}{(_{yy}r_{32})(_ya_1)}\right](_ya_1)(_{xy}r_{21})}{(_xa_1)(_{yy}r_{21})}$$

and from area y to area x

$$_{yx}e = \frac{\left[\dfrac{(_{xx}r_{31})(_xa_2) + (_{xx}r_{21})(_{xx}r_{32})}{(_{xx}r_{32})(_xa_1)}\right](_xa_1)(_{yx}r_{21})}{(_ya_1)(_{xx}r_{21})}$$

If the population sizes of each area $_xP_1$ and $_yP_1$ have been estimated from the mark-recapture survey, the number of individuals moving from area x to area y per day is $_{xy}E = (_xP_1)(_{xy}e)$, and from area y to area x is $_{yx}E = (_yP_1)(_{yx}e)$. Because Bailey's triple-catch method assumes that the population is closed, the calculated birthrate and death rate will be polluted with changes due to immigration and emigration.   Iwao (1963) has developed an estimate of the "survival factor" per time period.

Iwao applied his method to measuring the dispersal of the ladybird beetle, *Epilachna vigintioctomaculata*, between fields of cucumbers and kidney beans.   He found that the rate of immigration from cucumbers to kidney beans per day was .201.   The reverse emigration rate was .137.

11-6   SEQUENTIAL SAMPLING

A rather serious problem arises in estimating the parameter k of the negative binomial distribution in a population dispersed over a wide area.   If an area is divided into a number of subareas, k may be relatively constant from subarea to subarea, but may decrease if averaged over the entire area because of the aggregative pattern due to the heterogeneity of the environment.   This is a matter of relative scale.   Waters (1955) found that in a series of 50 plots the average k value of each plot was 7.20, but that the value of k for the entire area was 1.44.

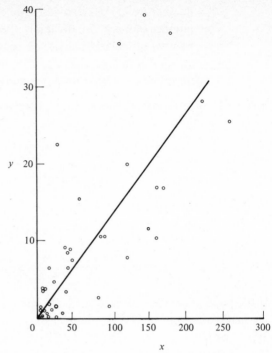

FIGURE 11-6
A method of determining a common value of k for a large area. The statistics x and y are plotted, and a regression line fit to the points. The common value of k, k_c, is equal to $1/\beta$, where β is the slope of the regression line. (*After Bliss, 1958.*)

To calculate a value of k representative of each of the subareas k_c, Bliss (1958) suggests the use of regression. Bliss analyzed Water's data on the number of spruce budworm larvae per twig in northern Maine. Two statistics, x and y, are calculated for each subarea, in this case a series of 50 plots, as

$$x = \hat{\mu}^2 - \frac{s^2}{N} \qquad y = s^2 - \hat{\mu}$$

where s^2 is the sample variance of each plot, and $\hat{\mu}^2$ is the sample mean number of larvae per twig in a given plot. In plot 1 the mean number of larvae per twig was 10.68, and the variance was 46.64. Therefore, $x_1 = (10.68)^2 - 46.64/25 = 112.20$ (there were 25 twigs counted per subarea), and $y = 46.64 - 10.68 = 35.96$. The values of x and y for each subarea are plotted against each other in Fig. 11-6. The slope of the regression line, b, is equal

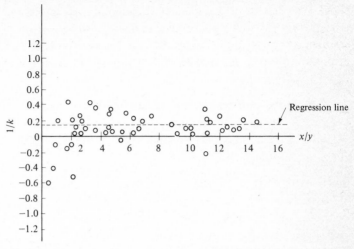

FIGURE 11-7
A plot of the value of the reciprocal of k of each subarea against x/y. The dotted line is the regression line. There is no significant clustering of points or trend toward the regression line, and it is concluded that a common value of k, k_c, is justified. (*After Bliss, 1958.*)

to $1/k_c$, where k_c is the common value of k for the entire area. Alternatively $k_c = 1/b$. The slope of the regression line in Fig. 11-6 is .138966. Therefore, the common value of k, k_c, is 7.106.

To check if the calculation of k_c is really justified, i.e., if the value of k within subareas is relatively constant, the values of x/y are plotted as in Fig. 11-7. If there is no marked trend or clustering of points, a common value of k can be assumed.

The value of a common k will become apparent in the following discussion of sequential sampling.

11-6.1   Sequential Sampling

If there is a large area where a pest species occurs, it may be necessary to estimate the density of the population at regular intervals of time without going through a strenuous sampling program using a large number of plots. However, it is not permissible to take only one or two samples to estimate the population density, because the population is not randomly dispersed. However, if (*1*) the subareas of the total area can be considered to have a common k, and (*2*) the dispersion pattern can be fit by the negative binomial distribution, then (*3*) sequential sampling can be used.

The first task in sequential sampling is to create two opposing hypotheses. Morris (1964) wished to distinguish between heavy and moderate infestations of *Abies balsamea* by the spruce budworm. There were two hypotheses termed H_0 and H_1 which should not be confused with null and alternative hypotheses:

H_0 = the number of egg masses is 100 or less per 100 square feet; moderate infestation.

H_1 = the number of egg masses is 200 or more per 100 square feet; heavy infestation.

Each hypothesis is subject to two types of sampling error: (*1*) α is the probability of accepting H_1 when H_0 is true, and (*2*) β is the probability of accepting H_0 when H_1 is true. Morris set both α and β equal to .1. If one hypothesis is accepted, there is 1 chance in 10 that the accepted hypothesis is wrong. The estimated common value of k was 5.232. Because the postulated density for hypothesis H_0 is 100 individuals, the values of p and q of the negative binomial based on hypothesis H_0 and $p_0 = 100/5.232 = 19.1131$ and $q_0 = 1 + p_0 = 20.1131$. Similarly for hypothesis H_1 the assumed density is 200 larvae per 100 square feet, so $p_1 = 38.2263$ and $q_1 = 39.2263$. Using these figures two lines are plotted as

$$d = sn + h_0 \qquad d = sn + h_1$$

where

$$s = \frac{\log(q_1/q_0)}{\log(p_1 q_0/p_0 q_1)} k$$

and

$$h_0 = \frac{\log B}{\log(p_1 q_0/p_0 q_1)} \qquad B = \frac{\beta}{1 - \alpha}$$

$$h_1 = \frac{\log A}{\log(p_1 q_0/p_0 q_1)} \qquad A = \frac{1 - \beta}{\alpha}$$

The number of samples taken in sequence is n. In the example $s = 138.791$, $h_0 = -87.257$, and $h_1 = 87.257$, if both α and β are equal to .1. The two points when $n = 1$, the first sample in a sequential series, are $d = (138.791)(1) - 87.257 = 43.166$ and $d = (138.791)(1) + 87.257 = 226.048$. If these two values of d are calculated for various values of n, two straight lines result (Fig. 11-8). The area between these two parallel lines can be termed an uncertainty zone. If a decision were made to spray the forest for the spruce budworm if the infestation were heavy, but not to spray if the infestation

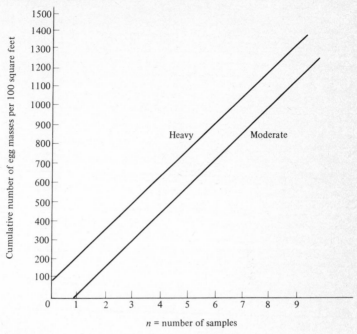

FIGURE 11-8

The two parallel lines enclose the uncertainty zone between a moderate and a heavy infestation of spruce budworm egg masses.   In sampling the first sample is taken, and if the number of egg masses falls between the two lines a second sample is taken, and added to the first; the process is continued until the cumulative number of egg masses falls outside the uncertainty zone.

were moderate or light, you would begin by taking one sample, $n = 1$.   If the number of egg masses is greater than 226, spray, but if less than 42, do not spray.   If the number of egg masses is between these two numbers, take a second sample, $n = 2$, add it to the first, and see if the cumulative number of egg masses is still inside the two lines or has moved outside.   The sampling is continued until either the cumulative number of egg masses has moved outside the two lines indicating either a moderate or a heavy infestation, or until some predesignated sample size has been reached.   If the cumulative sample remains between the two lines for some large sample size, the true infestation is probably somewhere between moderate and heavy.

PART FOUR

ASSOCIATIONS WITHIN PAIRS OF SPECIES

One of the most persistent beliefs in ecology is that the community of plants and animals in an area is an organized, distinct entity. This belief is reflected in a substantial literature on the naming or classifying of areas of vegetation. "Biocoenoses" of animals are also sometimes proposed. The great current interest in the measurement of species diversity also has a strong undercurrent of belief in the composition of the community as something more than just the sum of its parts, even though this feeling is rarely stated. This idea is also apparent in references to "energy flow between communities." The term *community* is a useful and necessary concept in delimiting the group of interacting species in an area. However, a community is, in large part, a subjective entity, and in specific studies its composition must be strictly defined.

The basic question, however, is, "Do communities represent changeable mixtures of species tolerating different environments and one another, or are communities constant quantities of the same species in the same proportions, each community being distinct and delimited from another?" Gleason, as early as 1926, came out against the concept of communities as homogeneous, well-defined assemblages of species. He held that a community of plants and animals was formed by the chance dispersal of propogules into an area and their subsequent differential mortality because of environmental factors. He also believed that plant species were distributed independently of each other, each species having its own characteristic tolerances to both the physical variables of the environment and the other species affecting its distribution. He stated that it was not possible to divide vegetation into clearly delimited areas of strongly associated species. Gleason's ideas at the time were for the most part summararily rejected by most plant ecologists. Forty years later most of Gleason's ideas became acceptable, although a compromise was reached between the community as a functional system of interacting species and the community as a collection of individually dispersed species. The plants and animals of a community in any given area respond to the pressures of the environment and interact with each other. The species and physical variables of an ecosystem form a functional system. However, individual species have different tolerances and optimal environments. Therefore, the dispersion and commonness of each species is, in general, different from the distribution and abundance of any other species. There is no conflict between species as parts of a functional ecosystem and species as adaptive organisms.

Not all species populations are independently distributed in a community. As we shall see, there are significant associations, both positive and negative, between many species of plants and animals due to interactions between the species or to similar responses of species to the same environmental variables. If two plant species both need selenium to grow, they will both be found on selenium-bearing soils, and, therefore, will be positively associated. Similarly, if two species of animals are in competition, and competitive exclusion is taking place, the two species will be negatively associated. In some cases one species will create conditions necessary for the survival of another. Common examples are many herbaceous plants found in forests, which are dependent on the trees for their existence. In some cases a herb may be dependent on a single species of tree. Associations between species are common, but clear-cut groupings of associated species corresponding to homogeneous, bounded communities appear to be rare or absent.

This chapter takes up the detection of association between two species, and its measurement. Chapter 13 discusses some of the techniques used to classify a set of quadrats into different "community types," or to ordinate them on a set of axes in an attempt to find the basic relationships between the species occurring in the quadrats. There is confusion in the literature between the measurement or detection of association between species and the classification of different sites into community types. The detection of association is clear-cut, and merely a statistical test; but the classification of sites is necessarily arbitrary and starts with the supposition, usually not stated, that classification is worthwhile.   Mathematically the two are closely inter-twined.   In fact, the measurement of association is so closely related to the methods of classification that classification almost always begins by measuring the degree of association between all possible pairs of species of a community.

Section 12-1 discusses the problem of detecting and measuring positive and negative association between a pair of species in a series of discrete units. Section 12-2 pursues the same subject when sampling is carried out in a continuous habitat rather than with discrete units.

12-1   ASSOCIATION BETWEEN TWO SPECIES OCCURRING IN DISCRETE UNITS

In the detection or measurement of association between a pair of species both species are assumed to occur in discrete units.   Quadrats placed in a continuous environment can be used, provided the considerations discussed in Sec. 12-2 are taken into account.

12-1.1   Detecting Association between a Pair of Species

A commonly used method of detecting association between a pair of species is to take a series of samples, such as ponds, quadrats, or trees, and count the number of units in which species A occurs but not species B, units with species B but not A, units with both species, and units with neither. The numbers of units in each group are listed as a 2×2 *contingency table.*

	Species B present	Species B absent	
Species A present	a	b	$a + b$
Species A absent	c	d	$c + d$
	$a + c$	$b + d$	n

We can also calculate the number of individual units in each class to be expected if the two species are distributed among the units independently of each other.   For example, the expected number of individual units in class a is $(a + b)(a + c)/n$.   The expected values based on a hypothesis of independence of the two species are compared with the observed cell totals a, b, c, and d by a chi-square test.   Suppose that in a sample of 500 logs the number of logs with the beetle species A but not species B is 50, both species are present in 200 logs, and so forth, giving a contingency table:

	Species B present	Species B absent	
Species A present	200	50	250
Species A absent	100	150	250
	300	200	500

If the two beetle species are independently distributed among the logs, a chi-square test on the 2×2 contingency table will be nonsignificant. The chi-square statistic for a 2×2 contingency table is

$$\chi^2 = \frac{(ad - bc)^2 n}{(a + b)(c + d)(a + c)(b + d)}$$

or in the example

$$\chi^2 = \frac{[(200)(150) - (50)(100)]^2 500}{(300)(200)(250)(250)}$$

$$= 83.33$$

This statistic is compared with a chi-square table with 1 degree of freedom. In this particular example the chi-square statistic is highly significant, and it is concluded that the two species of beetles are not independently distributed among the logs. The chi-square distribution is a continuous function. However, the data are discrete numbers. If n is small, the difference between the discrete distribution of the data and the continuous distribution of the chi-square distribution may be significant. If n is large, the differences due to the discreteness of the data are insignificant. If n is less than 30 or so, the *Yates correction* form of the chi-square test may be used.

$$\chi^2 = \frac{(|ad - bc| - n/2)^2 n}{(a + b)(c + d)(a + c)(b + d)}$$

The chi-square test has shown only that the two species are not distributed among the logs of the sample independently of each other. The test does not prove or disprove that the two species are positively associated or, conversely, negatively associated.

If n is small, less than 20, the corrected chi-square is not the most appropriate test for independence. If n is very small or if any of the cell totals are small, it is preferable, although laborious, to use the following exact test for independence in preference to the Yates correction for the chi-square. The test has a discrete probability distribution, unlike the chi-square distribution, and can detect either positive or negative association. In a contingency table the probability of obtaining an observed frequency of units with both species, cell a, given the fixed marginal totals $a + b$ and $a + c$ and sample size n, is

$$\Pr(a|n, a + b, a + c) = \frac{(a + c)!\,(c + d)!\,(a + c)!\,(b + d)!}{a!\,b!\,c!\,d!\,n!}$$

Consider the following numerical example:

	Species B present	Species B absent	
Species A present	2	13	15
Species A absent	10	3	13
	12	16	28

The probability of having two units with both species is

$$\Pr(2|28, 15, 12) = \frac{15!\,16!\,12!\,13!}{2!\,13!\,10!\,3!\,28!}$$
$$= .000987$$

Prior evidence leads one to suspect that the two species occur together too seldom, based on the supposition of independent association; i.e., they are negatively associated. It is possible to test for the significance of the deviation of a from its expected value based on the hypothesis of independent association by using a one-tailed test in a negative direction. If the hypothesis is that there are fewer joint occurrences of both species than expected on the basis of independence, the probability of a deviation as large as $E(a) - a$ or greater in the direction of negative association is

$$\Pr(\text{negative association}) = \sum_{i=a_{min}}^{a} \Pr(i|n, a + b, a + c)$$

The minimum value of a, of course, is zero, and a is the observed value of a. Similarly the one-tailed test for positive association is

$$\Pr(\text{positive association}) = \sum_{i=a}^{a_{max}} \Pr(i|n, a + b, a + c)$$

In this case a_{max} is 15. These calculations can be exceedingly tedious, although they can be easily carried out by using the tables in Beyer (1968) if both column or both row totals are less than 50. If the column totals are greater than 50, an approximate test of independence can be carried out by considering a to be a standardized normal variate X. The value of X is

$$X = \frac{a - E(a)}{[\text{var}(a)]^{1/2}}$$

where

$$E(a) = \frac{(a + c)(a + b)}{n}$$

$$\text{var}(a) = \frac{(a + b)(a + c)(b + c)(b + d)}{n^2(n - 1)}$$

Applying the test to the beetle example $E(a) = (250)(300)/500 = 150.00$, and $\text{var}(a) = (300)(250)(200)(250)/500^2(499) = 30.06$. The test statistic is $X = (200 - 150)/5.4827 = +9.1196$. From prior evidence one suspects that the two species are positively associated. With a one-tailed test in the direction of a positive association we find that the probability of a positive value of X this large, if both species are independently distributed among the logs, is infinitesimally small. Therefore, the two species of beetles are positively associated in the logs.

In the examples above, and in most real situations, a sample is treated as an entity, and the number of logs with species A and the number of logs with species B are set. Therefore, the marginal totals $a + b$, $a + c$, $b + c$, and $b + d$ are fixed. If a series of random samples were taken from a large population, not only would the frequencies of the cells vary but also the marginal totals. In using the above tests of independence the true hypothesis being tested is that the individuals in the sample are independently distributed among the units. This hypothesis is not equivalent to independence in the populations of both species. The generalization of the test on the sample to the populations of both species in most cases is justified but intuitive. The tests for contingency tables with random marginal totals, i.e., tests involving the population, not the sample, are complex and cannot be discussed in this text. If n is very small, the difference between fixed and random marginal totals may be important. The distinction between sample and population should be made, therefore, if n is small.

12-1.2   Cole's Measure of Association

The chi-square test only detects the presence or absence of some form of association; it is not a measure of the degree of association. Cole (1949) points out that a measure of association between two species has three desirable qualities:

1 The measure will be zero if the two species are distributed independently of each other.

2 The index will be $+1$ if both species are completely associated, and -1 if they are completely negatively associated.

3 The coefficient will vary linearly with a.

Cole (1949) has proposed a measure, actually two measures, satisfying conditions *1* and *2* with a proviso to be added later, but the index does not satisfy assumption *3*. Because of a lack of linearity of the measure with a, Cole created two coefficients, C_1 if the association is positive (*ad* greater than *bc*), and C_2 if the association is negative (*ad* less than *bc*).

$$C_1 = \frac{ad - bc}{(a + b)(a + c)} \qquad C_2 = \frac{ad - bc}{(b + d)(c + d)}$$

Suppose that the contingency table is

	Species B present	Species B absent	
Species A present	2	13	15
Species A absent	10	3	13
	12	16	28

The product (2)(3) is less than (10)(13), so $C_2 = (6 - 13)/(13)(16) = -.5962$. There is one rather serious defect in Cole's measure of association. If B is found only in those units where A is, the value of C_1 will be 1.00. But even though B is completely associated with A, A is not necessarily completely associated with B. There may be some units where A is, but B is not. For example, consider the two contingency tables:

50	40	90		50	0	50
0	20	20		0	20	20
50	60	110		50	20	70

Both tables give C_1 equal to 1, but it is clear that both species are completely associated only in the right-hand table. In the table on the left species B is completely associated with A, but not vice versa. The situation illustrated by the right-hand table is termed *complete association*. The table on the left illustrates *absolute association*.

12-1.3 The Point Correlation Coefficient

A measure taking into account the difference between absolute and complete association is the point correlation coefficient V. The coefficient V is the correlation between two variables X and Y, where X and Y can take only the values 1 or 0.

$$X = \begin{array}{l} 1 \text{ (species A present)} \\ 0 \text{ (species A absent)} \end{array} \qquad Y = \begin{array}{l} 1 \text{ (species B present)} \\ 0 \text{ (species B absent)} \end{array}$$

and is calculated as

$$V = \frac{ad - bc}{[(a + b)(a + c)(b + d)(c + d)]^{1/2}}$$

The value of V ranges from -1 to $+1$, and is equal to zero if the two species are distributed independently of each other. The value of V for the left contingency table above is $V = (1000 - 0)/[(50)(60)(90)(20)]^{1/2} = .4303$. For the right-hand table with complete association V is equal to $+1$.

The variance of the estimate of V is

$$\begin{aligned}
\text{var}(V) = V^2 \bigg| &-\frac{4}{n} + \frac{ad(a + d) + bc(b + c)}{(ad - bc)^2} - \frac{3[(a + b) - (c + d)]^2}{(4)(n)(a + b)(c + d)} \\
&- \frac{3[(a + c) - (b + d)]^2}{4n(a + c)(b + d)} \\
&+ \frac{(ad - bc)[(a + b) - (c + d)][(a + c) - (b + d)]}{2n(a + b)(c + d)(a + c)(b + d)} \bigg|
\end{aligned}$$

In this case the estimate of V and its standard error is $.4303 \pm .0514$.

12-1.4 Other Measures of Association

An incredible number of association measures have been proposed through the years. Goodall (in Tuxen, 1972) has written a comprehensive review of the subject, comparing the various measures and discussing the advantages and shortcomings of each. It should be mentioned that measures of association are available based on quantitative counts of one sort or another, rather than on simple presence-absence data. In some cases two or more species may be so common that they occur in all the quadrats or units, although they may be commoner in some of the units than in others. Some common types of quantitative measurements used in plant ecology are: (*1*) number of individuals; (*2*) percent of the total number of individuals that are species i; (*3*) percent of the total area covered by species i, termed cover; and (*4*) percent or total biomass of species i. Several other measurements of abundance are also used. Kendall's rank correlation coefficient is particularly appropriate for quantitative data in most, or at least many, situations. In order to test the significance of an ordinary correlation coefficient, both variables must be normally distributed. However, counts or percents are rarely, if ever, normally distributed. *Kendall's* τ does not depend on the distributions of the two variables; i.e., it is *nonparametric*. The measure is based on a comparison of the relative rankings of the observations on each species. The computations involved are discussed in most statistics texts and will not be presented. The measure ranges between $+1$ and -1, a value of 0 indicating independence of the two species.

12-2 ASSOCIATION BETWEEN TWO SPECIES IN A CONTINUOUS HABITAT

The methods of detecting and measuring association within a pair of species in a continuous habitat are generally the same as those used if the two species are dispersed in discrete units. Because of the arbitrary limits of the sample units, however, two serious sampling problems arise: (*1*) the spacing of the sample units, and (*2*) their size. The sample units in a continuous environment will be called quadrats.

In Fig. 12-1 the area marked A might be the top of a hill where two

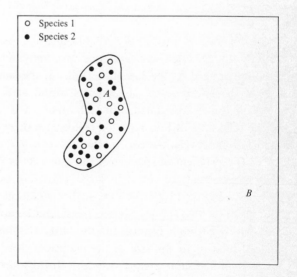

FIGURE 12-1
Two hypothetical grass species growing on top of a hill. If the sample area covers only the hill, the two species will appear to be dispersed independently of each other. If the total sample area B is sampled, the positive association between the two species will be detected.

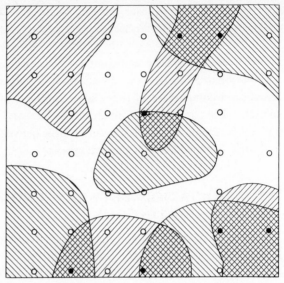

FIGURE 12-2
An illustration of the effect of sample spacing when the species occur as large, overlapping patches. Samples falling in an area of overlap between two species are marked black, other samples are open circles. If more than one quadrat falls in an area of overlap, the samples are not independent of each other, and the assumption of the chi-square test of independent samples is violated.

species of grass adapted to dry conditions live. Assume that these two species cannot survive elsewhere, and that the individuals of the two species are randomly mixed on the top of the hill. If the sample area covered only the top of the hill, a series of quadrats would show that the two species are not significantly associated with each other. If a series of samples were taken in the total area of Fig. 12-1, the resulting test would show a strong positive association between the two species.

The problem of spacing of the quadrats is often serious if each of the two species tends to occur in large, relatively well-defined aggregations or patches. Figure 12-2 represents such a situation. Because of their size the clumps of the two species overlap considerably, not for any biological reason, but purely through chance. If this area is sampled with quadrats closely spaced relative to the size of the clumps, several of the quadrats will tend to occur in the same area of overlap. If this occurs, the samples will not be independent, and one of the principal assumptions of the chi-square test will be violated. Therefore, the quadrats should be spaced far enough apart to prevent more than one quadrat from falling in the same area of overlap.

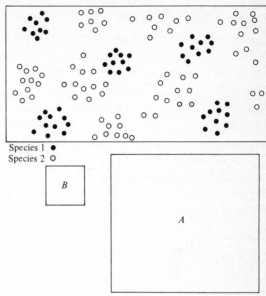

FIGURE 12-3
A hypothetical situation to illustrate the effect of the size of the sample quadrat on the detection of association between two species. A quadrat of size A indicates an independent dispersion of the two species, while a quadrat of size B detects the true negative association between the two species.

The problem of quadrat size can be illustrated by taking a patently absurd example. Suppose you wanted to measure the degree of association between saguaro cactus and ponderosa pine. Few persons will claim that the two species are associated, but it is possible by taking quadrat sizes of several thousand square miles scattered over the entire United States to come up with a significant association between the two. At the other end of the spectrum, if the quadrat is large enough to contain only one individual, any two species will show complete negative association. Figure 12-3 illustrates the effect of quadrat size for a more realistic situation. Suppose two species occurred in an area but each species was segregated from the other, perhaps because of an environmental mosaic or because of competitive displacement. If this habitat were sampled with a quadrat the size of A, the two species would prove to be independently dispersed relative to each other. Quadrats the size of B, however, would reveal a distinct negative association between the two species. Therefore, in creating quadrats to represent the discrete units of Sec. 12-1 care should be taken so that the sampling is not overly biased as a result of sample size, quadrat size, or sample spacing.

12-2.1 Plotless Sampling Methods

Because of the difficulties of sampling in a continuous habitat, Pielou (1961) developed a plotless technique utilizing the nearest-neighbor method to detect significant *segregation* between individuals of two species. In this case the pattern of one species relative to the other is measured and not the degree of association. In Fig. 12-4 species B is clumped relative to species A. Although the two species in Fig. 12-4 occur as distinct clumps and are distinctly positively associated, they are not segregated relative to each other.

If the populations of two species are small enough so that it is possible to tag all the individuals of both species, an individual is taken at random, its species noted, and also the species of its nearest neighbor. This process is carried out n times, and the data are compiled in a 2×2 contingency table

		Nearest-neighbor species		
		A	B	
Base plant	A	a	b	$a + b$
Species	B	c	d	$c + d$
		$a + c$	$b + d$	

A chi-square test is carried out to test if a is significantly greater than $(a + b)(a + c)/n$, the expected value of a based on the hypothesis of independent dispersion of the two species.

Pielou (1961) gives the following example for a woodland containing douglas fir, *Pseudotsuga menziesii*, and ponderosa pine, *Pinus ponderosa*. The contingency table with expected values in parentheses is

Base plant species	Nearest-neighbor species		
	Pine	Douglas fir	
Pine	30 (15.8)	23 (37.2)	53
Douglas fir	38 (52.2)	137 (122.8)	175
	68	160	

If the number of times a douglas fir has a pine as a nearest neighbor is indicated as F_{DP}, and the number of times a pine has a douglas fir as a

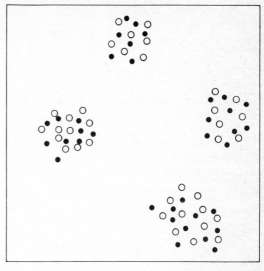

FIGURE 12-4
A hypothetical illustration of positive association between two species which are not segregated.

nearest neighbor F_{PD}, the test for the significance of the departure of F_{DP} and F_{PD} from the expected frequencies if the two species are not segregated, with a correction factor for small sample size, is

$$\chi^2 = \frac{(|F_{DP} - F_{PD}| - 1)^2}{F_{DP} + F_{PD}}$$

or in this case $\chi^2 = (|38 - 23| - 1)^2/(38 + 23) = 3.213$. With 1 degree of freedom, the probability of observing a larger value of chi-square by chance is .073. The chi-square statistic is not significant at the 5 percent probability level. There is no reason to reject the null hypothesis that the two species are not segregated.

Pielou (1961) has proposed as a measure of segregation

$$S = 1 - \frac{n(b + c)}{(a + b)(b + d) + (c + d)(a + c)}$$

The value of S is 0 if the two species are not segregated, and $+1$ if the species are completely segregated. If there is a negative segregation, i.e., the nearest neighbor is always the other species, S equals -1.

In the pine–douglas fir example

$$S = 1 - \frac{(38 + 23)(228)}{(53)(160) + (175)(68)}$$

$$= .318$$

Pielou (1962a) has extended her nearest-neighbor technique for measuring and detecting segregation between two species by utilizing transects.

13

COMMUNITY CLASSIFICATION AND ORDINATION

A pine forest is distinct from a prairie or a desert and is classified by a nonecologist as a "pine forest." One of the most extensive disciplines of ecology, particularly plant ecology, has been the classification of areas of vegetation into various "types." Around the turn of the century most classification was done by visual inspection, and usually with the assumption that communities of plants and animals were well-defined, integrated units which could be combined into well-defined real entities reflective of units in the real world (McIntosh, 1967). Vegetation was conceived of as consisting of a mosaic of discrete different community types. Later Gleason (1926), as discussed in Chap. 12, disputed this view, holding the opposing opinion that vegetation changes continuously and cannot be separated into distinct community types except arbitrarily. The first viewpoint can be referred to as the community-unit theory, and the second as the continuum concept. Most ecologists now accept the continuum concept, while recognizing also that there may be a good deal of discontinuity in vegetation in the field. Vegetation is expected to be continuous only if its environment is continuous and the vegetation undisturbed.

Even if communities are not viewed as discrete and natural units, it is sometimes necessary and useful to classify a series of sites or quadrats into admittedly arbitrary artificial units. An example might be a map of an area based on vegetation types for management purposes. A cornfield is distinctly separable from an adjacent forest, even if artificial and man-made. Even in a continuously changing area of vegetation such as the coniferous forests along an altitudinal gradient in western North America, it is still useful in some cases to divide the vegetation into forest types for purely practical purposes. A decision as to whether or not a classification should be carried out perhaps depends on whether the classification is useful or not.

The methods of community classification in this chapter are all based on "objective" statistical analysis of a set of data. A vast literature has accumulated on subjective ways of classifying and delimiting communities, either by visual inspection or by a series of simple criteria. An excellent summary of the early literature on community classification has been written by Whittaker (1962). In many instances a subjective division of the landscape is completely adequate. However, in objective classifications the criteria of classification are rigidly formulated, allowing two or more classifications to be compared. A multitude of classification techniques exist, and only a few methods can be discussed. For a complete discussion of classification and ordination the reader is referred to Tuxen (1972).

If a classification is attempted, the first decision is whether or not the units being classified (taken for simplicity to be quadrats) can be grouped into progressively larger units in a manner analogous to species, genera, families, and orders, e.g., several quadrats grouped into a pine forest, and several types of forests grouped together as forest as distinct from grassland. Another alternative is to plot the quadrats on a pair of coordinate axes and to look for clusters of points which can be delimited either by eye or by some form of statistical analysis. These clusters form a *reticulate classification*, as distinct from a *hierarchical classification*. Plotting a series of quadrats on a set of coordinate axes is termed *ordination*. Communities are often not clear-cut, but change continuously. Therefore, ordinated quadrats are widely scattered; they do not form clear-cut clusters. If the quadrats are widely scattered, hierarchical classification methods form groupings where groupings simply do not exist. Ordination methods are more noncommittal, and perhaps in many, if not most, situations are more appropriate.

In Sec. 13-1 hierarchical methods of classification will be taken up, including inverse classification, the grouping of species into a hierarchy rather than into quadrats. Section 13-2 discusses various methods of ordination, primarily polar ordination and principal components. In Sec. 13-3 some methods of cluster analysis are introduced for dividing ordinated quadrats into groupings. Section 13-4 discusses the continuum concept and examines the possibility of communities existing as real, integrated units in nature.

13-1 HIERARCHICAL CLASSIFICATIONS

If a hierarchical classification is used, three further decisions must be made: (1) whether the classification is to be divisive or agglomerative, (2) whether the classification is to be monothetic or polythetic, and (3) whether the data are to be qualitative or quantitative. In a *divisive* classification the whole population of quadrats is divided into smaller groups, usually two. These two groups are further divided until each group contains only a single quadrat. In contrast, *agglomerative* methods begin with individual quadrats, combining them into progressively larger groups until all quadrats belong to a single group. There are two advantages in using a divisive method:

1 The divisive method is usually faster because subdivision is usually not carried to the level of single quadrats. This cannot be true of agglomerative techniques, because the starting point is at the individual quadrats.

2 In agglomerative methods the chance absence or presence of species in the single quadrats from which the groupings begin may lead to bad initial combinations, inevitably resulting in a poor classification because of chance anomalies in a few of the quadrats (e.g., the chance absence of one of the characteristic species of a site in a quadrat).

In a *monothetic* classification two groups are combined or separated based on the absence or presence of a single attribute, usually a particular species. In a polythetic classification the two groups are joined or separated based on the overall similarity of the two groups. For this reason polythetic techniques have sometimes been termed similarity methods. Polythetic methods have the obvious advantage of considering all the information available in the sample, not just the information supplied by a single species. However, there are no feasible divisive-polythetic methods of classification as yet, and, as pointed out, divisive classifications may be preferred to agglomerative ones.

There is a multitude of different hierarchical classification methods available. The discussion in this chapter will be limited to two of the more commonly used techniques: association analysis and information analysis. The first is divisive-monothetic, and the latter agglomerative-polythetic. Pielou (1969) discusses other hierarchical classifications.

13-1.1　Association Analysis

Association analysis is a divisive-monothetic technique developed by Williams and Lambert (1959, 1960). In an association analysis the species present or absent in a series of quadrats are listed. If all the quadrats contain the same species, no classification is possible because all quadrats belong automatically to the same indivisible group. If there are differences in the species lists between the quadrats, the quadrats will be different, and a classification will be possible. The species in the quadrats may be positively or negatively associated, and an index of association within each pair of species is calculated. The most widely used index is V, the point correlation coefficient.

After an index of association is chosen, the index is calculated between all possible pairs of species over the series of N quadrats. These indices are listed in matrix form, and the absolute values of the indices (i.e., regardless of sign) in either the rows or the columns are summed. The association of a species with itself is defined as zero, as are all the associations of a species occurring in all the quadrats. Such a species is referred to as indeterminate. For example, a hypothetical case is

$$
\begin{array}{c}
\qquad\qquad\qquad \text{Species} \\
\qquad\quad 1 \qquad 2 \qquad 3 \qquad 4 \\
\text{Species}\begin{array}{c} 1 \\ 2 \\ 3 \\ 4 \end{array}
\left[\begin{array}{cccc}
\cdots & .82 & -.61 & .22 \\
.82 & \cdots & .53 & -.36 \\
-.61 & .53 & \cdots & .78 \\
.22 & -.36 & .78 & \cdots
\end{array}\right] \\[2mm]
\sum \quad 1.65 \quad 1.71 \quad 1.92 \quad 1.36
\end{array}
$$

If there are 100 quadrats, the first division into two groups is made by choosing the species with the highest column total of absolute V, here species 3 (1.92), and dividing the collection of quadrats into those with species 3 and those without species 3. The original collection of 100 quadrats now consists of two groups, one containing quadrats with species 3, and another containing quadrats without species 3. Each of these groups is treated as a collection unto itself, an association matrix is calculated for each, and the process is continued until no significant associations remain. This process is referred to as *normal association analysis*.

Williams and Lambert (1960) performed a normal association analysis on an area of heath bordering a bog in England. A total of 504 quadrats was taken at the site. The classification is shown in Fig. 13-1. Williams and Lambert used V as the index of association for dividing the sample. The stopping point of the division is set at a point where none of the

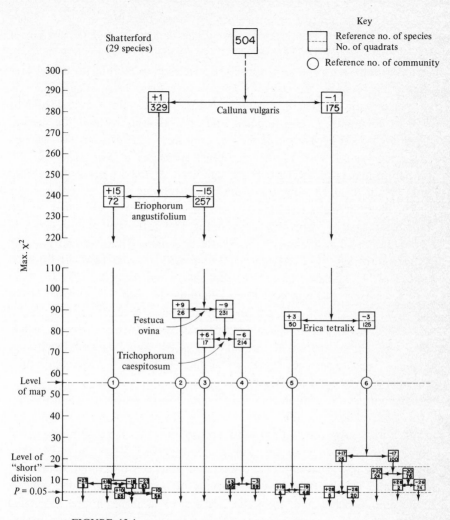

FIGURE 13-1
A normal association analysis of heath bog vegetation near Shatterford, England.
(*From Williams and Lambert, 1960.*)

associations in a group are significant at some preset significance level, say
$P = .05$. Williams and Lambert used the sum of chi-square as the scale of
the classification, referred to as maximum chi-square, but used V as their
index of association. This point is mentioned to avoid confusion.

The first division of the 504 quadrats was made on the plant species
Calluna vulgaris. Of the 504 quadrats, 329 contained the species and 175 did

not. The first group can be termed group A and the latter group a. Each of these two groups is treated as a separate collection, and the matrix of associations for each group calculated. The summed absolute V of $C.$ *vulgaris* in both groups is now zero, because the species is indeterminate in the first group and absent in the second. In group A *Eriophorum angustifolium* is the species with the highest summed absolute $V.$ Group A is divided into quadrats containing *E. angustifolium*, and into quadrats without the species. The next division of group a is on *Erica tetralix.* The division into groups is continued until none of the chi-square measures of association in the association matrix of each group are significant at the 5 percent probability level. Williams and Lambert also used as a stopping point the point where the maximum chi-square dropped below the single largest chi-square in the original data matrix. This is indicated in Fig. 13-1 as the level of short division.

Figure 13-2 is a map of the Shatterford area, with soil and contour profiles on the left. On the right-hand map Williams and Lambert list each quadrat as belonging to one of the first six groups formed. This stopping point is arbitrary and was kept low to simplify the map. In a classification the actual groups used and the level of groupings used ultimately depend more on what makes biological or practical sense than on any statistical test. The major division in this analysis is between bog and heath; i.e., $C.$ *vulgaris* is characteristic of heath in this part of England. A second division exists between wet and dry heath.

A normal association analysis divides a population of quadrats into successively smaller groups, until within a group there are no significant positive or negative associations between species. In a related process called *inverse association analysis* the groups are composed of species rather than quadrats. These groupings of species are abstractions of ecologically related species, and can be thought of as groups of positively associated species. In normal association analysis the indices of association are calculated between all possible pairs of species, depending upon the presence or absence of each species in each quadrat. In inverse association analysis the reverse is true. The quadrats are correlated in all possible pairs. An example is

	Species in quadrat 2	Species not in quadrat 2	
Species in quadrat 1	22	13	35
Species not in quadrat 1	11	6	17
	33	19	52

All possible correlations between quadrats are calculated as before and listed in matrix form with the absolute row or column totals.

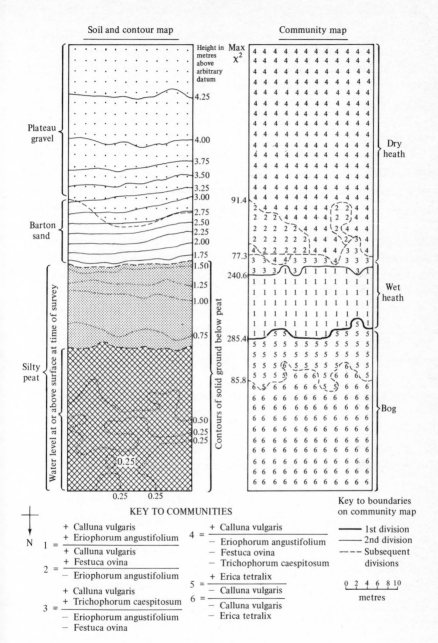

Shatterford
(29 species)

Soil and contour map

Community map

Plateau gravel

Barton sand

Silty peat

Water level at or above surface at time of survey

Height in metres above arbitrary datum

Contours of solid ground below peat

Max χ^2

91.4
77.3
240.6
285.4
85.8

Dry heath

Wet heath

Bog

KEY TO COMMUNITIES

Key to boundaries on community map

N

$1 = \dfrac{\text{+ Calluna vulgaris}}{\text{+ Eriophorum angustifolium}}$

$2 = \dfrac{\text{+ Calluna vulgaris} \\ \text{+ Festuca ovina}}{\text{− Eriophorum angustifolium}}$

$3 = \dfrac{\text{+ Calluna vulgaris} \\ \text{+ Trichophorum caespitosum}}{\text{− Eriophorum angustifolium} \\ \text{− Festuca ovina}}$

$4 = \dfrac{\text{+ Calluna vulgaris}}{\text{− Eriophorum angustifolium} \\ \text{− Festuca ovina} \\ \text{− Trichophorum caespitosum}}$

$5 = \dfrac{\text{+ Erica tetralix}}{\text{− Calluna vulgaris}}$

$6 = \dfrac{\text{− Calluna vulgaris}}{\text{− Erica tetralix}}$

——— 1st division
——— 2nd division
- - - Subsequent divisions

0 2 4 6 8 10
metres

FIGURE 13-2

A map of the Shatterford experimental area on the left, and a map of the vegetation types arrived at by the normal association analysis in Fig. 13.1 on the left. (*From Williams and Lambert, 1960.*)

The quadrat with the largest absolute row total is the critical quadrat, and the species are divided into those present in the quadrat and those not present in the quadrat. This division process is continued until the largest remaining row total falls below some level of significance, as in normal association analysis.

A simplified version of the Shatterford data using only 20 species and 56 quadrats is presented in Fig. 13-3. The quadrat with the largest absolute row total is quadrat 2. The species are divided into two groups, those present in quadrat 2 and those absent from quadrat 2. These two groups of species are treated as distinct collections of species, the correlations between the quadrats are recalculated for each group, and the analysis is carried out again. Four groups of species result from this analysis. Group A, consisting of *C. vulgaris*, *M. caerulea*, and *E. tetralix*, is a combination of species characteristic of heath in England. Group B contains species usually found in wet heath marginal to bog areas. These groupings contain positively associated species tending to occur under similar sets of environmental conditions, which may be useful as indicator species characteristic of certain types of vegetation and environmental conditions.

Association analysis usually uses presence-absence data. There are often cases in which one species occurs in all the quadrats, although it may be common in some quadrats and rare in others. Such a species is indeterminate, and if several indeterminate species occur, quantitative data are more correctly used than presence-absence records. Kershaw (1961) suggests using a covariance matrix rather than an association matrix, treating the covariance matrix as if it were an association matrix. The mathematical implications of this approach have not been investigated. A correlation matrix might also work.

13-1.2 Information Analysis

Information analysis is an agglomerative-polythetic method of classification developed by Williams, Lambert, and Lance (1966). Groups are formed by the initial combination of single quadrats into larger and larger groups based on the overall similarity of the groups. The similarity of any two groups is measured by the *information statistic I*. The concept of information content will be discussed in greater detail in Chap. 14. A group of quadrats contains "information," and the greater the differences between the quadrats making up the group, the greater the information content, i.e., the higher the value of I. Individual quadrats by definition have an information content

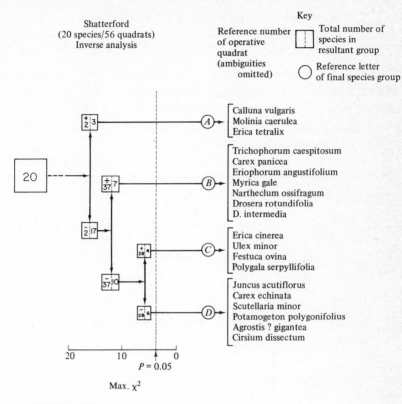

FIGURE 13-3
An inverse association analysis of the plant species occurring in the Shatterford experimental area. (*From Williams and Lambert, 1960.*)

of zero. If two quadrats are combined, the more closely the species compositions of the two quadrats agree, the lower the value of I of the combined group of two quadrats. If two groups of quadrats are combined, the information content of the new composite group is equal to the information content of the first group plus the information content of the second group plus an increment in information content caused by differences in species composition between the two groups. Symbolically

$$I_{\text{group 1}} + I_{\text{group 2}} + I_{\text{heterogeneity}} = I_{\text{new group}}$$

The procedure is to combine the two groups resulting in the least increase in total information content, i.e., those groups which are most similar. At the start the two quadrats giving a group with the smallest value of I are combined.

The information statistic I for a group is calculated as

$$I = \sum_j [n \log n - a_j \log a_j - (n - a_j)\log(n - a_j)]$$

where n is the number of quadrats, j is the species, and a_j is the number of quadrats in the group containing the jth species. The data are presence-absence figures. The base of logarithms is open to choice so long as one base is used consistently. Natural logarithms are most often used because of tables of $n \ln n$ in Kullback (1959).

To illustrate information analysis consider the hypothetical example below with three quadrats and three species. A plus sign indicates a species presence in a quadrat, and a minus sign indicates its absence.

	Species 1	Species 2	Species 3
Quadrat 1	+	+	−
Quadrat 2	−	−	+
Quadrat 3	−	+	+

The total information content of the group of three quadrats is

$$I = [3 \ln 3 - 1 \ln 1 - (3 - 1)\ln(3 - 1)] +$$
$$[3 \ln 3 - 2 \ln 2 - (3 - 2)\ln(3 - 2)] +$$
$$[3 \ln 3 - 2 \ln 2 - (3 - 2)\ln(3 - 2)]$$
$$= 5.729$$

The value of I for all possible pairs of quadrats is

	Quadrat 1	Quadrat 2	Quadrat 3
Quadrat 1	...	4.158	2.772
Quadrat 2	4.158	...	.693
Quadrat 3	2.772	.693	...

The net information gain in adding quadrat 3 to quadrat 2 is $0 + 0 + \Delta I = .693$ or $\Delta I = .693$. Clearly the smallest information gain is achieved by combining quadrats 3 and 2. The combination results in two groups: a group of quadrats 2 and 3, and a group containing only quadrat 1. These two groups will be referred to as groups 1 and 2, respectively. The index I is again calculated between all possible pairs of groups. In this case only one combination is possible, groups 1 and 2. The total information of this grouping is 5.729, and the net gain is $0 + .693 + \Delta I = 5.729$, or $\Delta I = 5.036$. If there were

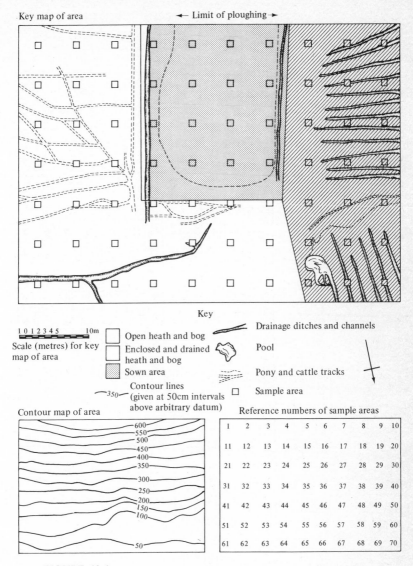

FIGURE 13-4

A map of the Markway experimental area. (*From Lambert and Williams, 1966.*)

354

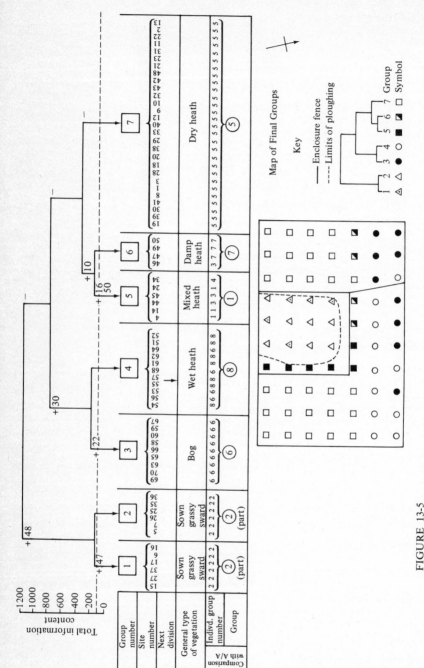

FIGURE 13-5

A normal information analysis of the vegetation of the Markway experimental area. *(From Lambert and Williams, 1966.)*

more than two groups remaining, the combination giving the smallest increase in I, i.e., the smallest value of ΔI, would be made.   In real situations the computations require the aid of a computer.

An agglomerative process must necessarily be completely carried out, but a divisive technique can be programmed to stop at some prearranged significance level.   However, the lower groupings, i.e., the smallest groups of quadrats, may not be particularly significant or biologically meaningful.   The level at which groupings become significant at some predetermined significance level, say $P = .05$, can be determined, because $2\Delta I$ is approximately distributed as chi-square with as many degrees of freedom as there are species in the group.

Lambert and Williams (1966) have carried out an information analysis on the Markway experimental area in England.   Figure 13-4 illustrates the area and the position of the quadrats, and Fig. 13-5 the resulting normal information analysis and the area mapped by groups.   There were 70 quadrats and 53 species.   The stopping point was at $P = .025$.   Lambert and Williams (1966) found little difference between the results of the normal association analysis and the normal information analysis.

An inverse information analysis can also be carried out.   Lambert and Williams found inverse information analysis gave more meaningful groups of species than inverse association analysis.

13-1.3   Some Comments

There is a voluminous literature on the hierarchical classification of communities.   Of the many techniques available, only two of the more widely used have been discussed.   There is no "best" method, and the selection of a method depends primarily on which gives results most consistent with the views of the researcher carrying out the study.   This problem will probably persist, because objective methods are being applied in an attempt to elucidate what is intrinsically a subjective subject.   Some of the difficulties of creating objective groups where subjective conditions exist can be eliminated by using the ordination techniques discussed in the next section.

13-2   ORDINATION

Ordination is the arrangement or plotting of either species or samples on an axis or axes.   The axes may represent environmental gradients or purely mathematical constructs derived from a matrix of similarities between the quadrats or species.   Ordination is an extensive subject recently reviewed by several authors in Tuxen (1972).

Whittaker and Gauch (in Tuxen, 1972) classify the plethora or ordination techniques according to the following criteria:

1 Based on what principle or model are the quadrats or species arranged?
2 Is ordination *direct*, using environmental gradients as the bases and axes, or is it *indirect*, using axes extracted from similarity measures applied to the quadrats or species?
3 Are samples or species ordinated?
4 What similarity index is used?
5 Are the axes of the ordination perpendicular to each other, or are they *oblique*?

Direct ordination is exceedingly useful in studying the relationships and distributions of the species in some area. Figure 13-13 is an example of a direct ordination of species along an environmental gradient. Whittaker (1967) plotted the importance values, importance being expressed as the percent of total stems over 1 centimeter in diameter, of five tree species along an elevation gradient in the Great Smoky Mountains. More will be said about the species distributions observed in Sec. 13-4. Suffice to say that the method is simple, direct, and often very useful.

In indirect ordination the axes of the ordination are mathematical constructs not corresponding directly to any quality of the environment. Two of the most commonly used indirect ordination methods will be presented in this section: polar ordination and principal components.

Table 13-1 PRIMARY MATRIX OF SPECIES COVERAGE PERCENTS IN NEW JERSEY SALT-MARSH SAMPLES

Sample number and species	1	2	3	4	5	6	7	8	9	10	11	12
1 *Spartina alterniflora*	75	30	5	20	5	1		10	1	2		
2 *Salicornia europaea*	5	10	2	1	1		2			2		
3 *Atriplex patula* var. hastata	1	10	2	1	1	2	5		1		5	2
4 *Distichlis spicata*		15	80	2	10	15	30	1	10	10	20	
5 *Suaeda maritima*				20	10							
6 *Salicornia virginica*				5	10							
7 *Juncus gerardii*			1			40	1					
8 *Scirpus olneyi*						5	20				1	
9 *Spartina patens*							20	10	50		2	5
10 *Iva frutescens*							5	1	2	1	20	10
11 *Phragmites communis*								1	10	20	5	30
12 *Solidago sempervirens*									1	5	1	2
Species cover total	81	65	90	49	37	63	83	23	75	40	54	49

13-2.1   Polar Ordination

Polar ordination was introduced by Bray and Curtis (1957). In a series of quadrats the importance of each species is recorded. In a study of a New Jersey salt marsh Cottam et al. (in Tuxen, 1972) recorded the percent of the area of the sample covered by each species in the salt marsh (Table 13-1). The data are then *doubly standardized*, first by dividing each number in a row by the sum of the row, and second by dividing each number in a column by the sum of the column (Table 13-2). Each sample quadrat is now represented by a column of relative importance values adding to 100 percent. After the double standardization has been carried out, the similarities between the samples are computed. The percentage similarity between two samples A and B is the sum of the smaller of the two values for each species. For example, the percentage similarity, PS, of samples 1 and 2 is $15.5 + 31.3 + 6.2 + 0.0 = 53.0$. The PS values are computed for all possible quadrat pairs (Table 13-3). The two most dissimilar samples, i.e., the sample pair with the lowest PS, are chosen as the end points of the first axis of the axes system. In this case this pair is quadrats 6 and 8. The length of the line between the end points is $PD = 100 - PS$. All the other sample quadrats are located along the axis between the end points, quadrats 6 and 8. The plotting of sample quadrat 7 is illustrated in Fig. 13-6. The PS of quadrat 7 with quadrat 8 is $D = 100 - 26.5 = 73.5$. Using the scale of the axis, take a compass and mark off an arc at a distance 73.5 units from the end-point quadrat 8. The distance from the end-point quadrat 6 is $D = 100 - 39.5 = 60.5$. Repeat the procedure with the compass. The intersection of the two arcs is a point. Drop a line from the point to the axis, and this is the

Table 13-2   PRIMARY MATRIX OF SPECIES COVERAGES IN NEW JERSEY SALT-MARSH SAMPLES AFTER DOUBLE STANDARDIZATION

ple number and species	1	2	3	4	5	6	7	8	9	10	11	12
Spartina alterniflora	62.5	15.5	4.5	13.4	3.5	0.8		31.1	0.7	1.3		
Salicornia europaea	31.3	38.6	13.4	5.0	5.3		7.3			9.7		
Atriplex patula var. *hastata*	6.2	38.6	13.4	5.0	5.3	12.1	18.2		5.3		22.7	9.1
Distichlis spicata		7.2	67.0	1.3	6.6	11.4	13.6	2.9	6.7	6.0	11.3	
Suaeda maritima				50.2	26.4							
Salicornia virginica				25.1	52.9							
Juncus gerardii			1.7			60.6	0.9					
Scirpus olneyi						15.1	36.4			2.3		
Spartina patens							14.5	46.6	53.4		1.8	4.5
Iva frutescens							9.1	11.7	5.3	2.4	45.3	22.7
Phragmites communis								7.8	17.8	32.2	7.6	45.5
Solidago sempervirens									10.7	48.3	9.1	18.2

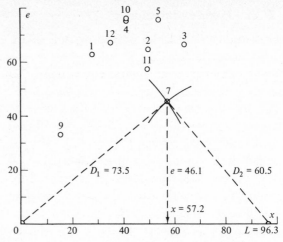

FIGURE 13-6
Ordination of the quadrats on the first axis in polar ordination. The quadrats are samples from a New Jersey salt marsh. (*From Cottam, Golf, and Whittaker, in Tuxen, 1972.*)

Table 13-3 SECONDARY MATRIX FOR SAMPLE SIMILARITIES, NEW JERSEY SALT MARSHES, PERCENTAGE SIMILARITIES OF SAMPLES AFTER DOUBLE STANDARDIZATION OF SPECIES COVERAGE DATA

Sample	1	2	3	4	5	6	7	8	9	10	11	12
1		53.0	24.1	23.4	14.1	7.1	13.5	31.1	6.1	11.0	6.2	6.2
2	53.0		38.5	24.7	20.7	20.2	32.7	18.4	12.7	17.0	29.9	9.1
3	24.1	38.5		15.8	20.7	26.0	35.2	7.4	12.7	17.0	24.7	9.1
4	23.4	24.7	15.8		66.4	7.1	11.3	14.6	7.0	7.6	6.3	5.0
5	14.1	20.7	20.7	66.4		12.7	17.2	6.4	12.6	12.6	11.9	5.3
6	7.1	20.2	26.0	7.1	12.7		39.5	3.7	12.7	6.9	25.7	9.1
7	13.5	32.7	35.2	11.3	17.2	39.5		26.5	31.9	15.7	42.7	22.7
8	31.1	18.4	7.4	14.6	6.4	3.7	26.5		63.3	14.4	23.9	24.0
9	6.1	12.7	12.7	7.0	12.6	12.7	31.9	63.3		37.7	35.8	43.7
10	11.0	17.0	17.0	7.6	12.6	6.9	15.7	14.4	37.7		25.1	52.8
11	6.2	29.9	24.7	6.3	11.9	25.7	42.7	23.9	35.8	25.1		50.2
12	6.2	9.1	9.1	5.0	5.3	9.1	22.7	24.0	43.7	52.8	50.2	
Mean	17.8	25.2	21.0	17.2	18.2	15.5	26.3	21.3	25.1	19.7	25.6	21.6

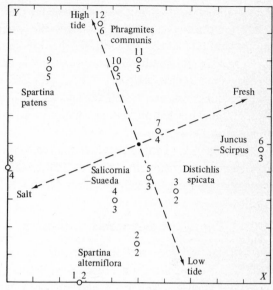

FIGURE 13.7
Ordination of the quadrats from a New Jersey salt marsh on two axes in a polar ordination.   The pair of dotted-line axes represents the general direction of tide level and salinity.   The species characteristic of each group of quadrats has also been added to the figure.   (*From Cottam, Golf, and Whittaker, in Tuxen, 1972.*)

position of quadrat 7 on the axis.   The procedure is the same for the remaining quadrats.   The distance x on the axis can be directly computed as $x = (L^2 + D_1^2 - D_2^2)/2L$, where L is the length of the axis; in this case $x = (.963^2 + .735^2 - .605^2)/(2)(.963) = .572$.   A second axis is now added by choosing another pair of quadrats as end points of the second axis.   These end points should be: (*1*) both among the samples in the middle part of the first axis, (*2*) close to one another in this position along the axis, but (*3*) more dissimilar (low PS) to each other than any other such pairs.   In the example quadrat pairs 1 versus 12 and 4 versus 10 are possible choices.   Quadrat pair 1 and 12 had the lowest value of PS, and the resulting axis was more interpretable in terms of factors of the environment, specifically tide level and salinity, than for quadrat pair 4 and 10.   The above procedure is repeated, and each quadrat is plotted on the second axis.   Each quadrat may now be plotted on both axes at X_1 and X_2 (Fig. 13-7).   In order to make the ordination ecologically interpretable, the rank of each quadrat site from lowest to highest in salinity is listed below the sample, and the rank in tide level is listed above the sample.   The dotted lines in Fig. 13-7 are drawn in the general direction of changes in salinity and tide level. Each quadrat in the ordination is now in a position relative to the two important physical variables: salinity and tide level. In addition, the dominant plant species occurring in clusters of sample points are also listed on the ordination. The relationships between the sample quadrats are now indirectly related by the ordination to salinity, tide level, and dominant species. Clusters of similar quadrats are evident, but no attempt has been made to group the quadrats into any sort of classification system.

Further axes could be included in the ordination, i.e., a third and higher axes, but the dimensionality of the ordination and a decrease in the importance of higher-order axes and their interpretability reduces their value.

13-2.2   Principal Components Ordination

One straightforward method of ordinating quadrats is to consider every quadrat an observation on 2 species variables (Fig. 13-9). The position of quadrat 2, say, is given by a pair of coordinates which are the quantities of the two species in the quadrat. If a third species is present, a third coordinate axis is added perpendicular to the other two axes, resulting in a three-dimensional set of coordinate axes.   A fourth species axis could be added, but the fourth axis would be in hyperspace. Needless to say, with more than three species the ordination becomes impossible to visualize. Some method of reducing the s original coordinates to new, transformed variables is needed, such that two or three of these mathematical variables account for so much of the variance of the original s variables, the species, that the quadrats can be plotted on these two or three new variables and still maintain positions relative to each other indicative of their similarity. Principal components ordination and its relative, principal coordinates ordination, are just such methods.

Suppose that the numbers of individuals of two species have been recorded in a series of n quadrats. The variances and covariances of these two variables, X_1 and X_2, species 1 and 2, can be listed in the covariance matrix S (see Chap. 9). These two species variables can be thought of as a pair of coordinates X_1 and X_2, and any observed number of the two species can be plotted on the coordinates.

The two species variables can be transformed into two new variables Y_1 and Y_2 (Fig. 13-8) by rotating the two coordinate axes X_1 and X_2 through some angle α. These two new variables are only mathematical and have no biological meaning. The two sets of coordinates are related by the angle between them. The variables Y_1 and Y_2 do not correspond to either X_1

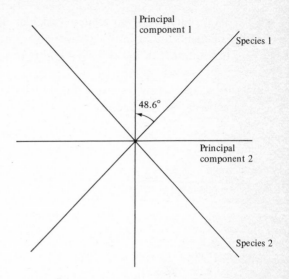

FIGURE 13-8
An illustration of the creation of principal components by the rotation of the coordinates representing the two original variables, in this case species, through an angle α of 48.6°.

or X_2, but each is, in a way, a combination of both. The variables Y_1 and Y_2 are derived so that:

1 The two variables are not correlated, termed *orthogonality*.
2 The first variable Y_1 is made as variable as possible.
3 The second new variable Y_2 is made as variable as possible under the condition of orthogonality to the first variable.

The importance of characteristics *2* and *3* can best be illustrated by supposing that 10 species are involved instead of only 2. If Y_1 is as variable as possible, and Y_2 as variable as possible, although less variable and orthogonal to Y_1, and so forth, for Y_3, Y_4, Y_5, etc., it is possible that the first two or three new variables, *principal components*, will account for most of the variance in the original variance-covariance matrix, even though the original species variables were of approximately equal variability. If the first three modified components account for 95 percent of the total variance of the total number of principal components, they will account for 95 percent of the total variance of the original variables. As successive components are calculated, the residual variance becomes smaller and smaller, usually at an approximately negative exponential rate. The basic purpose of principal components is to take a large number of correlated variables and reduce them to a small number of significant, uncorrelated variables, principal components.

The variances of the principal components are the latent roots of the variance-covariance matrix of the species. This is best illustrated by a numerical example involving only two species. Suppose that the covariance matrix between the two species is

$$\mathbf{S} = \begin{bmatrix} 17.903 & 19.959 \\ 19.959 & 23.397 \end{bmatrix}$$

The latent roots of the matrix $\mathbf{S}$, λ_1 and λ_2, are found from the characteristic equation (see Chap. 1) which is

$$\begin{vmatrix} 17.903 - \lambda & 19.959 \\ 19.959 & 23.397 - \lambda \end{vmatrix} = 0$$

or $\lambda^2 - 41.300 + 20.5148 = 0$. The two roots of the equation are $\lambda_1 = 40.797$ and $\lambda_2 = .503$. Note how the transformation takes two species variables of almost equal variance and converts them into two mathematical variables, where the first accounts for almost 99 percent of the total variance or the two species. The variable Y_2 can for practical purposes be ignored.

The components and the original variables are related by the angle of rotation α. The relationships between the two pairs of variables can be represented by a matrix $\mathbf{A}$.

$$\begin{array}{cc} & \text{Component 1} \quad \text{Component 2} \\ \mathbf{A} = \begin{array}{c} \text{Species 1} \\ \text{Species 2} \end{array} & \begin{bmatrix} \cos\alpha & \sin\alpha \\ -\sin\alpha & \cos\alpha \end{bmatrix} \end{array}$$

The elements of the $\mathbf{A}$ matrix correspond to the latent vectors of the latent roots. If the matrix $\mathbf{A}$ is represented as

$$\mathbf{A} = \begin{bmatrix} a_{11} & a_{12} \\ a_{21} & a_{22} \end{bmatrix}$$

the elements of the matrix $\mathbf{A}$, the proportional elements a_{ij}, can be found by the solution of the four equations leading to the latent vectors of the latent roots of a matrix (see Chap. 1). With more than three species these solutions can be conveniently found only with a computer. Therefore, the computations will not be carried out here. Several standard computer programs are available for the purpose. The principal axis method is perhaps the most commonly used. The matrix $\mathbf{A}$ must also satisfy the relationship $\mathbf{AA'} = \mathbf{I}$. The primary purpose of this condition is to provide a unique solution to $\mathbf{A}$, because there is an infinite number of possible latent vectors for every latent root. To fulfill this

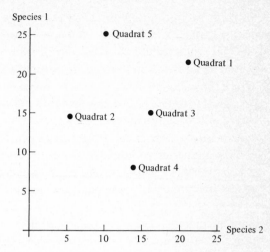

FIGURE 13-9
The ordination of five quadrats on two species.

condition the columns of **A** (the latent vectors) are calculated so that the sum of the squared elements of the column are equal to 1.0. This is called normalization. In the **A** matrix below note that $(.6571)^2 + (-.7578)^2 = 1$. You can verify that **A** fulfills the condition $\mathbf{AA'} = \mathbf{I}$. Normalization is carried out automatically by the principal axis computer program. The matrix **A** in the example is

$$\mathbf{A} = \begin{bmatrix} .6571 & .7538 \\ -.7538 & .6571 \end{bmatrix}$$

The parameter a_{ij} is the relationship between the ith species and the jth principal component. The relationship of the first species variable to the first principal component is $\cos \alpha = .6571$, and to the second component $\sin \alpha = .7538$. The relationship between the species and the two principal components is shown in Fig. 13-8. The angle of rotation is about 48.6 degrees.

Each quadrat is characterized and ordinated on a set of axes representing species according to the quantities of the species in it. In Fig. 13-9 quadrat 5 is ordinated at 10 individuals of species 1 and 25 individuals of species 2. Any quadrat is related to the new principal components through the linear combination

$$Y_i = a_{i1}X_1 + a_{i2}X_2 + a_{i3}X_3 + \cdots + a_{in}X_n$$

With only two species and the **A** matrix above a quadrat with 10 individuals of species 1 and 25 individuals of species 2 would be ordinated on the first principal component at $Y_1 = (.6571)(10) + (-.7538)(25) = -12.274$. The quadrat is ordinated on the second principal component at $Y_2 = (.7538)(10) + (.6571)(25) = 23.975$.

This two-species example is easily extended to three or more species. Instead of two coordinates at right angles to each other, three species would be represented by three lines at right angles in three dimensions. Fourth and subsequent species coordinates are in hyperspace. Although impossible to visualize, further dimensions are easily handled mathematically. Hopefully, the first two principal components account for such a large proportion of the variance of the species that the quadrats can be plotted on the principal components with relative positions indicative of the true relationships between them. Often, if not usually, a matrix of correlation coefficients is used instead of a covariance matrix. However, if a matrix of correlations is used in place of a covariance matrix, the principal components derived are usually not the same.

If a principal components analysis is used, the distances between the quadrats when the quadrats are plotted on the new set of coordinate axes are not necessarily proportional to the distances of the quadrats ordinated on the species axes. Gower (1966) has developed a technique called *principal coordinate* analysis whereby the original set of n coordinate axes is reduced to a smaller, significant set of axes as in principal components ordination and at the same time the proportionality of the distances between the ordinated quadrats from the old set of reference to the new is preserved.

13-2.3   Some Comments on Ordination

Although principal components ordination is far more elegant and formalized than either polar ordination or direct ordination, it is not necessarily the preferred method of ordination. One of the primary purposes of ordination is to create an ordination understandable in terms of the environment. Direct ordination of species along an environmental gradient is directly interpretable in terms of the environmental factors affecting the species. In polar ordination the end points of the axes are chosen by the investigator, and again the position of each quadrat in the ordination makes biological sense. In principal components ordination, however, the axes correspond to the purely mathematical criteria of maximum variability and orthogonality. Although mathematically succinct, the axes often have little or no interpretation in terms of biological factors.

Principal components and polar ordination both have one further weakness. In principal components ordination each component is postulated to be a linear function of the species. However, the species in a set of quadrats along an environmental gradient are not linearly related (Fig. 13-13); instead they are related in some complex curvilinear manner. Because the linear postulate of principal components ordination has been violated, the positions of the quadrats in the ordination are not indicative of the true relationships between the quadrats. For example, Whittaker and Gauch (in Tuxen, 1972) took samples along a simulated environmental gradient such as the one in Fig. 13-13. The sample quadrats from the simulated environmental gradient were subjected to a principal components ordination. Because the samples were taken along a gradient, the ordination should consist of a series of quadrats in a straight line when plotted on the component axes in order to make biological sense. However, because of the nonlinear distributions of the species, the ordinated quadrats formed curved line groupings, some of the curves becoming quite complex. Although this type of distortion is also present in polar ordination, the distortion is not as great. In addition, changes in species composition along the gradient also distort the principal components ordination, the distortion increasing as the change in species composition along the gradient becomes greater.

Whittaker and Gauch, after comparing several ordination methods, suggest the use of polar ordination. Polar ordination is relatively simple, and produces ordinations more interpretable in general than principal components ordination.

13-3 DISCRIMINANT ANALYSIS AND HOTTELING'S T^2

Let us consider a problem slightly different than those already discussed. Suppose there are two woodlots, both containing the same species of trees. The species of trees in the two woodlots, however, appear to differ in their abundances. Are the mean abundances of the trees in the first lot significantly different from those in the second?

To answer this question a series of samples is taken from each woodlot. The abundance of each species in each of the two woodlots is represented by its mean abundance, i.e., the average of the samples in each woodlot. For example, take two woodlots both containing the same four species of trees; 12 samples are taken in the second lot and 37 in the first. The data might be

	Woodlot 1 ($N_1 = 37$)	Woodlot 2 ($N_2 = 12$)
Species 1	$\bar{x}_{11} = 12.57$	$\bar{x}_{21} = 8.75$
Species 2	$\bar{x}_{12} = 9.57$	$\bar{x}_{22} = 5.33$
Species 3	$\bar{x}_{13} = 11.49$	$\bar{x}_{23} = 8.50$
Species 4	$\bar{x}_{14} = 7.97$	$\bar{x}_{24} = 4.75$

The mean abundances of the four species in the two woodlots can be listed as two mean vectors $\bar{\mathbf{x}}_1$ and $\bar{\mathbf{x}}_2$. Also, for each area calculate a matrix of the corrected sums of squares and cross products $\mathbf{A}_1$ and $\mathbf{A}_2$. The following two assumptions are now made:

1   Each species variable is normally distributed.
2   The covariance matrices of each set of data are equal (*homogeneous*).

If the second requirement is fulfilled (see below), a pooled covariance matrix for the two samples can be estimated as $\mathbf{S} = (\mathbf{A}_1 + \mathbf{A}_2)/(N_1 + N_2 - 2)$. The null hypothesis tested is that the mean vectors of the two woodlots are equal, i.e., $H_0: \boldsymbol{\mu}_1 = \boldsymbol{\mu}_2$, versus the alternative hypothesis $H_1: \boldsymbol{\mu}_1 \neq \boldsymbol{\mu}_2$. The test statistic is Hotteling's T^2

$$T^2 = \frac{N_1 N_2}{N_1 + N_2} (\bar{\mathbf{x}}_1 - \bar{\mathbf{x}}_2)' \mathbf{S}^{-1} (\bar{\mathbf{x}}_1 - \bar{\mathbf{x}}_2)$$

The quantity

$$F = \frac{N_1 + N_2 - p - 1}{(N_1 + N_2 - 2)p} T^2$$

has the variance ratio F distribution with degrees of freedom p, the number of variables, and $N_1 + N_2 - p - 1$.

In the hypothetical example suppose that the pooled covariance matrix is

$$\mathbf{S} = \begin{bmatrix} 11.2553 & 9.4042 & 7.1489 & 3.3830 \\ 9.4042 & 15.5318 & 7.3830 & 2.5532 \\ 7.1489 & 7.3830 & 11.5744 & 2.6170 \\ 3.3830 & 2.5532 & 2.6170 & 5.8085 \end{bmatrix}$$

then

$$\mathbf{S}^{-1} = \begin{bmatrix} .259064 & -.135783 & -.058797 & -.064719 \\ -.135783 & .186449 & -.038305 & .014382 \\ -.058797 & -.038305 & .159064 & -.016920 \\ -.064719 & .014382 & -.016920 & .211171 \end{bmatrix}$$

The statistic T^2 is

$$T^2 = \frac{(37)(12)}{(37 + 12)}[3.82 \quad 4.24 \quad 2.99 \quad 3.22]S^{-1}\begin{bmatrix} 3.82 \\ 4.24 \\ 2.99 \\ 3.22 \end{bmatrix}$$

$$= 22.05$$

The value of F is

$$F = \frac{37 + 12 - 4 - 1}{(37 + 12 - 2)4} 22.05$$

$$= 5.16$$

with 4 and 44 degrees of freedom. The probability of an F value this large or larger arising by chance if the null hypothesis is true is less than .005. The null hypothesis is rejected. The two areas are significantly different in the mean abundances of their species. The two woodlots can be considered to be two significantly different community types.

Before the T^2 test is performed the two covariance matrices should be tested for homogeneity. Suppose that

$$S_1 = \begin{bmatrix} 183.77 & 79.15 & 37.38 \\ & 50.04 & 21.65 \\ & & 11.26 \end{bmatrix} \quad S_2 = \begin{bmatrix} 451.39 & 271.17 & 168.70 \\ & 171.73 & 103.29 \\ & & 66.65 \end{bmatrix}$$

and

$$S = \begin{bmatrix} 295.08 & 175.16 & 103.04 \\ & 110.96 & 62.47 \\ & & 38.96 \end{bmatrix}$$

To test for the homogeneity of the two covariance matrices S_1 and S_2 compute the quantity MC^{-1} where $n_i = N_i - 1$,

$$M = (\sum n_i)\ln|S| - \sum_{i=1}^{k} n_i \ln|S_i|$$

and

$$C^{-1} = 1 - \frac{2p^2 + 3p - 1}{6(p + 1)(k - 1)}\left(\sum_{i=1}^{k} \frac{1}{n_i} - \frac{1}{\sum n_i}\right)$$

where there are p variables and k covariance matrices being compared. If only two covariance matrices are tested for homogeneity, $k = 2$. For the example matrices the determinant of S_1 is 794.05, the determinant of the second

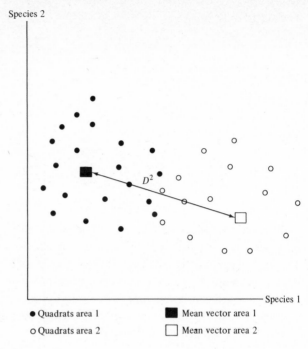

FIGURE 13-10
Two groups of quadrats from two areas plotted on two species to illustrate the concept of an average quadrat and the distance between two average quadrats.

covariance matrix is 12,639.89, and the determinant of the pooled covariance matrix is 5567.03. The quantity MC^{-1} is $(27.0586)(.9323) = 25.2265$. This quantity is approximately distributed as chi-square with $\frac{1}{2}(k - 1)(p + 1)p$ degrees of freedom. In this case there are 6 degrees of freedom. The probability of a chi-square this large or larger arising by chance if the two covariance matrices are homogeneous is exceedingly small. The two covariance matrices are not homogeneous, and the T^2 test of the mean vectors is not applicable.

If the observations in each of the two sample areas are quadrats, the quadrats can be ordinated on a p-dimensional set of coordinate axes as already discussed. In this case with four species we have a four-dimensional set of axes. The mean vectors $\bar{x}_1$ and $\bar{x}_2$ represent an average quadrat for each of the two areas, and the elements of the mean vector are the coordinates of the average quadrats. Figure 13-10 illustrates a possible case. The distance between these two average quadrats is a measure of their similarity; i.e., the closer they are the more similar they are. One measure of this distance is *Mahalanobis' D^2*.

$$D_{12}^2 = (\bar{\mathbf{x}}_1 - \bar{\mathbf{x}}_2)' \mathbf{S}^{-1} (\bar{\mathbf{x}}_1 - \bar{\mathbf{x}}_2)$$

Note that D^2 is equivalent to T^2 without the correction factor.

If the two mean vectors are not equal, it might be desirable to know which of the species are most responsible for the observed difference between the two areas. These species are found by using the *linear discriminant function*. The linear discriminant function is created by first calculating a column vector **a** of coefficients representing the importance of each of the species in causing the difference between the two areas. This vector **a** is

$$\mathbf{a} = \mathbf{S}^{-1}(\bar{\mathbf{x}}_1 - \bar{\mathbf{x}}_2)$$

or in the woodlot example

$$\mathbf{a} = \begin{bmatrix} .259064 & -.135783 & -.058797 & -.064719 \\ -.135783 & .186449 & -.038305 & .014382 \\ -.058797 & -.038305 & .150964 & -.016920 \\ -.064719 & .014382 & -.016920 & .211171 \end{bmatrix} \begin{bmatrix} 3.82 \\ 4.24 \\ 2.99 \\ 3.22 \end{bmatrix} = \begin{bmatrix} .030 \\ .204 \\ .010 \\ .443 \end{bmatrix}$$

The linear discriminant function is

$Y = .030$ species $1 + .204$ species $2 + .010$ species $3 + .443$ species 4

The most important species in terms of separating the two areas is species 4 followed by species 2.

It is possible that these two sample areas were contiguous, and that the initial division into two areas was more or less arbitrary. Near the boundary it may be problematical as to which community type a quadrat should be assigned. In Fig. 13-10 some of the quadrats can equally well be placed in either group. The discriminant function can be used to make this decision. In the initial analysis 37 samples were taken in area 1 and 12 in area 2. The probability of 1 of the total number of 49 quadrats being from the first area is $\frac{37}{49}$, and from the second area $\frac{12}{49}$. The first probability is designated h, and the second $1 - h$. Now suppose that it is no worse to classify mistakenly a quadrat really of community type 2 as community type 1 than it is to mis-classify a quadrat really of community type 1 as community type 2. If this is true, a quadrat represented by a column vector of the observed abundances of each species **x** will be classified as community type 1, area 1, if

$$\mathbf{x}'\boldsymbol{\Sigma}^{-1}(\boldsymbol{\mu}_1 - \boldsymbol{\mu}_2) - 1/2(\boldsymbol{\mu}_1 - \boldsymbol{\mu}_2)'\boldsymbol{\Sigma}^{-1}(\boldsymbol{\mu}_1 - \boldsymbol{\mu}_2) \geq \ln\left(\frac{1-h}{h}\right)$$

where Σ^{-1}, μ_1, and μ_2 represent the true population parameters estimated by S^{-1}, $\bar{x}_1$, and $\bar{x}_2$. Technically, this is valid only if the true population parameters are known. They rarely are. If the population parameters are estimated, classifying a quadrat into either of the two groups should be done with great care. For a discussion of the problem see Anderson (1958).

13-4   THE CONTINUUM CONCEPT

During the early history of plant ecology it was generally taken for granted that vegetation is made up of community types into which any stand of vegetation can be classified. These community types were thought to be well-defined natural entities, separated from other community types by narrow zones of contact called ecotones. A few persons disagreed with the prevailing view, in particular Gleason and Ramensky, who held that every plant species has its own characteristic tolerances to environmental gradients and, therefore, no two species have exactly the same spatial distribution. Gleason and Ramensky felt that communities occurring along continuous environmental gradients change continuously with gradual change in the density of each of the species. The latter view of vegetation is termed the continuum or individualistic concept.

If the densities of a series of plant species, or some other measure of a species importance, are plotted on a graph representing a transect along some environmental gradient, the curves representing each species will fall into well-defined groups if species do occur in well-defined community types (Fig. 13-11). Because of the sharp demarcation between communities, the curve for each species will be markedly platykurtic. If the species are independently distributed, indicating a continuous change in the composition of the community, a series of curves as in Fig. 13-12 will result. In a classic study Whittaker (1951, 1956) set out to test the two hypotheses by studying vegetation in the Great Smoky Mountains. He carried out his experiment by sampling along a transect up an elevation gradient. At each 100-meter interval of elevation five samples were taken and averaged. In each sample the percentage of the total number of stems over 1 centimeter in diameter at breast height made up by each of five species of trees was recorded. The transect covered an increase in elevation from 350 to 1,500 meters. The results of his observations are plotted in Fig. 13-13. Clearly this *gradient analysis* is like Fig. 13-12. Along the transect the species populations constantly and independently changed as elevation increased.

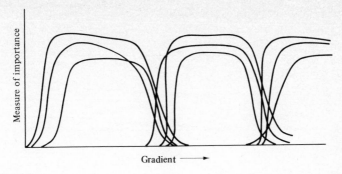

FIGURE 13-11
The distribution of populations of species along an environmental gradient if the community concept of the community is correct.

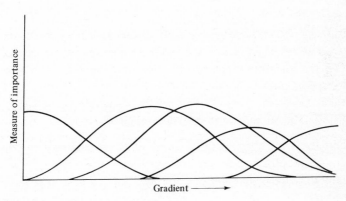

FIGURE 13-12
The distribution of populations of species along an environmental gradient if the individualistic concept of the community is correct.

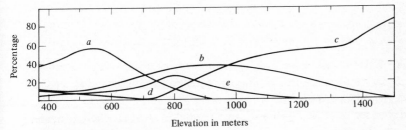

Elevation in meters

FIGURE 13-13
The importance of five species of trees along an elevation gradient in the Great Smoky Mountains. Importance is expressed as the percentage of the total number of stems over 1 centimeter in diameter at breast height. The four species were: (a) P. virginiana, (b) P. rigida, (c) P. pungens, (d) Q. marilandica, and (e) Q. coccinea. (From Whittaker, 1967.)

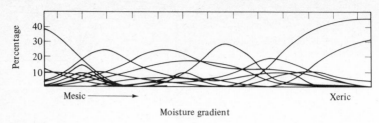

FIGURE 13-14
The population curves of several species of shrubs and trees plotted along a moisture gradient in the Great Smoky Mountains. Importance is expressed as the percentage of the total stem 1 centimeter or more in diameter at breast height. (*From Whittaker, 1967.*)

Elevation is only one possible gradient. Suppose that a series of samples is taken along a moisture gradient. Figure 13-14 represents another study by Whittaker along a moisture gradient with elevation kept relatively constant. The results are the same.

Whittaker (1967) has made the following general conclusions about communities of plants along environmental gradients:

1 Populations of species along continuous environmental gradients typically form bell-shaped curves of abundance, with densities declining gradually to scarcity and absence on each side of a central peak.

2 Species are not organized into groups with parallel distributions along the gradient. Each species is distributed in its own way, according to its own population response to environmental factors affecting it.

3 Because of the tapered form of the population gradient, the composition of communities changes continuously along environmental gradients if the gradients are uninterrupted and communities undisturbed. Disruptions will tend to produce distinct community types or groups, in the same way that a cornfield is distinct from a neighboring pasture.

Many other similar studies have been made, usually of a less direct and more complicated nature (e.g., see Brown and Curtis, 1952). These studies lead to the same general conclusions.

Gradient analysis does not have to be limited to a single environmental gradient. The density or some other measure of importance of a species in a sample can be plotted on a pair of coordinates representing two environmental gradients. In Fig. 13-15 four species of trees in the Great Smoky Mountains were studied, and the importance of each species expressed as the percentage of total tree stems present plotted on a pair of coordinates representing

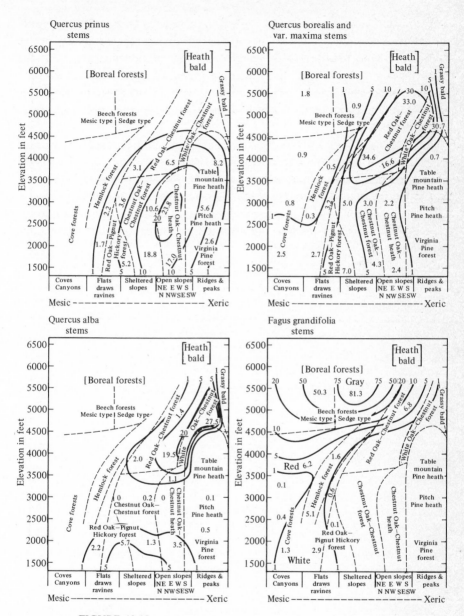

FIGURE 13-15

The importance of four species expressed as the percentage of the total stems over 1 centimeter in diameter at breast height as ordinated on two coordinate axes representing elevation and moisture. (*From Whittaker, 1967.*)

elevation and moisture. Quadrats in which the species are equally important are connected by lines. These lines are drawn over a schematic map of the "vegetation types" existing under given conditions of moisture and elevation. These lines of equal percentage form three-dimensional hills and valleys of a roughly bivariate normal distribution, but with many irregularities. The occurrence and commonness of each of the four species is independent of the other species.

Vegetation constantly changes along environmental gradients, and clear-cut, integrated community types do not exist except where they have been created by man or by some abrupt change in the environment. This does not lessen the practical value of delimiting such types for mapping purposes, so long as it is realized that the community groups arrived at by the hierarchical classification or cluster analysis are abstractions for convenience rather than real entities.

14

SPECIES-ABUNDANCE RELATIONS AND THE MEASUREMENT OF SPECIES DIVERSITY

The most obvious measure of species diversity is the number of species of some taxonomic group in an area. More species of butterflies occur in the Yucatan peninsula of Mexico than in the entire United States and Canada. Number of species is not the only way of looking at diversity, however. The abundance of each species can also be taken into consideration. In Fig. 14-1 the number of species of moths caught in a light trap at the Rothamsted Experiment Station in England with a given number of individuals, i.e., a given abundance, is graphed. Only the left side of the histogram is shown; the right tail of the distribution is very long. The importance of considering the relative abundance as well as number of species can be illustrated by two artificial "communities" both containing 10 species and 20 individuals (Fig. 14-2). Although both communities consist of 10 species, the relative abundances are different. The structures of the two communities are quite different. All the species in community 1 are equally common, but in community 2 some species are commoner than others. The species-abundance relationship in community 2 is graphed as a frequency distribution in Fig. 14-3.

Many diverse attempts have been made to fit different types of mathematical distributions to species-abundance relationships such as the one

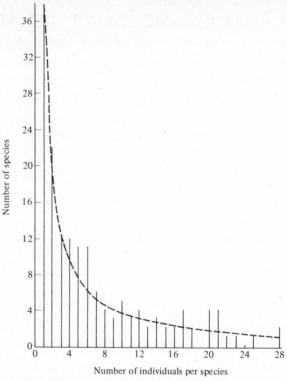

FIGURE 14-1

The frequency distribution of the abundance of moths in a light trap at the Rothamsted Experimental Station in England during 1935. The expected frequency distribution from the logarithmic series treated as a continuous function is shown as a dotted line. (*Data from Williams, 1964.*)

illustrated in Fig. 14-1. It is hoped that a distribution will be found fitting the data from many different types of communities and allowing the comparison of different communities through the parameters determining the two distributions. Perhaps these comparisons will reveal something about the underlying relations between the species in each of the two communities.

Some of these distributions were proposed as purely empirical fits to the data. Others, however, were derived from hypotheses about how the abundances of the species in the community should be related to each other. It had been hoped that, by specifying a set of conditions and deriving the distribution resulting from these hypotheses, specific conclusions about the interactions and relationships between the species and their environment could

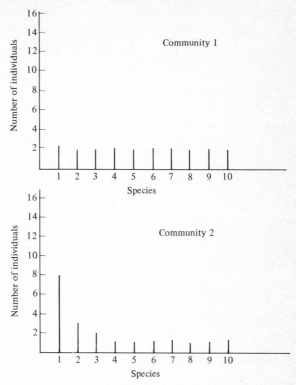

FIGURE 14-2
Two hypothetical communities to illustrate the importance of abundance as a determinant of species diversity.

be tested.   This approach has not been very fruitful, because the same distribution can often be derived from contrasting sets of initial premises.   In addition two distributions derived from conflicting postulates can sometimes both adequately fit an observed set of data.   The situation is analogous to the problem of fitting mathematical distributions to the observed spatial dispersion pattern of a species, as discussed in Chap. 5.   Therefore, even if the hypothesized distribution does fit the observed species-abundance relationship data, the fit neither proves nor disproves the postulates of the model. However, in terms of purely subjective value, the use of these models can help to summarize an observed species-abundance relationship, and to create heuristic hypotheses about interactions among the species in a community.

Because of the impossibility of determining the numbers of all the species in an area, species diversity studies are invariably carried out on

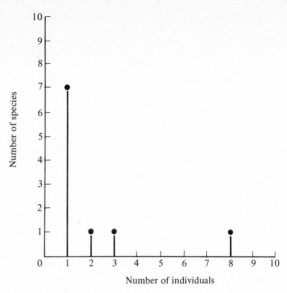

FIGURE 14-3
The frequency distribution of the abundances of the species in "community 2."

collections of particular taxonomic groups, such as the butterflies in a woods, the orbatid mites in 1 square foot of soil, or the ducks on Lake Cayuga. A community is necessarily arbitrarily limited both to area and taxonomy. Commonness is also a relative factor.   A mouse population of 100 individuals per acre is common, but a bacterial species with 100,000 individuals per acre is quite rare.   The effect of relative commonness can seriously bias measures of species diversity if disparate taxonomic groups are both included in the group of species taken to be the community.   For example, it is difficult to make meaningful comparisons between mice and bacteria purely on the basis of their abundances.

Species-abundance relationships have been rightly called "answers to which questions have not yet been found."   It is not likely that matching data and distributions will ever provide answers to questions concerning the basic relationships among species of a community.   However, the distributions do provide a convenient potential method of comparing two taxonomically limited groups of species.

Section 14-1 analyzes the relationships between area and the number of species found in the area.   Section 14-2 discusses some of the distributions which have been proposed to fit observed species-abundance frequency distributions.   Finally, Sec. 14-3 presents some empirical measures of species diversity.

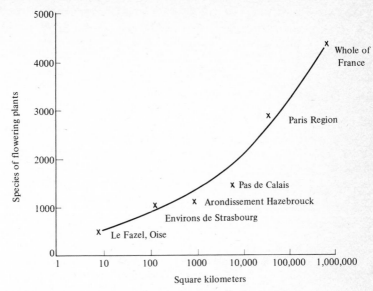

FIGURE 14-4
A plot of the number of species of flowering plants known from different-sized areas in France.   The area in square kilometers is plotted on a logarithmic scale. If the number of species were also plotted on a logarithmic scale, the relationship between area and the number of species would be roughly linear.

14-1   SPECIES-AREA RELATIONSHIPS

In Fig. 14-4 the number of species of flowering plants in increasingly large areas of land in France is plotted against the log of the area.   The shape of the curve drawn through the points by eye is roughly exponential.   This relationship between the number of species of a taxonomic group and an area of land or water turns out to be pretty general.   Preston (1962) presents some other examples.   In general data such as the plant data in Fig. 14-4 can be fit by the curve $S = CA^z$, where S is the number of species, A is the area, and C and z are two constants.   Transforming both the area and the number of species to logarithms results in the regression equation

$$\log S = \log C + z \log A$$

Given a series of values of S and A, a linear regression is calculated, and the two parameters C and z are estimated.

Kilburn (1966) investigated the relationship between number of species and area in several plant communities in Illinois and several other states. He utilized a 900-square-meter sample quadrat divided into areas of 100 square meters, 25 square meters, 4 square meters, 1 square meter, $\frac{1}{2}$ square meter,

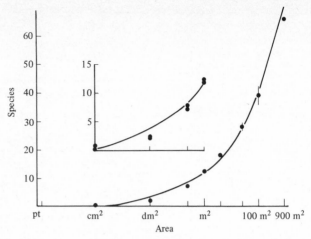

FIGURE 14-5
Species-log area data, shown by large dots, from an Illinois hill prairie, compared with $y = x^{.26}$, shown by solid line curves. Rooted data only plotted on lower complete graph. Rooted and covered data both plotted on smaller inset graph. Vertical lines represent double the standard error on either side of the point. (*From Kilburn, 1966.*)

1 square decimeter, and 1 square centimeter. In Fig. 14-5 the mean number of rooted species in each size quadrat is plotted for a hill prairie in Illinois. Transforming area and species to logarithms the regression equation relating area to number of species is $\log S = \log 12.4 + .26 \log A$. In nonlinear form the equation is $S = 12.4A^{.26}$. The parameter C is dependent on the area of the basic unit used, in this case a square meter, and represents the number of species in 1 square meter, 12.4. If the basic unit had been a square decimeter, C would have been 2.4. The parameter z, on the other hand, does not change and is constant for all sizes of basic units. The parameter z almost always lies in the range .25 to .35, even in situations as disparate as this plant community and the birds of the West Indies.

14-2 SPECIES-ABUNDANCE RELATIONSHIPS

The logarithmic distribution will be discussed first, because it is useful in providing empirical fits to observed species-abundance relationships. Following the discussion of the logarithmic distribution, three distributions derived from specific hypotheses of how the species of a community divide the resources available will be discussed: the niche preemption hypothesis, the log normal distribution, and the broken stick model.

14-2.1   The Logarithmic Series

In 1942 Corbert published an article on the relative abundances of 620 species of butterflies he had collected.   As in the moth data from the Rothamstad light traps, there were many rare species and relatively few common ones. The famous statistician Fisher decided that the frequency distribution of the number of species with a given number of individuals could best be fit by a logarithmic series

$$n_1, \frac{n_1 x}{2}, \frac{n_1 x^2}{3}, \frac{n_1 x^3}{4}, \ldots, \frac{n_1 x^{n-1}}{n} \qquad x < 1$$

where n_1 is the number of species in the sample represented by a single individual, n is the abundance class (e.g., $n_3 = 3$ individuals), and x is a constant related to the average number of individuals per species (Fisher, Corbert, and Williams, 1943).   Dividing n_1 by x and setting this equal to α, another constant, the series becomes

$$\alpha x, \frac{\alpha x^2}{2}, \frac{\alpha x^3}{3}, \ldots, \frac{\alpha x^n}{n}$$

The quantity αx is the number of species with one individual predicted by the logarithmic series, $\alpha x^2 / 2$ is the number of species with two individuals, and so on.   Adding all terms, the total number of species S is

$$S = \alpha[-\ln(1 - x)]$$

and the total number of individuals in the collection $N = \alpha x/(1 - x)$.   The parameter α is a measure of the diversity of the collection, and is low if the number of species is low in relation to the number of individuals, and high if the number of species is high.   The value of x, on the other hand, depends on the sample size.   To calculate the expected frequencies in each abundance class, x is first estimated by the iterative solution of

$$\frac{S}{N} = \frac{1 - x}{x}[-\ln(1 - x)]$$

If the ratio S/N is high, over 20, x is equal to .99 or more.   The parameter x cannot be more than 1.   In almost all cases x will be at least .9.   The value of α is found as

$$\alpha = \frac{N(1 - x)}{x}$$

Carrying out these calculations on Williams's moth data,

$$\frac{197}{6814} = \frac{1-x}{x}\left[-\ln(1-x)\right]$$

With $x = .994$, $.0289 = .0309$, and with $x = .9945$, $.0289 = .0277$. Using this latter value of x the parameter α is estimated to be $(6814)(.0055)/.9945 = 37.68$. The expected frequency of the first abundance class n_1 is $(37.68)(.9945) = 37.47$ species with one individual. The second class is $(37.68)(.9945)^2/2 = 19.13$ species with two individuals, and $n_3 = 12.35$, and so forth. The expected frequency distribution is plotted in Fig. 14-1. Technically a continuous curve should not be drawn through the expected values, because the logarithmic distribution is a series of discrete values, not a continuous function. However, in this case the curve has been drawn to illustrate the general shape of the predicted species-abundance relationship.

14-2.2   The Broken Stick Model

MacArthur (1957) was one of the first to attempt to develop a model based on a set of hypotheses of how the species in a community utilize the available resources, and to test these hypotheses by trying to match the observed abundances of the species in a sample with those expected if the model were true. For a variety of reasons this approach came to grief, although the model served as the stimulus for a great deal of research at the time.

The model is called the broken stick model, because in the most commonly used version MacArthur visualized the resources of the community as a line or stick. He postulated that this stick was randomly divided into s pieces. If each piece is thought of as the resources used by one species, its niche, the model postulates s species dividing the environment into s nonoverlapping niches of randomly allocated size. The expected abundance of the jth species N_j is a percent of the total number of individuals in the community. The expected percent abundance of the jth species is

$$E\,\frac{N_j}{N} = \frac{1}{S}\sum_{i=1}^{j}\frac{1}{S-i+1}$$

The only parameter of the distribution is s, the number of species. If s is known, the relative abundances of each species as ranked from the rarest to the commonest can be predicted. The relative abundance of the rarest species, $j = 1$, if there are 20 species, is

$$E\frac{N_1}{N} = \left(\frac{1}{20}\right)\left(\frac{1}{20 - 1 + 1}\right)$$

$$= \tfrac{1}{400} = .0025$$

or one individual in 400 will belong to the rarest species. For the second rarest species

$$E\frac{N_2}{N} = \frac{1}{20}\left[\left(\frac{1}{20 - 1 + 1}\right) + \left(\frac{1}{20 - 2 + 1}\right)\right]$$

$$= .0051$$

The expected number of individuals of the second rarest species, if there are 1,000 individuals, is $(.0051)(1,000) = 5.1$ individuals.

It turns out that the fit of the broken stick model to the observed abundances of each species is dependent on sample size (Hairston, 1969). More serious, perhaps, the same model can be derived from opposing sets of hypotheses. A second derivation was proposed by Cohen (1966). He postulated a multidimensional niche space available to s species, divided into s subniches. The rarest species occupies one subniche, the second rarest species two subniches, and so on, the commonest species occupying all the subniches. In contrast to MacArthur's formulation there is considerable overlap in the resources utilized by the species, although the species do not occupy exactly the same niches. The set of hypotheses leads to the same model as did the broken stick analogy, even though the situation is quite different. Cohen (1968) later derived the same model from yet a third set of assumptions. As in the case of the negative binomial distribution, discussed in Chap. 5, it does not appear that deriving a distribution based on a set of assumptions is a valid way of testing these assumptions in the field.

14-2.3 The Niche Preemption Model

Suppose the percent of the total available resources used by a species is determined by the success of a species in preempting for its own use part of the available resources. The less successful species occupy the resources left. The most successful, i.e., dominant, species occupies some fraction k of the resources or niche space. A second species is able to occupy a similar fraction k of the remainder left by the dominant species, and the third most important species k percent of the remaining resources left after species 1 and 2 have

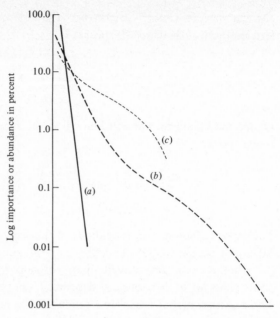

FIGURE 14-6

The general shape of three models expressing the manner in which the species of a community allocate the available resources. The species are ranked from commonest to rarest, and the rank of a species is plotted against the importance of the species in percent of the total number of individuals or biomass expressed in logarithms. The three hypotheses are: (a) niche preemption, (b) log normal distribution, and (c) broken stick model.

taken their share. If this set of postulates is true, the percent of the total number of individuals of all the species, or some other measure of importance such as biomass, ranked from commonest to rarest forms a geometric series

$$I_n = Nk(1 - k)^{n-1}$$

where I_n is the percent importance of the nth species, N is the total number of individuals, and k is equal to $1 - c$, where c is the ratio of the importance values of a species to that of its predecessor in the series of ranked species. Obviously, if the model is true, the ratio c must be constant between successive ranked species on the species list. If the species ranked from commonest to rarest are plotted against the logarithm of importance (Fig. 14-6), the predicted percent importance or abundance of the ranked species forms a straight line if the species-abundance relationship fits the niche preemption hypothesis.

Species-abundance relations similar to the predicted values of the niche preemption hypothesis are most often found in communities strongly dominated by a single species. Some plant communities, particularly those occurring in severe environments with a small number of species, are well fit by the geometric series of the niche preemption model.

14-2.4 The Log Normal Distribution

If some variable, such as fecundity, is determined by a large number of independent factors, the distribution of the variable will be normal. Therefore, if the niche of a species is dependent on a multitude of different factors, the sizes of the niches, i.e., the amount of resources used by each species, should be normally distributed. In a large and diverse community there should be many moderately abundant species and few common or rare species. This reasoning is the basis of the log normal distribution proposed by Preston (1948).

Plotting number of species against the number of individuals per species in steps of 1, 2, 3, etc., individuals is not the only way of relating abundances. It is just as reasonable to think of abundance as doublings of 1 to 2 individuals, 2 to 4 individuals, 4 to 8, 8 to 16, 16 to 32, and so on. Preston (1948, 1962) refers to each of these intervals as octaves. Note that using doubling intervals is equivalent to taking logarithms of the abundances to the base 2. If a species abundance falls on the line separating two octaves, e.g., an abundance of eight individuals, one-half of the species are assigned to the 4 to 8 individuals octave, and one-half of the species to the 8 to 16 individuals octave. The octaves are numbered; the octave with the greatest number of individuals is equal to zero, and the octaves are numbered in plus and minus directions from the mode of the curve. Preston (1948) postulates that data plotted in this manner can be fitted by the function

$$s = s_0 e^{-(aR)^2}$$

where s is the number of species in the Rth octave to the right and left of the mode (the curve is symmetrical), and s_0 is the number of species in the modal octave. The letter a is a parameter estimated from the data. The distribution of species into octaves is approximately normal.

The primary value of the log normal distribution is to summarize the observed abundance relationships in some types of communities. Therefore, the estimation of the parameters of the model will not be discussed in this chapter, and only some rather general comments will be made. Preston (1948) fit his log normal distribution to the Rothamstad moth data. The fitted function is $y = 35e^{-(.227R)^2}$. The observed number of species in each

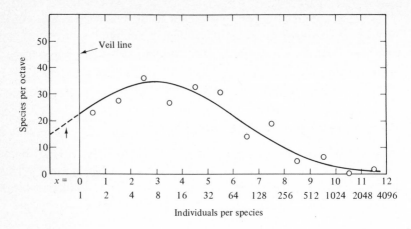

FIGURE 14-7
A log normal plot of the Rothamstad light trap moth catch.   The curve is discussed
in the text.   (*From Preston, 1948.*)

octave and the fitted curve are shown in Fig. '14-7.   The data seem to show
that there are many moderately common species and fewer rare or common
species.   The curve of the fitted function is truncated, with the left-hand limb
cut off.   The point of truncation is called the veil line.   Preston postulates
that the species theoretically belonging to the octaves to the left of the veil line
are so rare that they were missed by the sample.   As the size of the sample
increases, a larger percentage of the rare species will be caught, moving the
veil line to the left and revealing more of the curve.   However, it is difficult
to separate the effects of more complete sampling from the effects of environmental heterogeneity and changes in the species pool because of sampling
at different times and places.   There is a decided difference between the
logarithmic series and the log normal distribution.   If the distribution of the
number of species with a given number of individuals is best fit by a logarithmic
series, the same data plotted on an octave scale cannot show a mode to the
right of the veil line (see Pielou, 1969).   The presence of a mode to the right
of the veil line, as found in some of the examples analyzed by Preston (1948)
and others, seems to indicate that on the whole the log normal distribution
fits some situations better than the logarithmic series.   In homogeneous
samples taken at a single place and time the veil line usually falls very close
to the mode.   As a means of summarizing the abundance data in the latter
type of situation, there does not seem to be much to choose between the
two distributions.

The four distributions discussed in this section can be very useful in describing and summarizing the observed species-abundance relationships within a community. By fitting these models to the data, the abundance relationships of two or more communities can be compared. However, the models are heuristically motivated, and are of little value in determining the underlying interactions responsible for the observed abundance relationships among the species.

14-3 MEASURES OF DIVERSITY

One of the reasons for searching for a mathematical distribution fitting the observed frequencies of species with a particular number of individuals is to compare the diversity of two "communities." The parameter α of the logarithmic series is one possible measure of diversity. If it is certain that the relative abundances of the species in a sample or collection of individuals of some kind are fit by the logarithmic distribution, there is no reason why the parameter α cannot be used as a measure of the diversity of the group. It is clearly desirable, however, to have a more general measure of diversity that does not depend on the underlying frequency distribution of individuals among species. Some measures having this quality are discussed in this section: H and H', the information theoretical measures, and Simpson's index. Because H and H' are the most commonly used and widely accepted indices of diversity, they are presented first.

Diversity, as defined in this chapter, is dependent on not only the number of species in a collection but also on the relative abundances of each species. A community with all the species of about equal population density is more diverse than another community of the same number of species but with some species common and others rare. Community 1 in Fig. 14-2 is more diverse than community 2. The distribution of individuals among species is called *evenness*. Community 1 has a higher evenness than community 2. The problem remains of expressing the number of species and the relative abundances of each as a numerical measure.

Even though diversity has been defined as both number of species and the evenness of abundances, there are many situations in which simply the number of species is a more appropriate measure of diversity. In fact, number of species is the only truly objective measure of diversity. If the relative abundances of the species in a community are considered part of diversity, diversity will be dependent on how we define the relationship between relative abundance and diversity. The information-theoretic measures define the diversity of relative abundance as the evenness of the abundances. However, the definition is subjective. We can equally well define the diversity of relative abundance in exactly the opposite way. Therefore, if diversity is defined to be more than just number of species, the subjective nature of diversity must be borne in mind.

14-3.1   Information Measures of Diversity

The measures H and H' are taken from information theory. The two measures H and H' for the purposes of this book can be thought of as measuring the uncertainty of predicting the species of an individual drawn at random from the entire population of individuals of several species. In a collection or community of several species of about equal commonness it is very difficult to predict the specific identity of an individual drawn at random (i.e., more uncertain), but in another community with the same number of species and with one very common species and many rare ones, the probability of predicting the specific identity of a randomly drawn individual is much higher; i.e., the uncertainty is lower. Increasing the number of species makes prediction more uncertain. The uncertainty, the diversity, of a community can be increased either by increasing the number of species or by evening out the distribution of individuals among species. The measures H and H' indicate the average uncertainty per individual, called information in their context in information theory. The total information in a collection of N individuals is the average information per individual times the total number of individuals N.

The two commonly used information measures of diversity are Brillouin's H and the Shannon-Weaver index H'. The Brillouin measure is appropriate if all the individuals of a population are identified and counted. In contrast, the Shannon-Weaver measure assumes that a random sample is taken from an infinitely large population, and that all the species in the community population are represented in the sample. In both cases the assumptions of the index are quite restrictive. Of the two measures, the Shannon-Weaver H' is the more commonly used. Unfortunately it is rarely possible to take a truly random sample from an infinite population. Therefore, H' should be used with great care. Pielou (1966a, 1966b) discusses the uses of H and H' in greater detail than is possible in this chapter.

Brillouin's measure H   If all the individuals of a collection or community can be identified and counted, the diversity of the community will be measured by Brillouin's H

$$H = \frac{1}{N} \log \frac{N!}{N_1! \, N_2! \, N_3! \, N_4! \cdots N_s!} \tag{14-1}$$

where N is the total number of individuals, and N_1, N_2, ..., N_s is the number of individuals of each species.   The total information content of the collection, B, is $B = HN$.   The choice of the logarithmic base in Eq. (14-1) is arbitrary. If the base is 2, the unit of H will be called a bit.   If natural logarithms are used, the unit will be termed a natural bel (Good, 1950).   Various names have been applied to units of H if logarithms to the base 10 are used, including bel (Good, 1950), decimel digit (Good, 1953), and decit (Pielou, 1966a).

Because H measures the diversity of an entire population, it has no standard error.   Any two different values of H are, therefore, significantly different.   A possible example of a complete community for which H is appropriate is the community of insects found in the litter in a treehole.   The species of fish in a pond is another example.   If it is not possible to assume that individuals are drawn at random from a larger population of individuals, H will be the correct measure of diversity in the community. A collection of moths in a light trap is an example.   Brillouin's formula is the correct measure of diversity in the Rothamstad light trap sample of moths, because light traps are notoriously nonrandom collectors.   It cannot be assumed that the moth collection is a random sample from the total community of individuals.   It is, therefore, necessary to consider the collection of moths to be a population unto itself.   Inferences are made about the sample only, and we cannot extend the conclusions based on the sample to the population except subjectively.

In the calculation of H it is convenient to use the equivalent equation

$$H = \frac{c}{N} \left(\log_{10} N! - \sum \log_{10} N_i! \right) \tag{14-2}$$

where N_i is the number of individuals belonging to the ith species, and c is a constant for conversion of logarithms from the base 10 to the base chosen for the measure.   If the chosen base is 2, $c = 3.321928$, and if e, $c = 2.302585$. Equation (14-2) is convenient because Lloyd, Zar, and Karr (1968) provide a table of $\log_{10} n!$ for values of n from 1 to 1050.   For values of $\log_{10} n!$ greater than 1050 Sterling's approximation to the factorial is

$$\log_{10} n! \approx (n + .5)\log_{10} n - .434294482n + .39909$$

Logarithms of at least six-place accuracy should be used in the calculation of Sterling's approximation.

On July 26, 1966, a collection of caddisflies was caught in a black-light trap along a small stream in central Illinois.   The caddisfly collection from the light trap cannot be considered a random sample from the total fauna of caddisflies, and the only recourse is to treat the collection as a complete collection. The species and the numbers of each caught are listed in Table 14-1. By using Eq. (14-2) and the table in Lloyd, Zar, and Karr (1968), the calculated diversity of this collection in natural bels (i.e., using logs to the base e) is

$$H = \frac{2.302585}{802} \left[1982.6949 - (456.7265 + 416.6758 + \cdots) \right]$$

$$= 1.6530 \text{ natural bels per individual}$$

The total information content of the collection is $B = (1.6530)(802) = 1325.7060$ natural bels.

In a population the maximum possible diversity for a given number of species occurs if all the species are equally common. This is termed H maximum. A measure of the evenness of the distribution of individuals among

Table 14-1 THE NUMBER OF INDIVIDUALS OF SEVERAL SPECIES OF CADDIS-FLIES COLLECTED IN A BLACK-LIGHT TRAP ON JULY 26, 1966, IN A SMALL WOODS ALONG A STREAM IN CENTRAL ILLINOIS

Species	Number of individuals
Potamyia flava	235
Hydropsyche orris	218
Cheumatopsyche analis	192
Ocestis inconspicua	87
Hydropsyche betteni	20
Athripsodes transversus	11
Leptocella candida	11
Leptocella exquisita	8
Cheumatopsyche campyla	7
Polycentropus cinereus	4
Ocestis cinereus	3
Nyctiophylax vestitus	2
Cheumatopsyche aphanta	2
Neureclepsis crepuscularis	1
Triaenodes aba	1

Number of species = 15
Number of individuals = 802

SOURCE: Data from Poole, R. W., Temporal Variation in the Species Diversity of a Woodland Caddisfly Fauna from Central Illinois, *Trans. Ill. Acad. Sci.*, vol. 63, 1971.

species is the actual diversity of the collection as a percent of the maximum diversity, or $J = H/H$ maximum, where J is the measure of evenness. The maximum possible diversity is calculated as

$$H \text{ maximum} = \frac{1}{N} \log \frac{N!}{\{[N/s]!\}^{s-r}\{([N/s] + 1)!\}^r}$$

where $[N/s]$ is the integer part of N/s, s is the number of species, and $r = N - s[N/s]$. In the caddisfly example $N/s = 802/15 = 53.47$, so $[N/s] = 53$, and $r = 802 - (15)(53) = 7$. Then

$$H \text{ maximum} = \frac{1}{802} \log \frac{802!}{(53!)^8(54!)^7}$$

$$= \frac{1}{802} [\log_{10} 802! - (8 \log_{10} 53! + 7 \log_{10} 54!)]$$

$$= 1.1547 \text{ decits per individual}$$

For purposes of comparison the maximum diversity with logarithms to the base e is H maximum $= (1.1547)(2.302585) = 2.6589$ natural bels per individual. The purpose in calculating H maximum initially with logarithms to the base 10 is so the table in Lloyd et al. (1968) can be used. The calculated evenness of this collection is $J = 1.6530/2.6589 = .6217$.

H is dependent not only on the number of species and evenness, but the measure also increases with increasing N. Suppose that there are 10 species in equal proportions. With 100 individuals $H = 3.069$ bits per individual, but with 50 individuals $H = 2.903$ bits per individual. In contrast, H', the Shannon-Weaver index, is not dependent on sample size.

Shannon-Weaver measure H'   Suppose that it is possible to take a random sample from the entire pool of individuals available, and for mathematical reasons this pool can be considered indefinitely large. The number of species in the species pool is also known, and all species are represented in the random sample. Because the population is indefinitely large, it is not possible to count and identify all the individuals in the community, but it is possible to take a random sample from the population of all individuals. All species in the population must be represented in the sample, and s, the number of species, is known. An example might be all the species of Collembola collected in the top layer of soil in a swale. If the swale were of any size, it would be impossible to identify and count all the individuals, but in the laboratory it would be relatively easy to take a random sample by mixing the several thousand Collembola and drawing at random, say, 500 individuals. In this situation the correct measure of diversity is H', the Shannon-Weaver equation,

$$H' = - \sum_{i=1}^{s} p_i \log p_i \tag{14-3}$$

where s is the number of species, and p_i is the proportion of the total number of individuals consisting of the ith species. Unlike H, H' is an estimate of the diversity of the total population of individuals in the species pool. Again the base of the logarithms is open to choice, but in most cases the base e should be used. Equation (14-3) is a biased estimate of H'. The expected value of H', $E(H')$, can be found from the series (Hutcheson, 1970)

$$E(H') = \left[- \sum_{i=1}^{s} p_i \ln p_i \right] - \left[\frac{s-1}{2N} \right] + \left[\frac{1 - \sum p_i^{-1}}{12N^2} \right] + \left[\frac{\sum (p_i^{-1} - p_i^{-2})}{12N^3} \right] + \cdots$$

$$\tag{14-4}$$

Usually the terms past the first two are so small they are not worth calculating. Equation (14-4) must be calculated using natural logarithms.

The variance of the estimate of H' is found from the series

$$\text{var}(H') = \frac{\sum_{i=1}^{s} p_i \ln^2 p_i - (\sum_{i=1}^{s} p_i \ln p_i)^2}{N} + \frac{s-1}{2N^2} + \cdots \tag{14-5}$$

In large samples the first term is usually sufficient.

For purposes of illustration let us assume that the caddisflies caught in the light trap are a random sample of the total number of caddisflies present in the area, and that the fauna consists of the 15 species present in the sample. The estimate of H' from Eq. (14-3) is

$$H' = -[(.293 \ln .293) + (.272 \ln .272) + \cdots + (.001 \ln .001)]$$

$$= 1.6877 \text{ natural bels}$$

Correcting for the bias of the estimate

$$E(H') = 1.6877 - \frac{(15-1)}{(2)(802)} + \cdots$$

$$= 1.6791 \text{ natural bels}$$

Only the first two terms in Eq. (14-4) are used. The variance of the estimate of H' from only the first term in Eq. (14-5) is

$$\text{var}(H') = \frac{(3.7307 - 1.6877^2)}{802}$$

$$= .0011$$

The estimate of H' and its standard error are $H' = 1.6791 \pm .0332$. If H' has been estimated in two collections, the two diversities can be compared, with a t test, to see if they are significantly different.

$$t = \frac{H_1' - H_2'}{[\text{var}(H_1') + \text{var}(H_2')]^{1/2}}$$

The null hypothesis is $H_0: H_1' = H_2'$. The degrees of freedom of the test is

$$df = \frac{[\text{var}(H_1') + \text{var}(H_2')]^2}{\text{var}(H_1')^2/N_1 + \text{var}(H_2')^2/N_2}$$

where N_1 is the number of individuals in the first sample, and N_2 is the number of individuals in the second.

Hutcheson (1970) illustrated the use of this test by comparing the diversity of some bird casualties on January 1964 (57 individuals and 13 species) with the diversity of casualties at the same place on January 1965 (23 individuals and 11 species). Setting H_1' as the estimated diversity on January 1964, and H_2' as the estimated diversity on January 1965, $H_1' = 2.44155$, $H_2' = 1.80785$, $\text{var}(H_1') = .0148724$, and $\text{var}(H_2') = .0620399$. The t test is

$$t = \frac{2.44155 - 1.80785}{(.0148724 + .0620399)^{1/2}}$$

$$= 2.285$$

with 35 degrees of freedom. This value of t exceeds the 5 percent probability level, and it is concluded that the diversities of the bird casualties differed; i.e., $H_1 \neq H_2$.

In this type of collection the number of species in the entire community or population of species is known. Therefore, the evenness can be calculated as $J = H'/H'$ maximum, as before. The maximum value of H' is H' maximum $= \log s$, using the same base of logarithms used in the calculation of H'. In the caddisfly example H maximum $= \ln(15) = 2.70805$, and $J = 1.6791/2.7081 = .62004$.

If a species is not represented in the sample, $p_i = 0$ and $\ln p_i = -\infty$. Therefore, H' will be incalculable. If the sample contains the majority of the species available to be sampled, no great error results from carrying out the calculations as if all species were represented. If, however, there are many species in the sample represented by only one or two individuals, it is reasonable to assume that the true value of s is not known and that some species will not be represented in the sample. See Good (1953) for a discussion of this problem.

Diversity of vegetatively reproducing organisms   It is not possible to take a random sample of the individuals in a community of sessile, vegetatively reproducing organisms, because of the patchiness of the "individuals" and the impossibility of defining the individuals.   This situation is likely to occur most often in plant communities.   The procedure in such cases is to divide the area into a large number of quadrats, and then to take a series of quadrats at random.   In lieu of individuals, the weight of each plant species in each quadrat, or some other similar measure of importance, is used in the calculation, calling the weight per species in a quadrat N_i.   The diversity of one randomly chosen quadrat is calculated by using Brillouin's measure H (Eq. 14-1).   If there are k quadrats collected as a sample (say 100), 1 quadrat is taken at random and its diversity is measured.   Then a second quadrat is taken, the data from the first and second quadrats are combined, and the diversity of the combined quadrats is estimated.   A third quadrat is then taken, adding the data to that from the first and second quadrats, and the diversity is estimated.   This process is continued until all the quadrats have been combined, and the diversity of the combined 100 quadrats estimated.   As more and more quadrats are combined, the calculated diversity increases to an asymptote where the effect of adding new species to the sample is offset by the addition of more individuals of the common species (Pielou, 1966b).   In Fig. 14-8 the calculated value of H_k for the example discussed below, where H_k is the diversity of the combination of k quadrats, is plotted against k.   In this graph the value of H_k changes little for combinations of quadrats larger than 70.   This point is indicated as t, and the total number of quadrats sampled as z; in this case $z = 100$ and $t = 70$.   The values of H_k are calculated from $k = 70$ to $k = 100$.   A quantity h_k is calculated as

$$h_k = \frac{M_k H_k - M_{k-1} H_{k-1}}{M_k - M_{k-1}}$$

The quantity M_k is the total number of "individuals" (summed weight) of all the species in k quadrats, and M_{k-1} is the summed weight of all species in $k - 1$ quadrats.   If the h_k values from $k = t$ to $k = z$ are not serially correlated (see Sec. 10-4), an estimate of the diversity of the whole area from which the 100 random quadrats were taken is

$$H'_{pop} = \frac{1}{z - t + 1} \sum_{k=t}^{z} h_k$$

and the variance of the estimate of H'_{pop} is equal to $var(\bar{h})$ and

$$var(\bar{h}) = \frac{var(h)}{n}$$

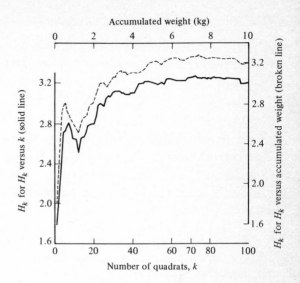

FIGURE 14-8
H_k versus k (solid line, left-hand ordinate scale); and H_k versus accumulated weight of plant material (broken line, right-hand ordinate scale).   (*From Pielou, 1966b.*)

where n is the number of h_k values, and var(h) is calculated in the usual way.

Pielou (1966b), who developed this method, has applied it to measure the diversity of the ground vegetation in a 3,000-square-meter tract of mixed woodland in Gatineau Park, Quebec. All herbaceous vascular plants and all woody plants and tree seedlings with a diameter of 2 centimeters or less were clipped at ground level in 100 randomly placed 1-meter-square quadrats. The fresh weight in decigrams for each species in each quadrat was measured, and used as N_i in Eq. (14-1). Figure 14-8 shows the results of the calculations of successive values of H_k. By inspection it was decided that by $k = 70$ the curve had leveled off. Values of h_k were, therefore, calculated for $k = 70$ to $k = 100$, rejecting two empty quadrats for which diversity was undefined. Using logarithms to the base e, the mean value of h_k and its variance were $\bar{h} = 3.056$ and var$(h) = .7638$. Therefore, for the quadrats beyond 70, var$(\bar{h}) = \frac{.7638}{29} = .026$. The estimate of the diversity of the area and its standard error are $H'_{\text{pop}} = 3.056 \pm .161$. A 95 percent confidence interval can be calculated as

$$H'_{\text{pop}} - 1.96[\text{SE}(H'_{\text{pop}})] \le H'_{\text{pop}} \le 1.96[\text{SE}(H'_{\text{pop}})] + \hat{H}'_{\text{pop}}$$

or in the example Pr$(2.740 \le H'_{\text{pop}} \le 3.372) = 95$ percent.

14-3.2 Simpson's Measure of Diversity

If two individuals are drawn at random from a population of N individuals, the probability that both individuals belong to the same species is

$$C = \sum_{i=1}^{s} \frac{n_i(n_i - 1)}{N(N - 1)}$$

where n_i is the number of individuals of the ith species. The higher this probability, the lower the diversity of the collection. Simpson (1949) was the first to suggest C as a measure of diversity. Sometimes $D = 1 - C$ is used as the index of diversity, because D increases with increasing rather than decreasing diversity. By applying this measure to the caddisfly data in Table 14-1, $C = .22909$ and $D = .77091$.

The Simpson measure of diversity expresses the dominance of or concentration of abundance into the one or two commonest species of the community. In contrast, the information-theoretic measures H and H' express the relative evenness of the abundances of all the species. Therefore, Simpson's index is most appropriate if we are most interested in the relative degree of dominance of a few species in the community, rather than the overall evenness of the abundances of the species.

14-3.3 Some Comments

The two information measures of diversity, H and H', are the most commonly used indices of diversity. However, both measures suffer from some inadequacies:

1 Both H and H' measure diversity in terms of number of species and absolute abundance. However, abundance is not necessarily a good indication of the importance of a species in a community. For example, a population of 1,000 individuals of a species of flagellate has a far more important effect on H' than 10 sea otters in the same area of water, even though ecologically the sea otters are far more important to the structure of the community than the flagellates. Therefore, H and H' are appropriate only in collections of species of approximately the same size and trophic level.

2 The diversity indices H and H' are logarithmically related to the number of species in the community of collection. If 10 species are added to a community of 20 species, the increase in H or H' will be greater than if 10 species are added to a community of 50 species.

In very rich communities H and H' rapidly approach maximum, asymptotic values, rather than increasing linearly with an increase in the number of species.

3   The Shannon-Weaver measure H' is independent of sample size, because it estimates diversity from a random sample containing all the species of the community.   Practically, this type of random sample may be impossible to take in diverse communities, because increasing sample size in a diverse community almost always turns up individuals of the rarer species.

4   The abundance relationships in a community are measured in terms of the evenness component of the diversity index.   Evenness expresses the degree of equality of the abundances of the species of the community.   In some cases, however, we may be more interested in the degree of dominance of the commonest species.   If dominance is of primary interest, Simpson's index of diversity is a more appropriate measure of diversity than either H or H'.   The two measures H and H' are most strongly affected by the abundances of the middle species of a community, rather than by the common or rare species.

15

INTERACTIONS AMONG THREE OR MORE SPECIES

A study of the interactions among three or more species introduces a whole new level of complexity into the simple two-species situations of Part II of this book. Although the classic deterministic models of competition and predation can be easily extended to three or more species, the models soon become untestable, because of the large number of parameters and interaction terms. As a consequence, perhaps, the majority of work on multispecies interactions has been observational and theoretical, almost always employing the heuristic niche concept.

This chapter begins by exploring the consequences of competition among several species, and the influence of competition on the number of closely related species occurring in the same habitat. Section 15-2 studies the interactions among a single predator or parasitoid species and its many host species. In addition, every species, except in rare cases, is fed upon by several predators, parasites, and diseases. Section 15-3 discusses population stability, diversity, and the community food web. Section 15-4 is about succession, and Sec. 15-5 takes up the subject of the formation and extinction of populations, particularly on islands.

15-1 COMPETITION AMONG THREE OR MORE SPECIES

In Chap. 6 a model of competition between two species was presented. An equivalent expression for some species i is

$$\frac{dN_i}{dt} = r_i N_i \frac{K_i - N_i - \alpha_j N_j}{K_i} \qquad j \neq i$$

where N_i is the density of the ith species, r_i is the intrinsic rate of increase of the ith species, K_i is the carrying capacity of the environment for the ith species in the absence of other competing species, N_j is the density of the jth species, and α_j is the effect of the jth species on the rate of population growth of the ith species.

The model can be easily extended to more than two species.

$$\frac{dN_i}{dt} = \frac{r_i N_i}{K_i} \left(K_i - N_i - \sum \alpha_{ij} N_j - \sum \beta_{ijk} N_j N_k - \cdots - \sum \omega_{ijk\ldots m} N_j N_k \cdots N_m \right)$$

$$(15\text{-}1)$$

Each α_{ij} represents the effect of the population density of the jth species on the ith species. If more than two species are competing, the possible interactions among three or more species must also be considered. The coefficients β_{ijk} represent the effect of the interaction between the jth and kth species on the rate of population growth of the ith species. As in the analysis of variance, the number of interaction terms can become overwhelming if more than a very few species are involved. In Eq. (15-1) $\omega_{ijk\ldots m}$ represents the interaction among species j, k, $\ldots$, m on the rate of population growth of the ith species. With four species Eq. (15-1) is, for the first of the four species,

$$\frac{dN_1}{dt} = \frac{r_1 N_1}{K_1} (K_1 - N_1 - \alpha_{12} N_2 - \alpha_{13} N_3 - \alpha_{14} N_4 - \beta_{123} N_2 N_3$$

$$- \beta_{134} N_3 N_4 - \beta_{124} N_2 N_4 - \gamma_{1234} N_2 N_3 N_4)$$

Because of the number of possible interactions Eq. (15-1) is difficult to apply, even to experimental situations. Therefore, it is exceedingly convenient to assume that the interaction terms (second-order and higher) are insignificant, reducing Eq. (15-1) to

$$\frac{dN_i}{dt} = \frac{r_i N_i}{K_i} \left(K_i - N_i - \sum_{j=1}^{m} \alpha_{ij} N_j \right) \qquad j \neq i \qquad (15\text{-}2)$$

(MacArthur and Levins, 1967; Vandermeer, 1969).

Vandermeer (1969) conducted an experiment utilizing four species of protozoans to investigate the importance of interaction terms in multispecies experimental situations. The four species were *Paramecium caudatum, P. bursaria, P. aurelia*, and *Belpharisma* sp. All species were raised in single-species cultures in order to estimate the parameters K_i and r_i (Table 15-1), and in two-species competition experiments to estimate the parameters α_{ij} and α_{ji}. These α_{ij} values are listed in Table 15-2. In this table α_{31} is the effect of the population density of *P. aurelia* on the population growth of *P. bursaria* (.50). This matrix is called a *community matrix* or a *competition matrix*.

Finally all four species were combined in several replicates to determine the effects of competition among four species on the population growth of each species. The results of the experiment are shown in Fig. 15-1. The fitted lines are the densities predicted by Eq. (15-2). The fits of the predicted and the observed results are quite good, although there are minor deviations, most noticeably in *Blepharisma* sp. In Vandermeer's experiment higher-order interaction terms were apparently unimportant and were ignored. Hairston et al. (1969), however, did find significant second-order interactions in experiments with three "species" of *Paramecium*.

15-1.1 Competition among Several Species in the Field

If second-order and higher interactions can be ignored, a community matrix can be used to predict roughly the number of species which might possibly coexist in the same habitat area. Estimating the parameters α_{ij} in any field

Table 15-1 THE INTRINSIC RATES OF INCREASE r, THE CARRYING CAPACITIES K, AND INITIAL DENSITIES N_0 OF THE FOUR SPECIES OF PROTOZOANS USED IN VANDERMEER'S FOUR-SPECIES COMPETITION EXPERIMENTS

Species	r	K	N_0
Paramecium aurelia	1.05	671	2.5
Paramecium caudatum	1.07	366	5.0
Paramecium bursaria	.47	230	5.0
Blepharisma sp.	.91	194	3.0

SOURCE: After Vandermeer, J. H., The Competitive Structure of Communities: An Experimental Approach with Protozoa, *Ecology*, vol. 50, 1969.

situation is likely to be impossible. Levins (1968) suggests as an approximation

$$\alpha_{ij} = \frac{\sum p_{ih} p_{jh}}{\sum p_{ij}^2} \qquad (15\text{-}3)$$

where p_{ih} is the percent of the total population of the ith species occurring in the hth environment, and p_{jh} is the percent of the jth species. In terms of a real situation, individuals of species i and j may be trapped in a series of areas, and the percent of the total of each species in each area expressed as p_{ij} and p_{jh}. If two species utilize approximately the same resources of the environment, and if the two species are associated strongly in the habitat, the parameters α_{ij} as calculated from Eq. (15-3) will be high. Unfortunately, it is impossible to separate the degree of association due to similar food sources from the responses of both species to other environmental characteristics such as temperature or humidity. Therefore, this approximation is only a crude one at best.

Levins (1968) determined the "competition coefficients" among five species of *Drosophila* at Mayagüez, Puerto Rico, by setting out a number of banana traps and calculating the percent of the total number of each species found at each trap. The coefficients in Table 15-3 were estimated from these data by using Eq. (15-3).

Table 15-2 THE COMMUNITY OR COM-
PETITION MATRIX BETWEEN
THE FOUR SPECIES OF
PROTOZOANS IN VANDER-
MEER'S FOUR-SPECIES COM-
PETITION EXPERIMENT (*PA*
= *P. AURELIA; PB* = *P. BUR-
SARIA; PC* = *P. CAUDATUM;
BL* = *BLEPHARISMA* sp.)

	jth species			
ith species	*PA*	*PC*	*PB*	*BL*
PA	1.00	1.75	−2.00	−.65
PC	.30	1.00	.50	.60
PB	.50	.85	1.00	.50
BL	.25	.60	−.50	1.00

SOURCE: After Vandermeer, J. H., The Competitive Structure of Communities: An Experimental Approach with Protozoa, *Ecology*, vol. 50, 1969.

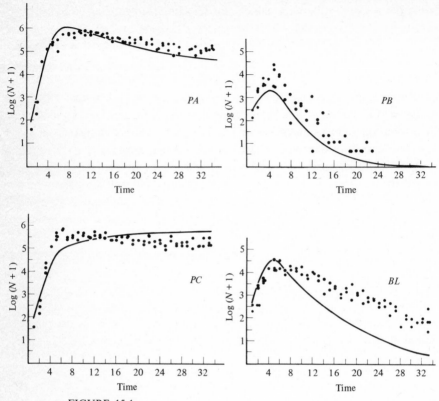

FIGURE 15-1

Changes in population density with time of four species of protozoans in a four-way competition experiment. Population densities have been transformed to logs. $PA = Paramecium\ aurelia$, $PB = P.\ bursaria$, $PC = P.\ caudatum$, $BL = Blepharisma$ sp. (*From Vandermeer, 1969.*)

If the changes in density of the population of each species of several competing species in a habitat are expressed in the form of Eq. (15-2), and if these equations are adequate representations of population growth in each species, the species populations will be in equilibrium when $dN_i/dt = 0$ for all the species, or

$$X_i = K_i - \sum \alpha_{ij} X_j \qquad i \neq j \qquad (15\text{-}4)$$

When in equilibrium, X_1, X_2, ..., X_m are all constants. To repeat, the values of K_i represent the equilibrium densities of the ith species in the absence of all other species. The values of X_i represent the population densities of every species, with all species present and all population densities constant.

You may remember, however, that in a two-species competition model an equilibrium between the two species can be stable or unstable. Therefore, even at equilibrium the population densities of any pair of species may be unstable, and the densities $X_1, X_2, \ldots, X_m$ may not be constant even though they are assumed to be constant in the method developed below. Therefore, the use of a community matrix in predicting the number of coexisting species is valid only if the equilibrium is completely stable, which is not very likely. The predictions of the model should be accepted with some reservations. Some other assumptions of Eq. (15-4) should be considered:

1 The competition equations adequately represent competition between two or more species.

2 The competition coefficients estimated by Eq. (15-3) really do represent the effect of competition between two species.

3 The competitive coefficients are constants.

4 It is possible to have an equilibrium population level for each species. When this is true, the density of each species is affected only by density-dependent factors, and stochastic events and population fluctuations due to physical factors of the environment are negligible or nonexistent.

Even though none of these assumptions is likely to be true, Eq. (15-2) is still a useful way of heuristically studying multispecies competition, because some insight is better than no insight at all.

Table 15-3   **THE COMMUNITY MATRIX FOR FIVE SPECIES OF *DROSOPHILA* FROM MAYAGÜEZ, PUERTO RICO. THE VALUES OF α_{ij} WERE ESTIMATED FROM EQ. (15-3)**

Species	Drosophila melanogaster	Drosophila latifasci- aeformis	Drosophila willistoni group	Drosophila dunni	Drosophila ananassae
Drosophila melanogaster	1.00	.30	.42	.61	.16
Drosophila latifasciaeformis	.72	1.00	.92	.72	.60
Drosophila willistoni group	.88	.81	1.00	.96	.47
Drosophila dunni	.90	.44	.67	1.00	.38
Drosophila ananassae	.18	.28	.25	.29	1.00

SOURCE: From Levins, R., "Evolution in Changing Environments: Some Theoretical Explorations," Monographs in Population Biology, Princeton University Press, Princeton, 1968.

The equations derived from Eq. (15-4) can be summarized in matrix form as $\mathbf{AX} = \mathbf{K}$, where $\mathbf{A}$ is the matrix of α_{ij} parameters, $\mathbf{X}$ is a column vector of the equilibrium densities, and $\mathbf{K}$ is a column vector of the carrying capacity of the environment for each species. Given $\mathbf{A}$ and $\mathbf{K}$, the equilibrium population of each species is $\mathbf{X} = \mathbf{A}^{-1}\mathbf{K}$. One must be able to measure the K_i densities, which in many situations may be impossible (see MacArthur, 1968). Levins (1968) estimated the values of K_i by using the frequencies of six species of *Drosophila* collected in a series of traps on December 16, 1965, as estimates of X_1, X_2, X_3, X_4, X_5, and X_6. He did not prove that these densities were equilibrium values. Because the $\mathbf{A}$ matrix and the $\mathbf{X}$ column vector were known, Levins estimated the $\mathbf{K}$ vector as $\mathbf{AX}$ (Table 15-4). If *Drosophila melanogaster* has an equilibrium level of three flies, and there is a total of 100 flies of all species, *D. melanogaster* will have a carrying capacity of 41 flies in the absence of the other five species.

MacArthur (1958) in a study of four warbler species in a spruce forest found that the four species, myrtle (N_1), black-throated green (N_2), black-burnian (N_3), and bay-breasted (N_4), divided the habitat (the spruce trees) among themselves by spending different amounts of time in different parts of the trees. The time spent by each species in each of the four parts of the tree is shown in Fig. 15-2. If the time spent by the ith species at the hth level is t_{ih}, and t_{jh} is the time spent by the jth species at the hth level, the

Table 15-4 ESTIMATE OF THE K VECTOR FOR SIX SPECIES OF *DRO-SOPHILA* FROM MAYAGÜEZ, PUERTO RICO. THE EQUILIB-RIUM VECTOR X WAS ESTIMATED FROM THE FRE-QUENCY OF EACH SPECIES IN THE TOTAL SAMPLE

Species	Frequency	K
Drosophila melanogaster	.03	.41
Drosophila willistoni group	.54	.89
Drosophila latifasciaeformis	.32	.92
Drosophila dunni	.01	.62
Drosophila ananassae	.08	.27
Drosophila "tripunctata"	.02	.35

SOURCE: From Levins, R., "Evolution in Changing Environments: Some Theoretical Explorations," Monographs in Population Biology, Princeton University Press, Princeton, 1968.

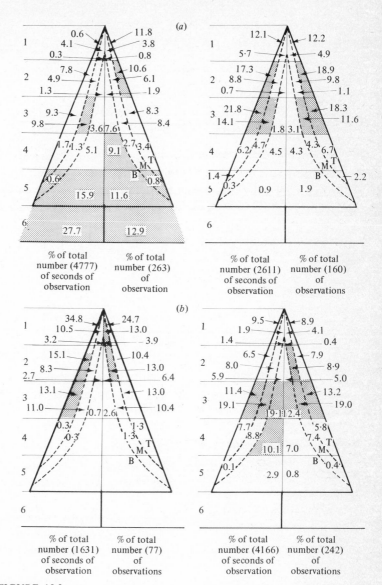

FIGURE 15-2
The figures are diagrammatic spruce trees subdivided into zones, with the time budgets of four species of warblers entered into the zones. From upper left to lower right, the species are myrtle, black-throated green, blackburnian, and bay-breasted. The shaded zones indicate the most concentrated activity. (*From MacArthur, 1958.*)

competition coefficients can be estimated by using these values in place of p_{ih} and p_{jh} in Eq. (15-3). The matrix $\mathbf{A}$ for the first three species of warblers is

$$
\begin{array}{c}
\qquad\qquad\text{Species 1}\quad\text{Species 2}\quad\text{Species 3} \\
\mathbf{A} = \begin{array}{c}\text{Species 1}\\\text{Species 2}\\\text{Species 3}\end{array}
\begin{bmatrix}
1.000 & .490 & .480 \\
.519 & 1.000 & .959 \\
.344 & .654 & 1.000
\end{bmatrix}
\end{array}
$$

In addition MacArthur estimated the competition coefficients of the fourth species to be $\alpha_{41} = .545$, $\alpha_{42} = .854$, and $\alpha_{43} = .654$. If K_1, K_2, and K_3 are the carrying capacities of each of the first three species when the other two are absent, the fourth species, the bay-breasted warbler, can invade and persist if

$$
K_4 > \begin{bmatrix} \alpha_{41} & \alpha_{42} & \alpha_{43} \end{bmatrix} \mathbf{A}^{-1} \begin{bmatrix} K_1 \\ K_2 \\ K_3 \end{bmatrix}
$$

In the example

$$
K_4 > \begin{bmatrix} .545 & .854 & .654 \end{bmatrix}
\begin{bmatrix}
1.000 & .490 & .480 \\
.519 & 1.000 & .959 \\
.344 & .654 & 1.000
\end{bmatrix}^{-1}
\begin{bmatrix} K_1 \\ K_2 \\ K_3 \end{bmatrix}
$$

By carrying out the indicated computations we find that species 4 is able to persist only if its carrying capacity K_4 is greater than the function $.1392K_1 + 1.0775K_2 - .4461K_3$. MacArthur (1968) estimated the four carrying capacities to be $K_1 = 6.190$, $K_2 = 9.082$, $K_3 = 6.047$, and $K_4 = 9.014$. Therefore, the inequality is $9.014 > (.1392)(6.190) + (1.0775)(9.082) - (.4461) \times (6.047) = 9.014 > 7.950$. The bay-breasted warbler can persist in the community because the inequality is fulfilled. This is a fortunate conclusion, because it does.

It must be reemphasized that these results are valid only if a truly stable equilibrium exists. The interactions among competing species in a community without static population densities remain to be studied.

15-1.2 Competition among Several Plant Species

Analysis of competition among plant species has usually revolved around the yield of a plant grown in monoculture as compared to its yield when grown in combination with another species. Therefore, competition is defined in terms of changes in biomass rather than changes in density. The method of analysis of multispecies competition presented below is taken largely from McGilchrist (1965) and Williams (1962).

The basic experiment is to grow each species by itself in r_1 replicates and measure the yield. Williams measured the weight of plant tops in grams per one-half pot (each pot was a replicate). The data were transformed to log(weight + 1) to stabilize variances in order to test the significance of the competition effects by an analysis of variance. In addition, each possible pair of species is grown in equal numbers in a pot. There are r_2 replicates of each species pair. The yield of each species in the two-species experiment is measured. If no competition is occurring, the yield of a plant in the two-species pot should be equal to the half-pot yield in the monoculture.

The transformed yield of the ith species grown in competition with the jth species in the kth replicate can be represented by the linear model

$$Y_{ijk} = \alpha_{ij} + \rho_k + T_{ijk} + e_{ijk}$$

where

α_{ij} = the mean for the ith species grown in competition with the jth

ρ_k = the differences between replicates $(\sum \rho_k = 0)$

T_{ijk} = pot effect, assumed to be an independent normal variable with zero mean

e_{ijk} = an error term.

There are two possible measures of competition between two species of plants. The first is the increase in yield of species i when it is grown with species j over species i yield in monoculture. The other measure is the depression species i causes in the yield of species j as compared to the yield of species j in monoculture.

McGilchrist (1965) defines the competitive advantage of species i over species j as an average of these two concepts. If this competitive advantage is denoted as γ_{ij}

$$\gamma_{ij} = \tfrac{1}{2}(\alpha_{ij} - \alpha_{ii}) + \tfrac{1}{2}(\alpha_{jj} - \alpha_{ji})$$

He goes on to define the competitive depression of species i and species j as one-half the decrease in the total yield of species i and j when grown in competition, as compared to their total yields in monocultures. The competitive depression of species i and species j can be represented as

$$\delta_{ij} = \tfrac{1}{2}(\alpha_{ii} + \alpha_{jj}) - \tfrac{1}{2}(\alpha_{ij} + \alpha_{ji})$$

The competitive advantage of species i can be averaged, and a general competitive ability over all p species is

$$k_i = \frac{1}{p} \sum_j \gamma_{ij}$$

The quantity k_i is called the *competition effect* of species i.   In the same way the average depression effect of species i is

$$\lambda_i = \frac{1}{p-2}\left(\sum_j \delta_{ij} - \frac{1}{p}\sum_i \sum_j \delta_{ij}\right)$$

There are also interactions between the competition effects and depression effects of each species.   The interaction between the competition effects of two species is

$$\theta_{ij} = \gamma_{ij} - k_i + k_j$$

The interaction between two depression effects is

$$\tau_{ij} = \delta_{ij} - \mu - \lambda_i - \lambda_j$$

where

$$\mu = \frac{1}{p(p-1)}\sum_i \sum_j \delta_{ij}$$

In the estimation of these quantities the following statistics are calculated

$$S_{ik} = Y_{iik}$$
$$C_{ijk} = \tfrac{1}{2}(Y_{ijk} - Y_{jik} - \bar{Y}_{ii.} + \bar{Y}_{jj.})$$
$$D_{ijk} = \tfrac{1}{2}(-Y_{ijk} - Y_{jik} + \bar{Y}_{ii.} + \bar{Y}_{jj.})$$

where Y_{ijk} is the yield of species i when grown with species j.   The number $\bar{Y}_{jj.}$ is the mean of all k replicates of the jth species in monoculture.   Dot notation is used to save space, and is discussed in Chap. 9.   In dot notation $\bar{D}_{...}$ stands for the mean of all the D_{ijk} values, and $\bar{D}_{i..}$ is the mean over all replicates and other species resulting in a mean value of $\bar{D}_{i..}$ for each species. If $\bar{S}_{i.}$ is represented as S_i and $\bar{D}_{ij.}$, and $\bar{C}_{ij.}$ as C_{ij} and D_{ij}, then

$$\hat{\alpha}_{ij} = S_i$$
$$\hat{\gamma}_{ij} = C_{ij}$$
$$\hat{k}_i = \frac{1}{p}\sum_j C_{ij} \quad i \neq j$$
$$\hat{\theta}_{ij} = C_{ij} - k_i + k_j$$
$$\hat{\delta}_{ij} = D_{ij}$$
$$\hat{\mu} = \frac{1}{p(p-1)}\sum_i \sum_j D_{ij} = \bar{D}_{...}$$
$$\hat{\lambda}_i = \frac{1}{p-2}\sum_j (D_{ij} - \bar{D}_{...})$$
$$\hat{\tau}_{ij} = D_{ij} - \hat{\lambda}_i - \hat{\lambda}_j - \hat{\mu}$$

The variances and covariances of each of these estimates may be found in McGilchrist (1965).

McGilchrist (1965) analyzed data collected by Williams (1962) on competition among seven species of plants, $p = 7$. The species mean, the competition effects, and the depression effects are listed in Table 15-5. The best competitor is species 1, and the two worst competitors are species 5 and 6. On the other hand, the species with the largest depressing effect is species 5. If species 5 were grown in mixed cultures with another species, it would tend to lower the overall yield more than any other species. Evidently species 2 is the best species to grow in mixed cultures. The interactions among competition and depression effects are given in Table 15-6. In choosing the number of replicates r_1 and r_2, the ratio r_1/r_2 should be approximately equal to $p/2$ where there are p species.

Table 15-5 ESTIMATES OF SPECIES, COMPETITION, AND DEPRESSION EFFECTS IN WILLIAMS' COMPETITION EXPERIMENTS AMONG SEVEN SPECIES OF PLANTS

Species	Species mean (S_i)	Competition effect (k_i)	Depression effect (δ_i)
1	1.495	.2850	.0347
2	1.790	.0361	−.0978
3	1.005	−.0189	−.0548
4	1.065	.0650	−.0143
5	1.010	−.2400	.0697
6	.750	−.2278	.0177
7	1.415	.1007	.0447
Standard error of entry	.052	.030	.029

SOURCE: From McGilchrist, C. A., Analysis of Competition Experiments, *Biometrics*, vol. 21, 1965.

Table 15-6 ESTIMATES OF THE INTERACTIONS BETWEEN COMPETITION EFFECTS AND DEPRESSION EFFECTS. THE INTERACTIONS BETWEEN COMPETITION EFFECTS $\hat{\theta}_{ij}$ ARE ABOVE THE DIAGONAL, AND THE INTERACTIONS BETWEEN DEPRESSION EFFECTS $\hat{\tau}_{ij}$ ARE BELOW THE DIAGONAL

Species	1	2	3	4	5	6	7
1	...	.0236	−.0189	.0175	−.0525	−.0803	.1107
2	−.1155	...	−.0300	.0014	.0264	−.0014	.0271
3	.0290	.0165	...	.0039	−.0536	.0561	−.0554
4	−.0089	−.0614	.0781	...	.0000	−.0228	.0457
5	.0920	.1045	−.0785	.0085	...	−.0053	−.0743
6	.0295	.0715	−.0040	.0605	−.2210	...	−.0540
7	−.0255	−.0155	−.0410	−.0765	.0945	.0640	...

SOURCE: From McGilchrist, C. A., Analysis of Competition Experiments, *Biometrics*, vol. 21, 1965.

Although the competition and depression effects and their interactions have been calculated, are there significant differences among them? Because the basic model of competition is a linear analysis of variance model, analysis of variance may be used to check for the significance of differences among the terms in Tables 15-5 and 15-6.   The reader is referred to McGilchrist (1965) for the methodology.

15-2   SEVERAL SPECIES OF PARASITES, PREDATORS, PREY, AND HOSTS

In spite of the importance of studying the interactions among groups of parasites, predators, hosts, and plants for purposes of biological control, the study of these interactions remains one of the more unexplored areas of ecology.   Although mathematical models are almost nonexistent, and experimental results are few, some good observational studies are available.   It is often impossible, however, to separate population interactions from the effects of other environmental factors, both physical and biological, in the field. Hopefully the hypotheses generated by the experimental and observational studies can be proved, and extended to more complicated systems in the not-too-distant future.

15-2.1   One Host and Several Parasites or Predators

Several species of predators and parasites may compete for the same host species population.   Utida (1957) has experimentally studied the numerical relationships between a population of the Azuki bean weevil, *Callosobruchus chinensis*, and two of its parasites, *Heterospilus prosopidis* and *Neocatolaccus mamezophagus*.   The three species were reared together in darkness at 30°C and 75 percent relative humidity.   The experiment was terminated after 4 years, about 70 generations, and all three species persisted the entire 4 years (Fig. 15-3). In contrast, when *C. chinensis* and *H. prosopidis* were reared together without *N. mamezophagus*, either the parasite or the host became extinct by the twenty-fifth generation in five of six replicates.   The two-species experiment is discussed in greater detail in Chap. 7.   The addition of the second parasite species seems to have decreased the probabilities of extinction of all three species, and increased the "stability" of the system.   There were also changes in the character of the fluctuations of the two parasite species when both were reared together with the host, as compared to when they were reared separately.   The fluctuations of *H. prosopidis* were lower in amplitude and

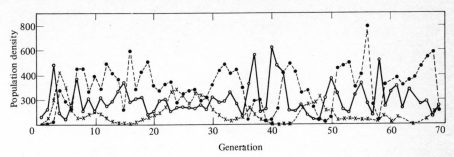

FIGURE 15-3
Population fluctuations in the interacting populations of a host, the Azuki bean weevil, *C. chinensis* (○), and its two parasites, *N. mamezophagus* (●) and *H. prosopidis* (×). (*From Utida, 1957.*)

longer in period when it was reared with *N. mamezophagus* (compare Fig. 15-4 with Fig. 15-3). The amplitude of the fluctuations of *N. mamezophagus* became broader, and the period longer (Fig. 15-4).

The fluctuations of the two parasite species in this three-species system tend to alternate, the highest densities of *H. prosopidis* occurring during low densities of *N. mamezophagus* (Fig. 15-5). Evidently *N. mamezophagus* is competitively superior to *H. prosopidis* at high host densities, but *H. prosopidis* is better able to find hosts at low host densities (Fig. 15-6). Differences in

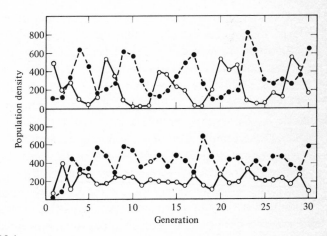

FIGURE 15-4
Population fluctuations in the population of the host, *C. chinensis* (○), and each of its two parasites (●). Upper figure with *H. prosopidis*, and lower figure with *N. mamezophagus*. (*From Utida, 1957.*)

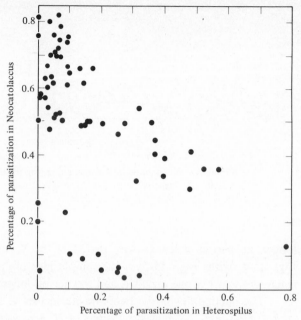

FIGURE 15-5

Relationship between the percentage of parasitism in the populations of *N. mamezophagus* and *H. prosopidis* in each generation when both are reared concurrently with the host *C. chinensis*. (*From Utida, 1957*.)

host-finding ability perhaps account for both the alternation of fluctuations and the continued coexistence of both parasite species. Evidently if both parasite species are to survive, differences in the exploitation patterns of the two parasites are necessary. In southern California the common hymenopterous parasite of the California red scale, *Aphytis chrysomphali*, was displaced from 1948 to 1958 by a second species of *Aphytis* introduced from China in 1948, *A. lingnanensis*. *Aphytis lingnanensis* was in turn displaced from the interior of southern California by a third species of *Aphytis* introduced from India and Pakistan in 1956, *A. melinus*. In laboratory experiments with two-species cultures of three species of *Aphytis* (*A. lingnanensis, A. melinus,* and *A. fisheri*) one species always eliminated the other (Debach and Sundby, 1963). The surviving species was almost always the species with the highest fecundity. No differences in exploitation pattern between the species were observed. Ecologically homologous parasites will apparently displace each other. The persistence of the two species of parasites on the Azuki bean weevil was probably due to differences between the parasites in exploitation rates and abilities at different host densities.

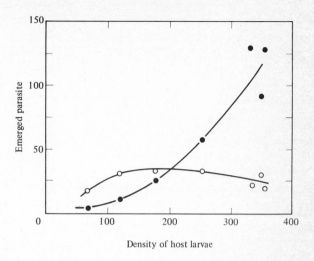

FIGURE 15-6
Relationship between the density of host larvae available for parasitism and the emerged number of parasites when both parasites attack simultaneously; *N. mamezophagus* ●, *H. prosopidis* ○. (*From Utida, 1957.*)

These experimental results may be compared with the field observations of Price (1970). Price studied the hymenopterous parasites of the Swaine jack pine sawfly, *Neodiprion swainei*, in Quebec. He initially divided the parasites into guilds. A *guild* is a group of functionally related species which are not necessarily taxonomically related (Root, 1967). The leaf miners of aspen in a given area represent a guild. The two guilds of parasites were those attacking sawfly cocoons and those attacking larvae. Five of the six species were specific parasites, attacking only *N. swainei*. The distributions of the five species were related to the availability of hosts. *Pleolophus basizonus* was dominant at high host densities, displacing *P. indistinctus*, dominant at low host densities. A similar relationship held between *P. basizonus* and *Endasys subclavatus* (Fig. 15-7). A fourth parasite, *Mastrus aciculatus*, occupied dry, open sites where other species were less numerous. The highest species diversity of parasites as measured by H' occurred at the ecotone around jack pine stands and at moderate host densities. As host density increased, *P. basizonus* became dominant, reducing species diversity (Figs. 15-8 and 15-9). Because of hyperparasitism by cocoon parasites larval parasites were common only at moderate host densities, and their populations were greatly reduced by the increase in *P. basizonus* at high host densities. These field studies are remarkably similar to the experimental findings of Utida (1957).

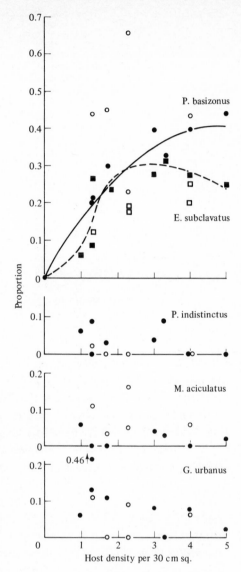

FIGURE 15-7
The proportions of *P. basizonus* and *E. subclavatus* in the total parasitoid complex, and of *P. indistinctus*, *M. aciculatus*, and *G. urbanus* in the cocoon parasitoid complex in plot 9. ● and ■, Proportions in increasing host populations; ○ and □, proportions in decreasing host populations. The regression line for *P. basizonus* proportions in increasing host population is $Y = .01720X - .0178X^2$, $R = .9748$. The regression is significant at the 5 percent probability level. (*From Price, 1970.*)

Some tentative hypotheses might be listed:

1 If several species of parasites attack a common host species, the exploitation patterns and abilities of the parasites must be different to allow coexistence.

2 Given differences in exploitation patterns, one parasite species may be favored at high densities of the host, reducing the host density and creating conditions favorable to the second species.

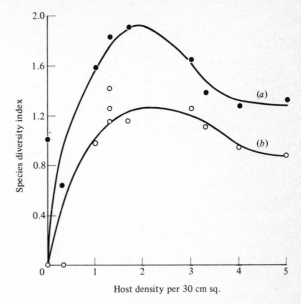

FIGURE 15-8
Change of species diversity index H' in increasing host cocoon populations at plot 9. (*a*) All parasitoids, (*b*) cocoon parasitoids only. The contribution to species diversity made by the larval parasitoid guild is the difference between (*a*) and (*b*). (*From Price, 1970.*)

3 Alternation of exploitation abilities may tend to decrease the fluctuations of a single parasite-host system, increasing the stability of the system and decreasing the chances of extinction of the host or parasite population.
4 At very high host densities, a parasite which is superior at high densities may cause the extinction of the other parasites.

15-2.2 One Parasite and Several Hosts

Specific parasites and predators are probably the exception and not the rule. A parasite or predator species usually feeds on several host species in different proportions. The proportion of each host species is dependent in part on the relative abundance of each species, and in part on the likes and dislikes of the predator. In studying the numerical changes in predator and host populations a distinction must be made between vertebrate and invertebrate predators, because of the learning capabilities of many vertebrate predators. The invertebrate parasite or predator species is considered first.

Utida (1953) performed an interesting experiment utilizing the two weevil species *C. chinensis* and *C. quadrimaculatus* and a parasite of both, the

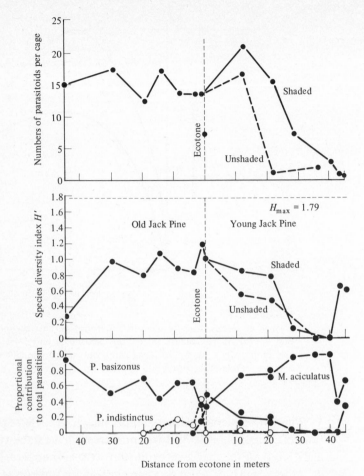

FIGURE 15-9
Transect at plot 3 across an old jack pine–young jack pine ecotone. Each point represents one sample point. Top, numbers of parasitoids per cage; middle, species diversity index, H', per cage; bottom, proportional contribution of *P. basizonus*, *P. indistinctus*, and *M. aciculatus* to total parasitism. (*From Price, 1970.*)

pteromalid wasp *N. mamezophagus*. When the two weevil species were reared together in the absence of the parasite in two replicates, *C. chinensis* became extinct in both cases by the fifth generation. In three replicates with the parasite, all three species persisted through the sixth generation (Fig. 15-10). Although the experiments were too short to be conclusive, the parasite may reduce the competitive advantage of one of the species over the other.

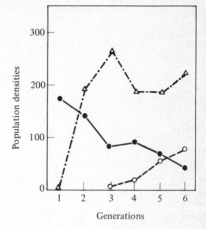

FIGURE 15-10
Competition between two species of hosts and their common parasite when one host, *C. chinensis* (●), is introduced after equilibrium is established between the other host, *C. quadrimaculatus* (○), and the parasite, *N. mamezophagus* (△). (*From Utida*, 1953.)

Holling (1965) has studied the change in the functional response of a vertebrate individual when primary and alternate food sources were both available.   In experiments on the functional response of the deer mouse *Peromyscus leucopus* to the density of cocoons of the sawfly *Neodiprion sertifer*, he always provided an abundant food supply of an alternate food source, usually dog biscuits.   The results of the experiment are illustrated in Fig. 15-11. The total amount of food eaten remained constant over changing densities of sawfly cocoons.   When a low-palatability alternate food such as dog biscuits was present at high sawfly densities, sawfly cocoons came to make up almost all of the food eaten.   In another experiment a high-palatability alternate food, sunflower seeds, was supplied in addition to the sawfly cocoons.   The diet, even at the highest sawfly densities, remained a mixture of the two food sources (Fig. 15-12).   The learning component of the vertebrate functional response makes the vertebrate predator more responsive to a sharp increase in the population density of one of its prey species (Chap. 7).   If one prey species becomes excessively common, the vertebrate predator has the ability to shift almost all of its predation to this species.   It has been postulated that in some cases a predator confronted with a very abundant food source may concentrate almost entirely on that food source, because it is easier to look for one particular type of prey than to search for several different types of prey at the same time.   If *search images* do exist, a predator may concentrate on an overly abundant or palatable species, taking the pressure off a less common species, which in general serves to damp the population fluctuations of the prey species.

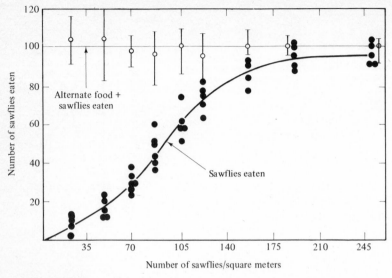

FIGURE 15-11
Effect of density of prey, *N. sertifer* cocoons, on number of prey eaten by a deer mouse, *P. leucopus*, when an alternate, low-palatability food, dog biscuits, is present. (*From Holling, 1965.*)

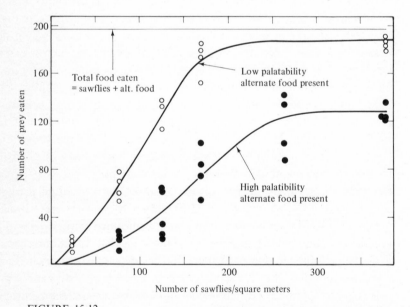

FIGURE 15-12
Effect of prey density and palatability of alternate food on number of prey eaten by a deer mouse, *P. leucopus*. Low-palatability alternate food = dog biscuits; high-palatability alternate food = sunflower seeds. (*From Holling, 1965.*)

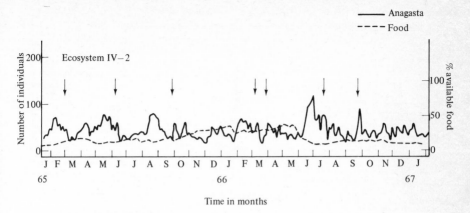

FIGURE 15-13
Changes in the density of the meal moth *A. kuehniella* when reared in the absence of a predacious mite. (*From White and Huffaker, 1969.*)

15-2.3 Three-Level Interactions

A three-level interaction might be: plant-herbivore-predator. Some insight into this three-species interaction can be gained by studying an experiment of White and Huffaker (1969). In this experiment meal moths, *Anagasta kuehniella*, were reared in cages containing 25 salve tins filled with 8 grams of rolled wheat. One tin of food was replaced each week, subjective measures of the amount of food available as a percentage of the maximal amount possible were kept, and the number of adult moths was counted. The results of one replicate are shown in Fig. 15-13. Both the number of moths and the amount of food available were fairly constant, as would be expected if the moths were growing logistically. The amount of unused food remained at a constant low level. When a mite predacious toward the eggs of the moth was added, the situation was radically changed (Fig. 15-14). Some of the most noticeable effects were:

1 Instead of small, continual, random fluctuations in moth density about a mean value, there was a pronounced pattern of large, repeated population cycles.

2 Although no distinct population generations of the moth were present in the controls, there were frequently rather distinct generations in the presence of the predator.

3 In the controls the amount of available food was generally constant and small, but in the presence of the mite the amount of food fluctuated and was small only during one short period of the cyclic behavior of the system.

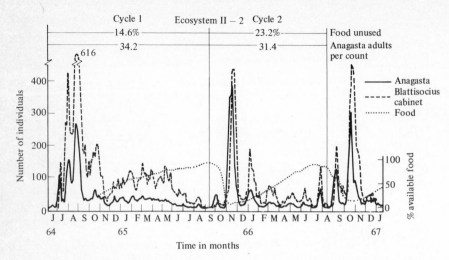

FIGURE 15-14

Indian meal moth–egg mite ecosystem replicate, series II: the population numbers of dead adult *A. kuehniella* and the recorded population samples of living *B. tarsalis* (the mite) observed on cabinet surfaces in ecosystem II-2 plotted twice weekly. The visual estimation of the concurrent availability of food in the ecosystem is expressed as a percentage of the maximal level and plotted weekly, beginning October 1964. (*From White and Huffaker, 1969.*)

15-3   THE FOOD WEB, POPULATION STABILITY, AND SPECIES DIVERSITY

In a community of plants and animals, plants are eaten by herbivores, herbivores by carnivores, and carnivores by higher carnivores in a fleas-upon-fleas manner. Each step in this *food chain* is called a *trophic level*. At each step in the food chain some individuals or parts of some individuals go to a decomposer group

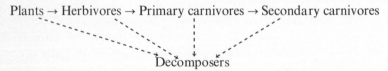

Plants → Herbivores → Primary carnivores → Secondary carnivores

Decomposers

This simplistic viewpoint, although generally correct, is usually unsatisfactory because secondary carnivores eat primary carnivores and herbivores, omnivores eat everything, and so on. The food chain is really a *food web*. Figure 15-15 is one discrete part of the littoral food web off the northern coast of the Gulf of California. At the bottom of this food web are the "herbivores," taking

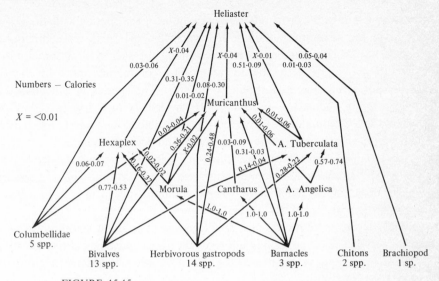

FIGURE 15-15
The feeding relationships by numbers and calories of the *Heliaster*-dominated subweb off the northern coast of the Gulf of California. (*From Paine, 1966.*)

herbivore in a broad sense, and above the herbivores are four levels of predators, the species of each level feeding on the species below it, but not on those above. In a terrestrial community the web is usually much more entwined. Rather than trying to peg each species into a distinct trophic level, it is more sensible to categorize each species by its food species and the relative abundance of each food source eaten. In Fig. 15-15 the number of individuals of each prey species eaten by the starfish *Heliaster* sp., the top carnivore, is listed, as well as the number of calories derived from each food source. In terms of number of individuals barnacles make up the majority of the diet, but in terms of calories the starfish derives its food primarily from 13 species of bivalves and 14 species of herbivorous gastropods.

The food web is the natural manifestation of the species interaction patterns discussed in Secs. 15-1 and 15-2. Each of the species in the food web affects the population fluctuations of every other species to one degree or another. The sum total of these interactions constitutes the food web. The food web can be thought of as a system in a sort of quasi-equilibrium. The species in a community are determined not only by interactions among themselves, but also by the adaptation of each species to different environmental conditions. Therefore, the existence of a species in the food web depends upon whether or not the species can successfully survive the physical conditions of the area, and whether or not the interactions of the species with the other species in the community allow it to invade and persist in the community. Chance also plays a large part in determining the species of a community. A species may be perfectly able to survive in a community, but if by chance it does not invade the area, it will be absent. The species in a community are constantly fluctuating in density. If the density of a species is low or becomes low at some time, the species may become extinct purely by chance. A species may invade a community, increase and decrease in density, and then become extinct. This invasion-extinction cycle may occur repeatedly (see Sec. 15-5). In many communities there are usually several common species which persist almost indefinitely. These common species may form a basic community with relatively stable densities. However, the species of a community are in continual flux, and when we talk about stability, we refer to the persistence of species and not so much to the constancy of the densities.

If we sample in an area at a given time, we will observe a group of species, the community. However, the constituency of the community changes with time, as well as distance. In terms of gradient analysis, time is just one more gradient we may use to ordinate species. In summary, the species of a community and their abundances are determined by: (1) the pattern of interactions among the species, (2) the physical variables of the environment, and (3) the vagaries of pure chance. However, the community does represent a partially balanced system. Why, therefore, do some of the species neither overexploit nor outcompete the other members of the community? In other words, why is the community food web at least partially stable?

15-3.1 Community Stability

Paine (1966) performed an experiment on the littoral community of the shore-line of Mukkaw Bay, Washington. In the middle intertidal zone there was a conspicuous zone of three species: the mussel *Mytilus californianus*, and the two barnacles *Balanus cariosus* and *Mitella polymerus*. Below this band was a zone of greater diversity, including scattered immature individuals of the above three species, *Balanus glandula*, anemones, two species of chitons, two limpets, four macroscopic benthic algae, a sponge, and the nudibranch *Ansodoris* sp. The stable position of the upper band was due to predation by the top predator, *Pisaster ochraceus*. In experimental areas the shore was kept free of this starfish. In these areas *B. glandula* rapidly set throughout the entire area, and was itself displaced by rapidly growing *Mytilus* and *Mitella*. Eventually the area was dominated by *Mytilus* and small patches of *Mitella*. Removal of the top predator caused the extinction of several species, and reduced the species diversity of the community to only one or two very common species.   In this case the presence of the top predator was apparently necessary to prevent the exclusion of the majority of the species in the food web.

Some tentative conclusions about the relative stability of different types of species interactions reached in this book are:

1   In competition between two species, one species almost always appears to cause the extinction of the other in the absence of mitigating factors, such as fluctuating environment and spatial heterogeneity.

2   There is not enough information available on multispecies competition to make any conclusions about its stability.

3   In single-species cultures with a constant food supply, the population reaches a maximum determined by the food supply, and then fluctuates around the maximum in an erratic way.   If there are significant time lags in the response of birth and death rates to an increase or decrease in density, the population may begin to fluctuate periodically, sometimes severely.   If the carrying capacity is small or the fluctuations are large, the chance extinction of the population is a good possibility.

4   Parasite-host and predator-prey relationships appear to be inherently unstable.   Spatial heterogeneity is apparently necessary if both species are to survive.

5   Disease-host interactions also appear to be unstable, except when there is a large host population or spatial heterogeneity.

6   A one-host, two-parasite system appears to be more stable than a single-parasite system.

7   If there are two hosts and one parasite, there is circumstantial evidence indicating that the presence of the parasite allows both host species to coexist where otherwise only one would survive.

8   Spatial heterogeneity and temporal heterogeneity appear to decrease the probability of a population becoming extinct.

These conclusions are only hypotheses to be tested, and may depend on the biological characteristics of the species involved in the interaction.

MacArthur (1955) hypothesized that community stability is a function of food web complexity.   The greater the number of links in the food web, the more stable the community.   Insect populations on *Brassica oleracea* are apparently subject to more violent fluctuations if the collards are grown in pure stands, than if the plants are mixed in with many other species of plants (Pimentel, 1961).   It seems reasonable to expect that in the more complicated food webs a large system of checks and balances will occur.   However, greater complexity can lead to less rather than more stability.   Zwölfer (1963) studied the population dynamics of six species of Lepidoptera and their parasites.   In the four cases in which there were few predators and parasites other than one or two dominant parasites, the parasites were able to keep the moth population from reaching outbreak levels.   In two cases in which there were many predators in addition to the dominant parasite species, the parasites were not able to prevent the moth species population from reaching outbreak proportions.

Watt (1965) analyzed data from the Canadian Insect Forest Survey of the numbers of 552 species of Lepidoptera collected from several species of trees and shrubs.   An analysis of the data yielded the following two generalizations:

1   The degree of fluctuation of individual species populations was less the greater the number of species competing for the tree species.

2   The degree of fluctuation of individual species populations was greater the larger the number of species of trees fed upon by that species.

Proposition *2* leads us to believe that more generalized herbivores tend to be more variable in density, either because of the biological characteristics of the species or because of some unknown set of interactions between the species and the rest of the community.   Proposition *1* suggests that the greater the diversity of competitors for the same food plant, the greater the stability of the populations of the herbivores.   Although an increase in the number of links in the food web may increase the stability of a community, the presence of a top, dominant predator or parasite may be the determining factor in maintaining this stability.   This was true in the studies of Paine (1966) and Price (1970) reported above.   Price found that the diversity and stability of the populations of the cocoon parasites on the sawfly population were determined in large part by the fluctuations of the dominant parasite, *Pleolophus basizonus.*

Although it is possible, as I have done, to make general statements about the conditions favoring the stability of species in a community, every community is unique.   The stability or lack of stability of a community is due to a unique set of interactions among the species and between the species and the physical environment.   What may be true in one community may not be true in another.   There is a great need for careful, empirical studies on the changes in the population of each species in a food web with time, taking into consideration both biotic and physical environmental factors.

15-3.2   Species Diversity

The stability of the species in a community and the diversity of species in an area are two sides of the same coin.   There is a great deal of interest in species diversity.   This interest has been generated largely by the remarkable differences in number of species found in the temperate and tropical regions of the world.   At a field station in the northern cordillera of Venezuela I collected more species of geometrids (inchworms) in 3 months than are known in the entire United States and Canada.   In temperate forests a community of trees may consist of only a few species.   Tropical forests, on the other hand, may contain several hundred species.

The number of species in an area may be due in general to two factors: (*1*) the evolutionary history of the geographic region, and (*2*) the interactions of the species and their relationships to the physical environment.   These factors are not mutually exclusive.   Evolution affects the pattern of interactions in a community, and the pattern of interactions in a community determines to a great extent the course of evolution.   It is impossible within the limitations of this book to discuss evolutionary ecology except in a general way. Geographic isolation allows the evolution of specific differences between two isolated populations of the same species.   If one of the two new species invades the habitat of the other, both species will most likely compete for the same food resources.   Competition may cause the extinction of one of the two species.   Therefore, there may be a strong selective pressure for the two species to evolve more specialized food habits in order to divide the food resources between them and avoid competition.   Evolution should, therefore, lead to greater numbers of more specialized species.   If specialization evolves on one trophic level, this specialization leads to a greater diversity of the food resources available to a higher trophic level, allowing greater numbers of species in successively higher trophic levels.   For example, the larger the number of food plant species available, the larger the number of specialized herbivores that can exist.   The degree of specialization is, of course, limited by the minimum level of resources needed by a species to persist. Evolution by itself should automatically increase the number of species in an area up to a limit imposed by the environment.   Therefore, if geographic areas have the same environmental conditions, the area which has had the same conditions longest in geologic time should have the greatest number of species.   The tropical forests of South and Central America may have been largely unaffected by the glaciations of the Pleistocene.   The North American continent was extensively glaciated during the Pleistocene, and the weather of the unglaciated areas greatly modified.

MacArthur and MacArthur (1961) noted that in a temperate forest the division of resources among several bird species took the form of a separation of the habitat into areas or levels in the forest.   The more "diverse" the foliage (the larger the number of habitat types), the greater the number of species of birds.   MacArthur and MacArthur studied this relationship by censusing the birds in several areas just large enough to hold 25 pairs of birds.

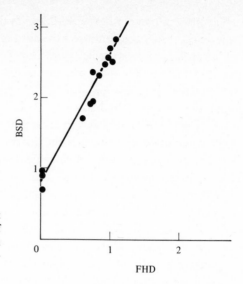

FIGURE 15-16
Bird species diversity plotted against
foliage height diversity for bird censuses of
a variety of habitats in temperate North
America. (*From MacArthur and Mac-
Arthur, 1961.*)

They calculated bird species diversity by using the Shannon-Weaver measure H'

$$\text{Bird species diversity} = -\sum_{i=1}^{s} p_i \ln p_i$$

where p_i is the proportion of the ith species. Assuming three levels to the
forest, ground to 2 feet, 2 to 25 feet, and 25 feet and above, they also
calculated a foliage height diversity as

$$\text{Foliage height diversity} = -\sum_{i=1}^{3} p_i \ln p_i$$

where p_i is the proportion of the total foliage in the ith layer. Calculating
the bird species diversity and the foliage height diversity for each area, and
plotting each area using foliage height diversity and bird species diversity
as axes, results in a linear relationship, indicating that the number of species
of birds can be predicted from the complexity of the habitat (Fig. 15-16).
MacArthur, Recher, and Cody (1966) also calculated bird species diversity at
several sites in Panama and Puerto Rico. When three foliage levels were
assumed, all the Panama points fell above the line in Fig. 15-16 (Fig. 15-17).
They found, by assuming that the birds at the Panama sites recognized
four levels instead of three, that the discrepancies between the points and the
regression line disappeared. In contrast all the Puerto Rican data points fell
below the line, and MacArthur et al. postulated that the birds in Puerto Rico

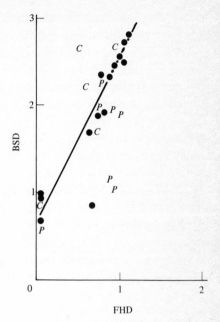

FIGURE 15-17
Bird species diversity plotted against foliage height diversity for censuses from Puerto Rico (points marked *P*), areas in or near the Canal Zone (points marked *C*). and temperate North America (solid points and regression line). Foliage height diversities were calculated assuming three layers of vegetation. (*From Mac-Arthur et al., 1966.*)

recognized only two habitat levels in the forest.   These experiments suggested the following hypotheses to MacArthur and his coworkers:

1   More species of birds are found in diverse habitats.
2   The bird species of South and Central America are more specialized in their habitat preferences than are temperate American or insular fauna.   In other words, the Panamanian species of birds divided the habitat into four vegetation layers, the American bird species into three layers, and the Puerto Rican birds into only two.

In the birds studied by MacArthur, heterogeneity of the habitat evidently allowed more species to coexist than if the habitat were homogeneous.   A pond with fallen logs, rocks, and mud contains more potential microhabitats than a pond with only a muddy bottom.   Habitat complexity, however, does not always increase the number of species in an area.   Paine (1966) found no relationship between the habitat complexity of the benthos off the coast of western North America and the number of species in the community.

Stability of weather conditions may also increase the number of species coexisting in a community.   If weather conditions remain generally constant throughout the year, and if these conditions are conducive to primary production by plants, primary production will be relatively great and constant.

It has been postulated that uniformity of production leads to greater stability in herbivore populations and allows greater specialization. Unfortunately there is little empirical evidence to prove that herbivore populations in tropical regions are less variable in density than are temperate species. Whittaker (1970) has noted that in general the number of species in a plant community increases toward lower elevations and warmer climates. In temperate plant communities areas with intermediate moisture conditions are usually richer in species than are areas with very wet or very dry conditions. It is possible that it is difficult for a species to adapt to extremes of temperature and moisture; the less optimal the conditions, the fewer the species that can evolve and successfully coexist in an area.

Although some of the determinants of species diversity are fairly obvious, there is no simple answer to the species diversity problem. The number of species in an area is determined by (1) the evolutionary history of the area, (2) the pattern of interactions among the species of the community, (3) the fluctuations of the physical variables of the environment, and (4) the spatial heterogeneity of the habitat. All these factors interact, and no single factor is responsible for the number of species occurring in an area. Within a community trends in diversity are often clear. However, it is impossible to separate the effects of these four factors from each other. It is possible to generalize about diversity, but a community is probably too complex to understand well enough to explain or predict diversity with any precision.

15-4   SUCCESSION

If an area of beech-maple forest in upstate New York were completely stripped of all vegetation and the denuded area allowed to go fallow, eventually the area would be colonized by weedy species of plants. With time these weeds would be replaced by other species, and in turn these species would be replaced by still others. Given time (perhaps 100 years), the original beech-maple forest would return; it would persist almost indefinitely, barring further disruption or changes in climate. This Clementsian view of *succession* has played a dominant role in American plant ecology. Clements and others believed that succession consists of a series of well-defined *seres* of definite duration, leading inevitably to an unchanging *climax association* adapted to the general climate of the area, e.g., beech-maple forests in the Finger Lake region of New York. The capriciousness of nature, however, does not allow for such nice, neat, deterministic stages, and the concepts of succession and climax have since been recognized as variable, overlapping, nondeterministic, and largely subjective, as a result of the work of Gleason, Tansley, Whittaker, and others. But in spite of the weaknesses of the fundamentalist approach to succession, there is no doubt that succession does occur, and as such represents one of the more interesting subjects in ecology. Probably the most lucid and complete discussion of succession and climax has been written by Whittaker (1953). The general discussion and the basic definitions of this section are taken largely from his article.

In spite of the enormous interest in directional changes in plant communities, little quantitative work has been done on succession. Some exceptions are an article by Watt (1962) on the effects of excluding rabbits from grassland, and the article by Williams et al. (1969) discussed later in this chapter.

Quarterman (1957) studied the early stages of plant succession on abandoned croplands in the central basin of Tennessee. The importance values of different species of herbaceous plants in the same field 1 and 3 years after abandonment are shown in Table 15-7. With increasing time the species which had originally colonized the field were replaced by other species. There was a decided shift from almost all forbs to predominantly grasses. By studying a series of fields of different ages Quarterman found the general sequence of dominant plants was:

1 Year 1: *Erigeron strigosus, E. canadensis, Ambrosia artemisiifolia*
2 Year 2: *A. artemisiifolia, E. strigosus, E. canadensis*
3 Year 3: *Aster pilosus, Solidago altissima, Bromus japonicus*
4 Years 4 to 8: *Andropogon virginicus, A. pilosus*
5 Year 12: *S. altissima, A. virginicus, Aster pilosus*
6 Year 15: Open woods of *Ulmus* and *Celtis*; herb layer dominated by *A. virginicus, S. altissima, A. pilosus,* and *Panicum* sp.
7 Year 25: *Ulmus* and *Celtis*; herb layer chiefly *Sanicula trifoliata, Geum virginianum,* grasses.

Even though there is a definite succession of plant species, there are no clear-cut boundaries by which seres may be defined; some species persisted through several years, and species appeared and disappeared at least partially independently of each other. The similarity of Quarterman's observations on succession to gradient analysis is apparent. In fact, time in succession may be thought of as a gradient, and the pattern of species replacement studied by plotting the importance values of each successional species along the time gradient. There is a bias in Quarterman's observations because, as we shall see later, each field's succession is dependent on the complex of factors unique to each field. A true picture of succession can be found only by following succession in the same field over a period of years. Each field is a case unto itself, and there is no guarantee that successive stages will be the same in species or duration from field to field.

Succession is not confined to plants. Succession in animal populations in feces or carrion is well known. As the population densities of plant species change during succession, so will the densities of the animal species

Table 15-7 RELATIVE IMPORTANCE OF HERBACEOUS SPECIES IN THE SAME FIELD 1 AND 3 YEARS AFTER ABANDONMENT

Species	1 year	3 years
Erigeron canadensis	103.8	
Gnaphalium obtusifolium	101.8	
Aster pilosus	88.1	88.9
Oxalis stricta	70.5	41.1
Sida spinosa	67.8	31.0
Ambrosia artemisiifolia	51.3	
Lespedeza striata	50.8	21.5
Grass seedlings	34.1	
Polygonum pennsylvanicum	33.8	
Solidago altissima	31.8	110.0
Bromus japonicus	30.6	226.5
Acalypha virginica	21.4	
Solanum carolinense	20.4	10.1
Erigeron strigosus	10.4	10.3
Cirsium sp.	10.2	
Desmodium paniculatum	10.1	
Draba sp.	10.1	
Elymus sp.	10.1	
Euphorbia sp.	10.1	
Lactuca scariola	10.1	
Oenothera laciniata	10.1	
Physalis heterophylla	10.1	
Torilis japonica	10.1	
Chaerophyllum sp.	...	181.3
Andropogon virginicus	...	85.2
Geranium sp.	...	62.7
Hieracium sp.	...	44.0
Ipomea sp.	...	40.6
Allium sp.	...	30.7
Number of species	23	14

SOURCE: From Quarterman, E., Early Plant Succession on Abandoned Cropland in the Central Basin of Tennessee, *Ecology*, vol. 37, 1957.

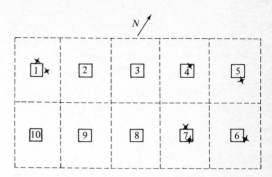

FIGURE 15-18
Layout of the experimental area in Queensland, with the positions of relevant tree stumps noted as stars. (*From Williams et al., 1969.*)

associated with particular plants or habitat types.   Pearson (1959) studied the changes in small mammal populations with old field succession, and found significant changes in these populations as the fields underwent succession.   Not all succession is necessarily directional.   The end result of succession in carrion is not a climax community, but rather nothing. Kershaw (1964) presents an excellent summary of cyclic vegetation changes. One of the more interesting examples of cyclic vegetation change is the hummock-hollow cycle studied by Watt (1947).   In some habitats subjected to severe environmental stress the pattern of succession may reverse itself. If a grassland is severely overgrazed by cattle, the most palatable plants will rapidly disappear.   As grazing continues, the grass cover is progressively reduced, and in the open spots weeds characteristic of the first stages of succession may begin to appear.   Reverse succession under environmental stress is called *retrogression.*

Williams et al. (1969) have carried out an interesting experimental analysis of succession.   In their experiment a 40 × 20 meter area in a moist, subtropical rain forest in Queensland was cleared of all vegetation, and the litter was scraped off by using a bulldozer, except where tree stumps prevented its removal.   The area was divided into 10 quadrats (Fig. 15-18), and the plant populations in each quadrat sampled on 12 occasions: February 4, 1958; February 28, 1958; May 6, 1958; July 9, 1958; October 2, 1958; March 31, 1959; December 9, 1959; March 7, 1960; December 14, 1960; November 30, 1961; May 13, 1963; and June 30, 1964.   The sampling program resulted in 120 samples for 10 quadrats on 12 dates.   A total of 118 species of plants was recorded, but in the analysis only the 32 commonest species were used.   Each quadrat differed slightly from every other quadrat in environmental conditions because of the stumps, packing of the soil by the bulldozer, and shade from the surrounding trees.

Williams and his coworkers performed a normal association analysis on the 120 quadrat-period observations (see Chap. 13) in order to investigate the course of succession in each of the 10 quadrats. The analysis resulted in seven groups, A to G. The quadrat periods belonging to each group are marked on seven 10 × 12 grids in Fig. 15-19. Group A was essentially temporal, representing the initial stages of succession spread over all the quadrats, if somewhat erratically. Group B was also temporal, and represented the main influx of colonizing species during sample periods 3 to 5 in all quadrats. At sample period 6 spatial heterogeneity appeared. Group C represented a temporal period in quadrats 6 to 7, but thereafter, until the observations terminated, the area behaved as four distinct site groups: 1, 2, 3, 5; 4, 6, 7; 8; and 9, 10. Because of small environmental differences in the 10 quadrats, succession took four directions after the demise of the first ubiquitous colonizing species. The four groups were all in a 20 × 40 meter area, indicating the importance of edaphic factors in the course of succession.

Each of the plant species appeared and disappeared independently of the others (Fig. 15-20). There was no indication of clear-cut groupings of species corresponding to seres. An inverse normal analysis of the data matrix resulted in six groups of species (Table 15-8). Group I, exemplified by *Solanum mauritianum*, colonized the area immediately and persisted over almost the entire area throughout the period of observation. Group II also contained initial colonizer species, mostly short-lived perennials and some mature forest species whose colonization had evidently been premature. The species in Group II colonized and then disappeared. Group II differed from Group I in that its members persisted longer at certain sites. Groups IV and V represented the second phase of succession. However, the species in Group V persisted and those of Group IV did not. Group VI, represented by *Eugenia corynantha* and *Melodinus acutiflorus*, consisted of mostly rain forest trees appearing later in succession and at only a few of the sites.

Three major conclusions can be drawn from the rain forest experiment:

1 There was considerable heterogeneity among quadrats. The course of succession was determined by edaphic and environmental conditions intrinsic to each quadrat. The heterogeneity increased with time, and during the period of observation there was no convergence to a common climax type.

2 There were no clear-cut groups of species replacing each other as succession progressed.

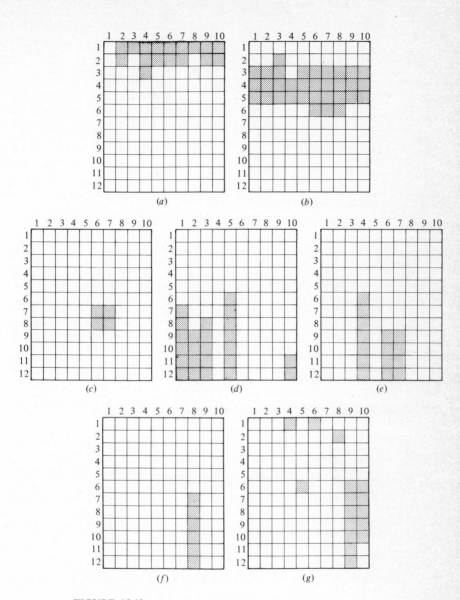

FIGURE 15-19
The division of the 120 quadrat-period observations into one of the seven groups A to G from the normal association analysis. (*From Williams et al., 1969.*)

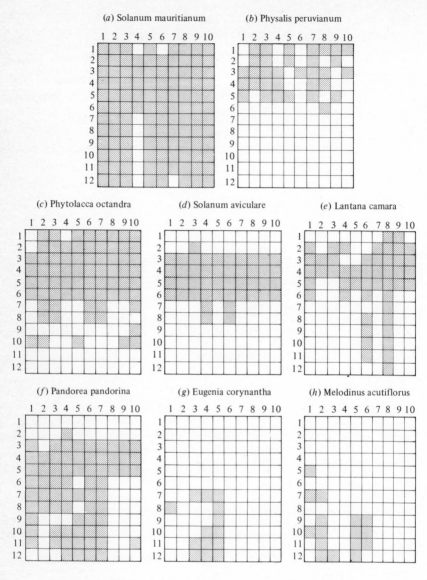

FIGURE 15-20
The quadrat-period distribution of selected species representing the six species groups resulting from an inverse association analysis of the data matrix. (*From Williams et al., 1969.*)

3   There was a definite break at observation period 5 from the original ubiquitous colonizing species to the species later making up the vegetation of the area.

These observations are in agreement with Whittaker's (1953) view of climax vegetation as "a pattern of populations corresponding to the pattern of environmental gradients and more or less diverse according to diversity of environments and kinds of populations in the pattern. Succession may be defined as a directional change in the densities of species population making up the community and the climax as a fluctuation around an average."

Whittaker postulates that a climax may be interpreted as a partially stabilized community steady state adapted to maximum sustained utilization

Table 15-8   **GROUPS OF SPECIES ARRIVED AT BY THE INVERSE ASSOCIATION ANALYSIS**

Group I
  Omalanthus populifolius
  Solanum mauritianum

Group II
  Cayratia clematidea
  Cucurbitaceae, unidentified
  Legnephora moorei
  Muehlenbeckia cunninghamii
  Parsonsia ventricosa
  Passiflora alba
  Physalis peruvianum
  Solanum nigrum

Group III
  Acacia melanoxylon
  Phytolacca octandra
  Solanum sporadotrichum

Group IV
  Duboisia myoporoides
  Erigeron canadensis
  Lantana camara
  Solanum aviculare
  Trema aspera

Group V
  Dioscorea transversa
  Pandorea pandorina
  Rubus rosifolius
  Urtica incisa

Group VI
  Archontophoenix cunninghamii
  Cryptocarya erythoxylon
  Eugenia brachyandra
  Eugenia corynantha
  Eugenia luehmannii
  Hibbertia scandens
  Lonchocarpus blackii
  Melodinus acutiflorus
  Panicum pygmaeum
  Sambucus australasica

SOURCE: From Williams, W. T., Lance, G. N., Webb, L. J., Tracey, J. G., and Dale, M. B., Studies in the Numerical Analysis of Complex Rain-Forest Communities.   III.   The Analysis of Successional Data, *J. Ecol.*, vol. 57, 1969.

of environmental resources in biological productivity. Whittaker has enumerated a number of general trends in directional succession:

1 Progressive development of the soil, with increasing depth, increasing organic content, and increasing differentiation of layers or horizons toward the mature soil of the climax.

2 The height, massiveness, and differentiation into strata of the plant community increases.

3 Productivity, the rate of formation of organic matter per unit area in the community, increases with increasing development of the soil and increasing utilization by the community of environmental resources. The microclimate within the community is increasingly determined by characteristics of the community itself.

4 Species diversity increases from the simple communities of early succession to the richer communities of late succession.

5 Populations rise and fall, and replace one another along the time gradient in a manner much like that of stable communities along environmental gradients (see Chap. 13).

6 Relative stability of communities increases. Early stages are in some cases of little stability, with populations rapidly replacing one another. The final community is usually stable, dominated by longer-lived species which maintain their populations, community composition no longer changing directionally.

Figure 15-21 illustrates the increase in soil nitrogen with increasing time from early alder successional stages to a climax spruce forest at Glacier Bay, Alaska. Olson (1958) has studied successional stages in the Lake Michigan area, and discusses the development of the soil with succession. It appears, in fact, that the early colonizers are adapted to rapid colonization and dispersal, and in their haste cause changes in the soil favoring the slower-dispersing secondary colonizers. These secondary colonizers themselves change the habitat, creating conditions favorable to the later species of the succession.

Observation *3* will be deferred to Chap. 17, but item *4* can be tested with the rain forest succession experiment. Using the Brillouin information measure *H*, Williams et al. (1969) calculated the overall diversity of the site for each of the 12 observation periods and the mean diversity of the 10 quadrats for each of the 12 observation periods (Fig. 15-22). The diversity of the site as a whole rose rapidly until period 6, dropped at the break between the colonizer and builder stages, and then rose to a plateau during the later stages of succession. On the quadrat level the results were different.

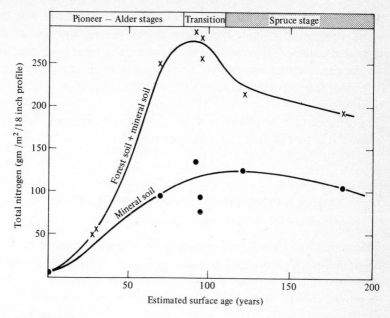

FIGURE 15-21
Change in total nitrogen content of soils on surfaces of varying ages in succession from a pioneer alder stage to a later spruce stage. (*From Crocker and Major, 1955.*)

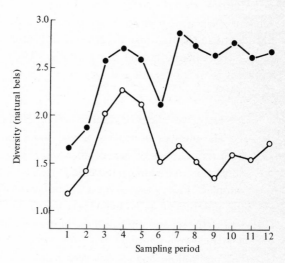

FIGURE 15-22
Changes in species diversity in the Queensland succession experiment. ●, Overall site diversity; ○, mean diversity of individual sites. (*From Williams et al., 1969.*)

The peak was at period 6, the diversity falling sharply after that and never really rising much again.   The greater diversity of the site as a whole during the later successional stages was apparently a result of the increasing heterogeneity of the community.   On the local level diversity may not increase. Therefore, it is important to specify the area involved when making statements on diversity.

Statement 6 is generally true of plant communities, but we do not know if it is true of animal populations as well.

15-5   THE EXTINCTION AND FORMATION OF POPULATIONS

This subject was developed primarily as a study of the establishment of communities of plants and animals on islands.   There is a very real and obvious similarity between an empty island and a recently abandoned field or burned forest.   In both cases new communities are formed by the influx of colonizing species.   Even in the absence of succession only a few of the original organisms dispersing onto the island can be expected to survive and produce sizable populations.   The populations of the species that do survive may later become extinct, because they are not well adapted and cannot successfully compete with other species, or merely by chance.

Patrick (in MacArthur and Wilson, 1967) suspended glass slides of different surface areas in Roxborough Spring in Pennsylvania.   She counted and identified the diatoms attached after 1 week in one set of replicates, and after 2 weeks in another set of replicates.   These slides can be considered to represent islands of small and large size.   Some conclusions reached by this experiment were:

1   The larger slides held more species of diatoms than the smaller slides.

2   The number of species on the slides after the first week was larger than the number of species observed the second week, even though the density of individuals increased greatly from the first to the second sample.   For example, on a 25-millimeter-square slide in experiment 2 the number of individuals increased from 29,600 to 466,665 from the first to the second week, but the number of species dropped from 47 to 29. The probable cause of the drop in number of species was increased competition among species with increasing density.

3   In the drop of species number from the first to the second week, it was generally the rarer species which became extinct.

15-5.1   Extinction and Formation of Single Populations: Theory

The formation of a population is a complex subject, because it is possible to consider population formation as either the spread of a population throughout a homogeneous area or the spread of a population among a series of discrete units.   It is perhaps easiest to study the formation and growth of a population in a discrete unit with immigrants arriving at a constant rate v.

If a population is absent from a favorable area and immigrants are arriving at a constant rate v, the expected number of individuals n at time t will be

$$E(n, t) = \frac{v}{b - d} \left[e^{(b - d)t} - 1 \right]$$

(Bailey, 1964).   If $b = .9$, $d = .2$, $t = 1$, and $v = .1$, the mean number of individuals in the area at $t = 1$ is $E(n, 1) = (.1/.7)(e^{.7} - 1) = .1448$.   The expected numbers of individuals for different values of t are graphed in Fig. 15-23.   The expected value increases exponentially, because v is small relative to the intrinsic rate of increase r.   If b is greater than d, the growth of the population, once established, will tend to make immigration unimportant.   However, if $b = d$, the expected number of individuals at time t will be $E(n, t) = vt$.   The immigrants tend to accumulate, and the growth of the population will be due entirely to the addition of immigrants.   If b is less than d

$$\underset{t \to \infty}{E(n)} = \frac{-v}{b - d}$$

The population exists only because of immigration from the outside.   **The greater the discrepancy between the death rate and the birthrate, the smaller the population.**

Because of the relative unimportance of immigration unless v is very high or r is very low, it is possible to view the formation of populations as the average number of days it takes the initial founder to arrive (see Chap. 4), followed by the exponential growth of the population.   If the growth of the population is density dependent, the addition of immigrants is important only if the carrying capacity of the environment is small.

It is almost always assumed that a logistically or exponentially growing population once established will persist indefinitely, provided the environment continues to be favorable.   This may be true of large populations, but small populations may become extinct because of random fluctuations, even if the intrinsic rate of increase r is greater than zero.

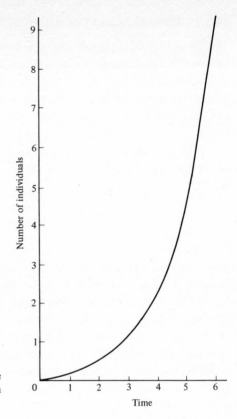

FIGURE 15-23
Growth of a population as function of time started with a single propagule with parameters $b = .9$, $d = .2$, and $v = .1$.

The probability of an exponentially growing population becoming extinct by some time t, $P_0(t)$, is

$$P_0(t) = \left| \frac{d(e^{(b-d)t} - 1)}{be^{(b-d)t} - d} \right|^a$$

where a is the initial size of the population. If the population is founded by a single immigrant, the probability that it will be extinct at $t = 1$, if $b = .7$ and $d = .2$, is

$$P_0(1) = \left| \frac{.2(e^{.5} - 1)}{.7e^{.5} - .2} \right|^1$$

$$= .1360$$

The probability of the extinction of the population with passing time is shown in Fig. 15-24. The probability of extinction is asymptotic if b is greater

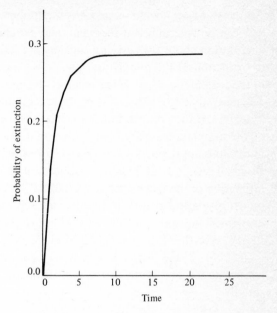

FIGURE 15-24
The probability of extinction of a population founded by a single propagule as a function of time. The parameters are $b = .7$ and $d = .2$.

than d, and the asymptote is equal to $(d/b)^a$. As the population grows, the chance of extinction because of random fluctuations becomes less and less. With only one initial founder, the limiting probability is .2856. With one propagule the probability of ultimate extinction is almost 29 percent, but with two propagules this probability is reduced to $.2856^2 = 8.16$ percent. With four propagules the probability of ultimate extinction is only .6 percent. If the birthrate is less than the death rate, the population will always become extinct, the time to extinction depending on the values of b, d, and a.

If the growth of the population is density dependent, the theoretical probability of its becoming extinct becomes difficult to determine (e.g., see Bailey, 1964). However, if the carrying capacity of the environment is large, the probability of extinction will be almost equal to the probability of extinction of an exponentially growing population. If the carrying capacity of the environment is large, the distribution of densities about the carrying capacity after the population has reached the carrying capacity will be approximately normal (see Chap. 3). The mean of the normal distribution can be related to the standard deviation as $m = ks$, where m is the mean,

s is the standard deviation, and k is a parameter (Bartlett, 1960). The value of k can be determined if the mean and standard deviation of the frequency distribution about the carrying capacity are known (see Chap. 3). The time to extinction of the population after it has reached the carrying capacity is of the magnitude $e^{k^2/2}$. If the mean is 48.5 and the variance is 75, $k = 5.639$. The mean time to extinction is of the magnitude $e^{15.902}$, a very large number. Therefore, if the population has reached the carrying capacity, and if the size of the population is large enough so that the probability of there being n individuals approximates a normal distribution, the possibility of extinction is so small that for all practical purposes it is not worth considering. This estimation of the mean time to extinction is valid only if Bartlett's stochastic model of logistic growth is an adequate representation of density-dependent growth in the species. If time lags are important, the oscillations induced in the growth of the population can severely reduce the mean time to extinction. If the oscillations are very severe, the population can become extinct at the low point in the oscillation by chance, and may do so quite rapidly.

15-5.2   The MacArthur-Wilson Theory of Biogeography

MacArthur and Wilson (1963, 1967) reasoned that, on an island separated from the mainland and the main pool of potential colonizer species, the number of species is determined by a balance between the rate of arrival of propagules of new species and the rate of species extinction. On a defaunate island next to a mainland area with P species there are P potential species which may become part of the developing community on the island. Some species are more likely to disperse to the island than others. In the beginning these rapidly dispersing species will reach the island quickly, and the rate of immigration of new species to the island will initially be large, tapering off with time. As the number of species on the island increases, the number of new species immigrating becomes less, partly because there are few new species left to immigrate, and partly because the remaining species are those least likely to disperse. If all P species are found on the island, the rate of immigration of new species is obviously zero. For heuristic reasons the rate of immigration of new species is postulated to decrease exponentially as the number of species on the island increases.

As the number of species on the island increases, the rate at which species become extinct must also increase. With more species there are more species to become extinct. In addition, it is reasonable to assume that, with an increase in the number of species, competition between species will become

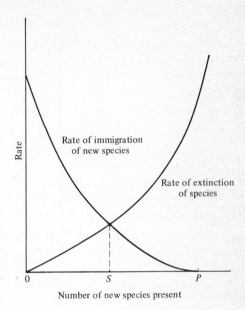

FIGURE 15-25
A graphic representation of the Mac-Arthur-Wilson model of the equilibrium number of species on an island. Where the immigration curve and the extinction curve intersect, the theoretical numbers of species on the island are at equilibrium.

important, increasing the rate of extinction.   Again the rate of extinction may be postulated to be exponentially related to the number of species on the island.

Both the immigration curve and the extinction curve may be plotted on the same pair of coordinates (Fig. 15-25).   The intersection of the two curves is a stable equilibrium point referred to as the equilibrium number of species. The immigration of new species to the island at this point is compensated for by the extinction of species already there.   Before the equilibrium number of species is reached the rate of new immigrants is greater than the rate of extinction, and the number of species on the island increases.   As the distance of the island from the mainland increases, the immigration rate of new species may be expected to drop.   Therefore, near islands will have a steeper immigration curve than far islands.   The rate of extinction for some given number of species is less on large islands than on small islands, because large islands will be able to support more species.   A series of immigration and extinction curves representing different distances of an island from the mainland, and island size, are drawn in Fig. 15-26.   The equilibrium number of species on an island depends on the size of the island and its distance from the species pool.   If the islands are of the same size, those nearer the mainland or source area will have more species than those farther away, other things being equal.   The species-area relationship of the number of land

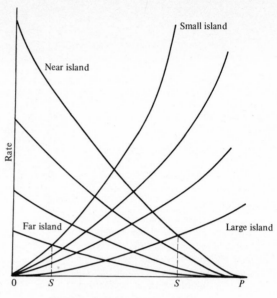

FIGURE 15-26
A graphic illustration of the effect of changing island size or distance from the source area.

and freshwater bird species on various islands and archipelagos of the Moluccas, Melanesia, Micronesia, and Polynesia is shown in Fig. 15-27. The islands farthest from the mainland are enclosed by squares, and those closest to the mainland by circles. A line has been drawn through two of the islands nearest the source area, and with the highest species densities, to give an idea of the degree of departure of the other islands from the potential densities of the archipelagos with high immigration rates. In each case the far islands had relatively fewer species for a given area than did the near islands.

Another interesting prediction of the model is that the number of species increases more rapidly with area for far groups of islands than for near groups of islands (Fig. 15-28). This is always true if the logarithms of the numbers of species are used to plot the two curves. If a line is drawn through the far islands in Fig. 15-27, its slope is far greater than a line drawn through the near islands, suggesting that this prediction is true.

In summary the hypotheses of the model are:

1 The number of species on an island is a balance between the rate of immigration of new species and the rate of extinction of species already there.

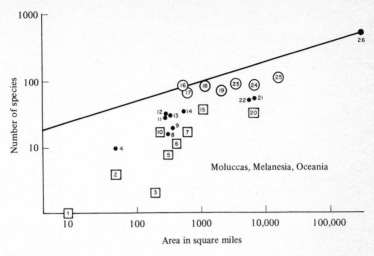

FIGURE 15-27

The numbers of land and freshwater bird species on various islands and archipelagos of the Moluccas, Melanesia, Micronesia, and Polynesia. Near islands are enclosed in circles, far islands in squares, and intermediate islands are left unenclosed. A line is drawn through two of the islands nearest the source regions and with the highest species densities in order to give a clearer idea of the degree of departure of the other islands from the potential densities of archipelagos with high immigration rates. Wake, 1; Henderson, 2; Line, 3; Kusaie, 4; Tuamotu, 5; Marquesas, 6; Solomons, 7; Ponape, 8; Marianas, 9; Tonga, 10; Carolines, 11; Palau, 12; Santa Cruz, 13; Rennell, 14; Samoa, 15; Kei, 16; Louisiade, 17; D'Entrecasteaux, 18; Tanimbar, 19; Hawaii, 20; Fiji, 21; New Hebrides, 22; Buru, 23; Ceram, 25; Solomons, 25; New Guinea, 26. (*From MacArthur and Wilson, 1967.*)

2   The number of species on an island depends on the size of the island and its distance from the source area.

3   Far islands have fewer species for a given size area than close islands.

4   The logarithm of the number of species on an island increases more rapidly for distant islands than for near islands.

MacArthur and Wilson (1967) have developed the model mathematically. An alternate formulation used by Simberloff (1969) has been developed by Bossert (1968) and Holland (1968).

15-5.3   An Experiment

To test the predictions and equations of the MacArthur-Wilson model Wilson and Simberloff (1969) performed a rather unusual experiment. Seven red mangrove, *Rhizophora mangle*, islands in the Florida Keys, of varying

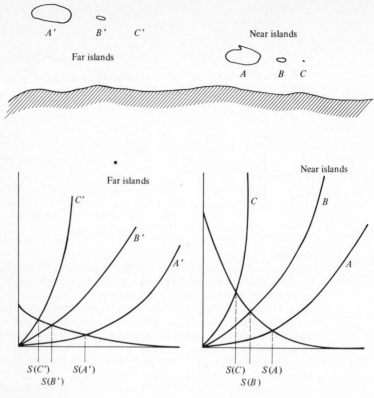

FIGURE 15-28
A diagrammatic illustration to show how species diversity as represented by the equilibrium number of species increases faster with size on far islands than near islands.

distance, size, and direction from the source areas, were covered with plastic and fumigated to remove all arthropods. The terrestrial fauna of these islands is composed almost entirely of arboreal arthropods, with from 20 to 50 species present on an island at any given moment. The potential species pool of the source area was estimated to consist of 600 to 700 species. After fumigation each of the islands was censused every 18 days, up to day 280. The islands were subsequently sampled at day 360 and day 720.

The *colonization curves*, indicating the increase in number of species with time, for four of the mangrove islands are shown in Fig. 15-29. The curves rise rapidly to a rough plateau at about the *two-hundredth* day. There is also an indication that the curve first peaks and then plateaus at a slightly

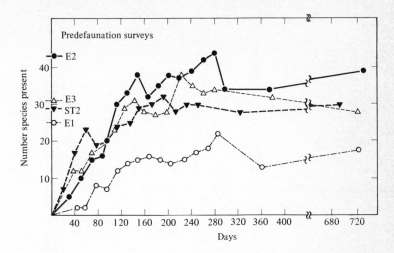

FIGURE 15-29
The colonization curves of four small mangrove islands in the lower Florida
Keys whose entire faunas were exterminated by methyl bromide fumigation. The
figures shown are the estimated numbers of species present by the criteria utilized
by Simberloff and Wilson (1966) and Simberloff (1969). (*From Simberloff and
Wilson, 1970.*)

lower number of species.   The original number of species on each of the
four islands is indicated by the hash marks on the ordinate.   The number
of species on each island appears to plateau near the number of species
originally on the island before fumigation.   This is evidence that an equilibrium
number of species exists.   Although there appears to be an equilibrium
number of species on an island, the species before and after fumigation were
not necessarily the same (Table 15-9).   However, as time passed, the species
composition of each island became more and more like the composition
before fumigation, indicating, perhaps, some sort of successional process.

15-5.4   Some Characteristics of Colonizing Species

Not all the P species in the source area are equally likely to colonize an
island whether that island is an island, lake, mountain range, or ploughed
field.   In fact, it is often easy to eliminate the majority and with some
accuracy predict the initial species composition of the islands.   In an
abandoned field the first colonizers are easy to predict; they are invariably
a few species of weeds.

Table 15-9 PERCENTAGES OF SPECIES THAT WERE PRESENT AT BOTH OF TWO GIVEN CENSUSES ON FOUR EXPERIMENTAL ISLANDS

Name of experimental island	A. Censuses: just before defaunation and 1 year later			B. Censuses: just before defaunation and 2 years later			C. Censuses: 1 and 2 years after defaunation		
	No. species in common	Total no. in both censuses	Percent in common	No. species in common	Total no. in both censuses	Percent in common	No. species in common	Total no. in both censuses	Percent in common
E1	2	29	6.9	5	26	19.2	7	18	38.9
E2	10	54	18.5	13	51	25.5	16	34	37.2
E3	8	40	20.0	7	35	20.0	16	31	51.6
ST2	11	37	29.7	17	31	54.8	12	34	35.3

SOURCE: Simberloff, D. S., and Wilson, E. O., Experimental Zoogeography of Islands. A Two-Year Record of Colonization, *Ecology*, vol. 51, 1970.

The most obvious characteristics of a colonizing species are rapid dispersal and the ability to survive in disturbed or marginal habitats. In terms of the probability of successful colonization, the probability of a group of propagules forming a constant population is approximately $1 - (d/b)^a$. The probability of survival is increased if a or b is increased, or d is decreased. Except among social animals not much can be done about a, but it might be postulated that successful colonizers have relatively high instantaneous birthrates. Lewontin (1965) and Cole (1954) have shown that it is generally easier to increase r and thus the birthrate, given a constant death rate, by decreasing the developmental time rather than by increasing fecundity. The fact that almost all early colonizing species of plants in succession are forbs is probably no fluke. In fact, it has been postulated that one of the factors contributing to the demise of early successional plants is their rapid growth and subsequent self-poisoning. The analogy between weeds and man is a rather painful one.

In conclusion we might postulate that colonizing species: (1) are highly dispersive, (2) become adapted to disturbed or marginal habitats, and (3) have high birthrates and rapid development.

15-5.5 Some Concluding Remarks

In spite of the interest and work in community ecology, not much is really known about the ebb and flow of or the formation and extinction of communities of populations, particularly in terms of relative probabilities. Our knowledge of the relationships between competitors, predators, parasites, and environmental factors in any large group of species is appallingly limited. Community ecology remains perhaps the least known, most difficult, and certainly one of the more interesting areas of ecology.

PART FIVE

16

PRODUCTION, BIOMASS, AND ENERGY IN SINGLE POPULATIONS

Despite the emphasis throughout the first four parts of this book on changes in numbers of individuals, changes in the organic matter stored in the population and the formation of new matter are also legitimate studies. Beef is sold by the pound, not by the cow, and fish production is measured in tons, not in millions of fish. Plants use sunlight as a source of energy to synthesize various organic compounds, and herbivores obtain nourishment by eating plants. In the transfer of energy from one trophic level to the next part of the food is wasted, part of the organic compounds are broken down, releasing the energy needed in metabolism, and another part may be converted to new organic compounds during the growth of the animal's body.

This chapter begins by presenting the basic concepts of the energetics of animal populations. In Sec. 16-2 production in plant populations is described. Section 16-3 discusses production in one particular exploited population.

16-1   PRODUCTION IN ANIMAL POPULATIONS

A piece of meat or a tomato can be expressed in terms of the number of calories of energy it contains.   Therefore, organic matter represents a potential source of energy to anything that eats it.   An individual animal feeds on some food source.   Of the food consumed by the individual, part will not be digested and will be excreted as feces.   Part of the assimilated food energy will be metabolized in respiration, and part will be incorporated into the body of the animal as produced body tissue or offspring.   This very simple relationship can be symbolized as

$$C = P + R + F + U$$

where the symbols, following the definitions of Petrusewicz and Macfayden (1970), are:

C = consumption: total intake of food by heterotrophic organisms during a defined period of time.

P = production: the net balance of food transferred to the tissues of a population during a defined time period, i.e., the net balance between assimilation and respiration.

R = respiration: that part of the food intake converted to heat and dissipated in life processes (metabolism) during a defined period of time.

F = egesta: that part of consumption rejected as feces or regurgitated.

U = excreta: material derived from assimilation and excreted in urine or through the skin.

Biomass is the mass of living organisms present in a population at some given moment.   Biomass may be expressed in a weight measure, such as dry weight in grams, or in calories.   If biomass is measured at two times, say 1 week separates the second sample from the first, production is equal to the change in biomass ΔB plus the tissue lost either through emigration, death, exuviae, or predation, or through respiration (weight loss), or $P = \Delta B + E$, where E represents the various losses.   Production consists of two components: (*1*) production due to reproduction, and (*2*) production due to body growth.

A rate called *production turnover* can be defined for some given time period as the ratio of the production during this time period to the average biomass $\overline{B}$, or $\theta = P/\overline{B}$.   If the average biomass for a period of 1 month

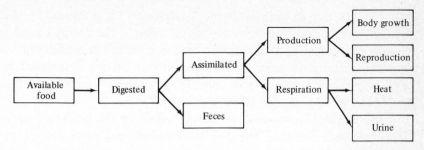

FIGURE 16-1
An animal viewed as a series of components, and the transfer of energy through the components.

were 100 kilograms per square meter and the production during the month were 10 kilograms per square meter, the turnover rate per month would be $\theta = \frac{10}{100} = .1$.

16-1.1   Efficiencies and the Energy Budget of an Animal Population

An animal can be viewed as a series of components. Energy, represented by food, flows through the components, and at each transfer part of the energy is lost. Figure 16-1 is a diagrammatic representation of a series of components. Some of these transfers are of more interest than others. The ecologist may ask, "What percentage of the energy ingested as food will be assimilated?" or perhaps, "What percentage of the energy ingested will be incorporated into body growth?" Both questions represent *efficiencies*; i.e., the first question concerns the *assimilation efficiency*, or the percent of the ingested energy assimilated by the animal. Two types of production efficiencies are commonly used. The first, *gross production efficiency*, represents the percent of calories ingested incorporated into new protoplasm. *Net production efficiency* is the percent of calories assimilated incorporated into new protoplasm. Production efficiencies can be further subdivided into reproduction efficiency and growth efficiency. The possibilities are almost endless. An intellectual dung beetle would probably be most interested in defecation efficiency. An incredible nomenclatorial tangle has evolved in the naming of efficiencies. Therefore, the reader must be careful to divine exactly the pathway of energy transfer involved, when interpreting efficiencies listed in articles on energetics.

The first step in the calculation of efficiencies for a laboratory animal population is the construction of an *energy budget*. The energy budget lists the calories ingested, egested, respired, used in growth, and used in reproduction. The calories in ingested food and production are determined by a bomb calorimeter. Several types of bomb calorimeters have been developed. Petrusewicz and Macfayden (1970) discuss their uses.

Paine (1965) constructed an energy budget for the predatory opistho-branch *Navanax inermis*. This predatory animal was particularly easy to work with because it ingests its food whole. In some species it is difficult to determine how much food is ingested. Paine determined the calories per gram dry weight for several prey species with a bomb calorimeter. When the length of the shell of two of the most common prey species, *Bulla gouldiana* and *Haminoea virescens*, was plotted against the logarithm of caloric content, linear relationships resulted (Fig. 16-2). Therefore, the caloric content of an ingested prey was accurately determined by measuring the length of its shell, and converting to calories by using the regression equations of the linear relationship. The calories egested were determined by collecting the feces, drying them, and determining their caloric content by using a bomb calorimeter.

The calories assimilated are simply the calories ingested minus the calories egested. Of the calories assimilated some will be respired; some will go toward growth and some toward reproduction. The calories allocated to reproduction can be estimated by gathering up the reproductive products of one kind or another, drying them, and determining their caloric content with a calorimeter. The calories of growth are determined by weighing the animal before and after the experiment, and converting from wet weight to caloric equivalents. The calories expended in respiration can be determined by using a respirometer. Petrusewicz and Macfayden (1970) discuss respiro-meters and their use. Paine (1965) used a polarographic oxygen electrode at 17°C. Freshly collected animals were permitted to clear their guts for 12 to 24 hours. The microliters of oxygen consumed by an animal per hour were determined with a respirometer. When the logarithm of the oxygen consumed was measured in a series of animals and plotted against the logarithm of the dry weight of the animal at the end of the experiment, the relationship in Fig. 16-3 developed. This linear relationship was fit by the regression equation $\log O_2 = \log 5.934 + .885 \log W$, where O_2 is the oxygen consumption, and W is the dry weight of the animal. The sum of the calories used in respiration, growth, and reproduction is a check on the accuracy of the calories assimilated as determined by calories ingested minus calories egested. The total energy budgets of five *N. inermis* individuals are

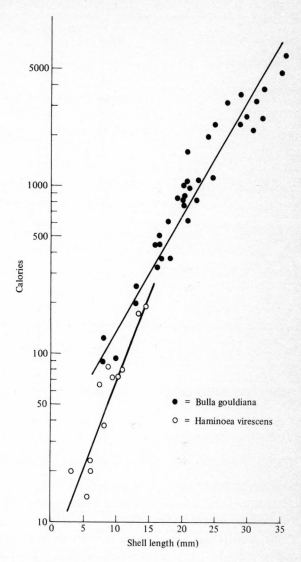

FIGURE 16-2
Relationships between shell length and the logarithm of caloric content of two common prey species of *N. inermis*.   Sample sizes are *B. gouldiana*, *n* = 36, and *H. virescens*, *n* = 12.   (*From Paine, 1955.*)

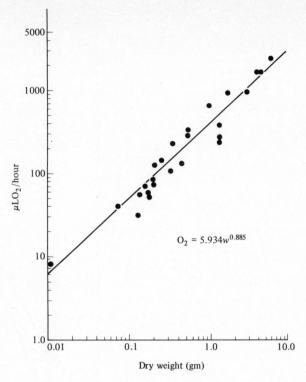

FIGURE 16-3

A log-log plot of microliters of oxygen consumed per hour, and dry weight of 25 individuals of *N. inermis*. The equation of the regression line is included on the graph. (*From Paine, 1965.*)

given in Table 16-1. The assimilation efficiency is determined by dividing the calories assimilated by the calories ingested, or for individual 1: assimilation efficiency = $\frac{1427}{2107}$ = .68. The gross production efficiency is equal to the sum of the growth and reproductive calories total divided by the calories ingested. For individual 1: gross production efficiency = $(181 + 617)/2107$ = .38. The net production efficiency is the total of reproductive and growth calories divided by the assimilation calories; net production efficiency = $(181 + 617)/1427$ = .56. These efficiencies are also listed in Table 16-1.

16-1.2   Measurement of Ingestion Using Radioactive Tracers

The *Navanax* example is, in many ways, fortuitous, because the number and size of the prey could be directly observed. In many cases, particularly in field experiments, the amount of food consumed cannot be directly determined.

However, the amount of food ingested per unit of time can be estimated by using radioactive tracers. Several different radioactive isotopes are available. Cesium 134 is often used because its half-life is long, 2.07 years, and the loss of radioactivity due to decay can be ignored. Another commonly used isotope is phosphorus-32. The half-life of this isotope is only 14.2 days, and if phosphorus-32 is used, all data should be corrected for the effects of the decay of the isotope. The basic experiment is to tag the food with a known amount of radioactivity. After the isotope has been ingested by the animal from the food, measurements of activity, usually in nanocuries, are made at intervals of time. With time the radioactivity of the animal will fall off. The radioactivity at some time t is expressed as a percentage of the activity at time zero, the beginning of the experiment. If the log of the percent activity is plotted against time, an exponentially shaped curve usually results. Figure 16-4 illustrates these activity curves for the isopods *Armadillidium vulgare* and *Cylisticus convexus*. The time at which the curve drops below the 50 percent mark is the *biological half-life* of the isotope; the animal in one way or another eliminated one-half of the isotope from its body. The exponentially shaped curves in Fig. 16-4 are typical of two-component systems. Of the isotope ingested part will be lost almost

Table 16-1   ENERGY BUDGETS FOR FIVE *N. INERMIS* INDIVIDUALS FED ON *H. VIRESCENS*. THE INGESTION MINUS EGESTION COLUMN WAS USED TO CALCULATE ASSIMILATION AND ASSIMILATION EFFICIENCIES

Individual	Days	Ingestion calories	Egestion calories	Respiration calories	Growth calories	Reproduction calories	Assimilation calories
1	8	2,107	680	491	181	617	1,427
2	8	3,697	1,118	1,112	1,034	0	2,579
3	16	449	190	216	108	0	259
4	16	986	490	290	310	0	496
5	7	2,597	907	643	0	758	1,690

Individual	Assimilation efficiency	Gross production efficiency	Net production efficiency
1	.68	.38	.56
2	.70	.28	.40
3	.58	.24	.42
4	.50	.31	.62
5	.63	.29	.45

SOURCE: After Paine, R. T., Natural History, Limiting Factors and Energetics of the Opisthobranch *Navanax inermis*, *Ecology*, vol. 46, 1965.

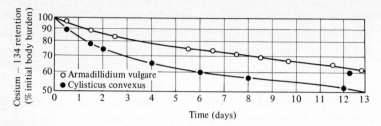

FIGURE 16-4
Cesium-134 retention curves for two isopod species. Results based on laboratory populations at 20°C. Data points are means of 25 individuals. (*From Reichle, 1967.*)

immediately through defecation. The remainder will be assimilated and lost more slowly through metabolism. Therefore, the curves in Fig. 16-4 can be separated into two components, one component representing the loss of activity of the assimilated isotope, and the other component representing the loss of activity of the nonassimilated isotope. To find the relative importance and biological half-life of each component, the level part of the activity curve is extrapolated back to time zero as a straight line (Fig. 16-5).

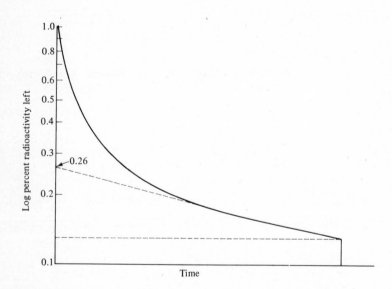

FIGURE 16-5
A diagrammatic of the method used to separate the two components of a radioactive retention curve. The method is described in the text.

The intercept of this line represents the proportion of the total activity going through the slower component, i.e., the assimilated isotope. In this case the assimilated isotope represents 26 percent of the total activity. Therefore, 74 percent of the isotope has been lost through egestion. To find the biological half-life of the assimilated isotope find one-half of 26 percent on the ordinate, .13, and draw a line parallel to the abscissa. The point where the straight line representing the assimilated component crosses the line is the biological half-life of the assimilated isotope. To find the first component, the defecated isotope, subtract the activity represented by the second component, the straight line, from the observed activity, and replot these corrected values. The point where the line representing the first component crosses the 50 percent mark is the biological half-life of the defecated part of the ingested isotope. Table 16-2 lists the relative proportions and biological half-lives of these two components for the two isopod species in Fig. 16-4.

An elimination rate can be calculated from these statistics as

$$k' = \frac{.693}{p_1 T_{b1} + p_2 T_{b2}}$$

where p_1 and p_2 are the proportions of activity of the first and second components, and T_{b1} and T_{b2} are the respective biological half-lives. For example, the value of k' for *A. vulgare* is

$$k' = \frac{.693}{(.14)(1.04) + (.86)(26.50)}$$

$$= .030 \text{ nanocurie per day}$$

Table 16-2 PARAMETERS OF THE RADIOCESIUM RETEN- TION CURVES FOR TWO SPECIES OF ISOPODS (p = PROPORTION WHOLE BODY ACTIVITY; T_b = BIOLOGICAL HALF-LIFE OF RESPECTIVE COMPONENTS; k' = WEIGHTED AVERAGE ELIMINATION RATE)

Species	p_1	T_{b_1}, days	p_2	T_{b_2}, days	k' per day
Armadillidium vulgare	.14	1.04	.86	26.50	.030
Cylisticus convexus	.30	.96	.70	26.50	.037

SOURCE: From Reichle, D. E., Radioisotope Turnover and Energy Flow in Terrestrial Isopod Populations, *Ecology*, vol. 48, 1967.

If the animal is allowed to continue to feed on the tagged food, an equilibrium will eventually be reached between the radioactivity ingested and the radioactivity eliminated. At equilibrium the input in nanocuries per day, or some other period of time, can be calculated as

$$I = k'Q_e M$$

where I is the input, k' is the elimination rate, M is the dry weight of the animal, and Q_e is the equilibrium activity density. In the isopod *A. vulgare* Reichle (1967) found that $Q_e = 16.93$ nanocuries per milligram dry weight of isopod. The average dry weight was $M = 22.41$ milligrams per isopod. As already shown, $k' = .030$. The input is $I = (.030)(16.93)(22.41) = 11.38$ nanocuries per day. The activity of the food was 22.16 nanocuries per milligram of dry weight food. Therefore, the uptake in dry weight food eaten per day is $\frac{11.38}{22.16} = .51$ milligram of food dry weight per day per isopod.

The consumption of an animal in the field can be estimated by measuring its weight and activity density, if the population is feeding on tagged food. The elimination rate k', however, is dependent on temperature (Fig. 16-6). Reichle (1967) found that a regression of the form

$$\log k' = a + b \text{ temperature}$$

could be used to predict k' for a given temperature. For *A. vulgare* $\log k' = -2.38 + .039$ (degrees centigrade).

16-1.3   Estimation of Production in the Field

In estimating the production, ingestion, and respiration of an animal population in the field, the most common approach is to determine average energy budgets of individuals under a variety of conditions in the laboratory. The population density of the animal in the field is determined, and the average energy budget per individual multiplied by the number of individuals in the population. Some sort of average temperature and humidity conditions in the field must be assumed during the period of time of the study. If production is estimated over a long period of time, say a year, it will be necessary to calculate budgets for smaller intervals of time, say a week, and sum the production figures over the year. The population should also be divided into age classes, and energy budgets constructed for each age class. Some methods of estimating production and respiration under specific conditions are presented below, after the growth curve is discussed. Estimates of energy flow in natural populations are usually only crude estimates, because of the various errors and approximations used. Usually the relative magnitudes of the transfers are of more interest than the estimates themselves.

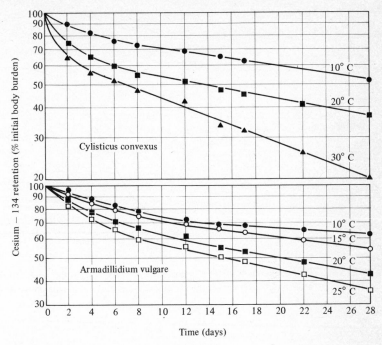

FIGURE 16-6
Cesium-134 retention curves for *A. vulgare* and *C. convexus* acclimated to several different temperatures.   Data points are means of 10 individuals.   (*From Reichle, 1967.*)

16-1.4   The Growth Curve

If the weight of an individual is plotted against its age, a sigmoid curve similar to Fig. 16-7 is often the result.   Figure 16-8 illustrates the relationship between age and weight in plaice.   The curve has an upper asymptote W_∞, the maximum weight obtained by an individual, and an intercept on the weight axis W_0, the weight of an individual at birth.   Several equations have been proposed to model the relationship between weight and age. Bertalanffy (1934) postulated that the rate of weight gain can be thought of as the synthesis of tissue minus its breakdown.   For our purposes the Bertalanffy equation can be modified to the recurrence relation

$$w_{t+1}^{1/3} = W_\infty{}^{1/3}\left(1 - e^{-K}\right) + w_t{}^{1/3}e^{-K}$$

where w_t is the weight of an individual at age t, and w_{t+1} is its weight one age period later.   The model presumes that the animal grows isometrically

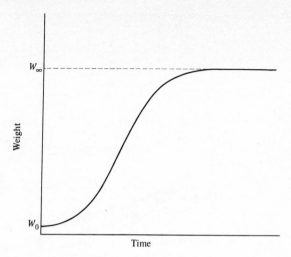

FIGURE 16-7
Diagrammatic figure of a typical growth curve. W_0 is the weight of the animal at birth, and W_∞ represents the asymptote of the curve or the maximum weight attained.

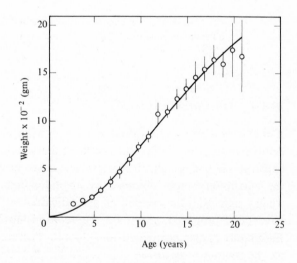

FIGURE 16-8
Growth curve of plaice. Data points are the average weight of fish in each age group from Lowestoft and Grimsby market samples from the years 1929–1938. The fitted curve was obtained with the Bertalanffy growth equation. The parameters were $W_\infty = 2867$ grams, $K = .095$, and $t_0 = -.815$ years. The length of each vertical line is equal to six times the standard error of the corresponding mean. (*From Beverton and Holt, 1957.*)

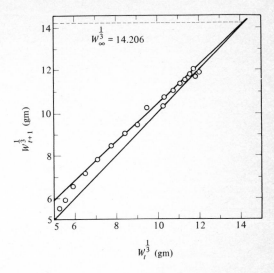

FIGURE 16-9
Fitting the Bertalanffy growth equation to data. First stage: plot of $w_{t+1}^{1/3}$ against $w_t^{1/3}$. Where the best straight line through these data cuts the bisector through the origin there is an estimate of $W_\infty^{1/3}$. The slope of the line is an estimate of e^{-K}. (*From Beverton and Holt, 1957.*)

and maintains a constant specific gravity. The parameter K is equal to $k/3$, where k is the rate of destruction of mass per unit mass. To estimate the two parameters W_∞ and K experimentally derived values of $w_t^{1/3}$ and $w_{t+1}^{1/3}$ are first plotted against each other as in Fig. 16-9, and a straight line is fitted to the points by least-squares regression (see Chap. 9). Second, a line is drawn bisecting the angle at the origin. The intersection of the two lines is the value of $W_\infty^{1/3}$. In this example the two lines intersect at 14.206, so W_∞, the maximum weight attained by plaice, is 2867 grams. The slope of the regression line is an estimate of e^{-K}, and therefore indirectly of K. The weight at birth W_0 may be estimated from the equation

$$W_0 = W_\infty(1 - e^{Kt_0})^3$$

where t_0 is an imaginary age representing the age at which the weight of the animal is zero. To estimate t_0 and K, W_∞ is first estimated. From experimental values of w_t the quantity $\log(W_\infty^{1/3} - w_t^{1/3})$ is plotted against t (Fig. 16-10). A line is fitted to the points by regression. The slope of the line is equal to K. In the example $K = .095$. The value of t_0 is the point where the regression line intersects a line drawn horizontal to the t axis

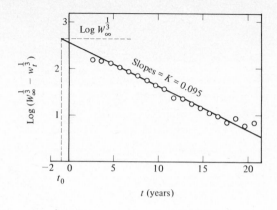

FIGURE 16-10
Fitting the Bertalanffy growth equation to data. Second stage: plot of $\log(W_\infty^{1/3} - w_t^{1/3})$ against t. The slope of the line is an estimate of K, and the value of t where it has an ordinate of $\log W_\infty^{1/3}$ is an estimate of t_0. (*From Beverton and Holt. 1957.*)

representing $W_\infty^{1/3}$. In this case $t_0 = -.815$. Predicted values of the weight of an individual at some age t may be calculated as

$$w_t = [W_\infty^{1/3} - (W_\infty^{1/3} - W_0^{1/3})e^{-Kt}]^3 \qquad (16\text{-}1)$$

The Bertalanffy curve fitted to the plaice data is shown in Fig. 16-8.

In some situations the Bertalanffy equation may not adequately represent the relationship between age and weight. For example, Fig. 16-11 shows the relation between weight and age in the flour beetle *Tribolium castaneum*. Instead of an asymptote there is a distinct drop in weight from prepupal and pupal stages to the adult stage. There may also be differences in rates of growth between males and females. Perhaps one of the more vexing problems is the relationship between the rate of growth and population density.

The maximum weight attained by an individual is sometimes dependent on the density of the population. If butterfly larvae are reared in crowded cages, the adults are often smaller than wild-caught specimens or specimens reared singly. Beverton and Holt (1957) discuss several examples in fish in which growth rate and maximum weight attained are functions of density. They conclude that the decrease in weight and growth with increasing population density is primarily a matter of decreasing food supply. Adult fish, at least, have a great plasticity of growth rates, and the response of a fish population to changes in density will be primarily a change in growth rate,

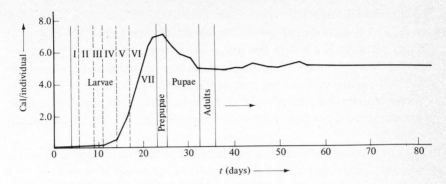

FIGURE 16-11
Growth curve of one individual of the grain beetle *T. castaneum.* Weight is expressed as calories. (*From Klekowski, Prus, and Zyromska-Rudzka, 1967.*)

not an increase in mortality. This is also true of many insects and of the cladoceran genus *Daphnia.* Interference may also be a significant factor in reducing the growth rate and the maximum weight attained by an individual. Species may respond to greater density by decreasing the rate of growth and the average weight of an individual, rather than by changes in the birthrate and death rate.

Because of the limitations of the Bertalanffy growth equation a more general but purely empirical fit of a growth curve may be obtained by using a polynomial regression equation (Chap. 9) of the form

$$w_t = a + b_1 t + b_2 t^2 + b_3 t^3 + e_t$$

16-1.5   Estimation of Production by Using the Growth Curve

Production in field studies is perhaps most easily estimated from the equation $P = P_r + P_g$, where P_r is the production of the population channeled into reproduction, and P_g is the production going into growth. The amount of production due to reproduction is

$$P_r = v_r W_r$$

where W_r is the average weight of a newborn individual, and v_r is the number of newborns added to the population at time T, the period over which production is being measured. The value of v_r is estimated by the methods and equations of Chaps. 1 and 11.

The relative importance of P_r to P_g varies from species to species. Phillipson (1967) found that in the wood louse *Oniscus asellus* the yearly production of a single individual under natural conditions could be divided: (1) $P_r = 8.477$ calories per gram live weight per year, and (2) $P_g = 2.718$ calories per gram live weight per year. The production due to reproduction accounted for 76 percent of the total production per individual per year. In some very rapidly reproducing species the relative importance of P_r may be so great that productivity may be due almost entirely to reproduction. In other animals, such as large mammals, production may be due almost entirely to growth.

The methods of estimating production due to growth depend upon the situation. Production is most easily measured in a group of animals all born at the same time, a cohort. Barring this fortuitous occurrence P_g can be calculated from the growth rate of an individual, or from the length of the stages in the life history of an animal with distinguishable life history stages.

16-1.6 Estimation of P_g Based on a Cohort

Both a survivorship and a growth curve can be drawn by following the mortality and growth of individuals in a cohort. The growth curve represents the average weight of an individual at time t, and may be computed by fitting either the Bertalanffy growth equation or a polynomial regression to the raw data. The production of the cohort due to growth during the period t_0 to t_1 (period T_1) is equal to

$$P_g(T_1) = (N_{t_0} - N_{t_1})\left(\frac{\Delta W_t}{2}\right) + N_{t_1} W$$

Because of the curvilinearity of the growth curve the quantity $\Delta W_t/2$ is not equal to $\Delta W/2$ (Fig. 16-12). Having calculated $P_g(T)$ for every time interval from the formation of the cohort to the death of the last individual, the total production of the cohort due to growth is simply the sum of the $P_g(T_i)$. Petrusewicz and Macfayden (1970) demonstrate other methods of estimating production in a cohort.

16-1.7 Estimation of Production Based on the Growth Rate

If the population does not consist of a single cohort, the procedure is to break it down into small enough age groups so that the effects of the curvilinearity of the growth curve are minimal. The relationship between time and weight within each age group is assumed to be linear. Provided the age groups are small enough, the approximation is usually satisfactory. The weight of an individual in each of the age groups is measured twice, and the weight at each measurement indicated as w_t and w_0, where t is the number of time units, say days, separating the weighings. Within the age group the

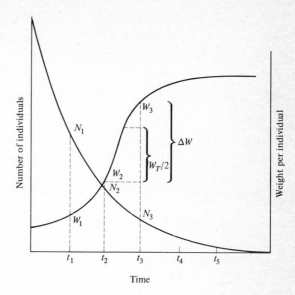

FIGURE 16-12

The estimation of production due to growth in a cohort of individuals. The exponential curve is the survivorship curve, and the sigmoid curve is the growth curve. The parameters and method of estimation are discussed in the text.

weight gain per time period $(t = 1)$ per individual of the sth age group is $v_s = (w_t - w_0)/t$. The total production of the sth age group due to growth is

$$P_g(s) = v_s N_s T \qquad (16\text{-}2)$$

where N_s is the number of individuals in the age group, and T is the number of time periods covered by the sth age group. Equation (16-2) is valid only if no deaths occur during period T. If deaths do occur, as they surely must, some approximation to N_s will be needed. The mean number of individuals in the age group during time T might serve as an approximation, although this assumes a linear relationship between age and death. If the mean number of individuals in the age group is $\bar{N}_s$, Eq. (16-2) becomes $P_g(s) = v_s \bar{N}_s T$. The total production to growth in period T is the sum of the quantities $P_g(s)$.

16-1.8 Estimation of Production Based on the Longevity of a Series of Age Groups

In insects and many marine animals the population of individuals of a species can be divided into easily identifiable stages such as egg, larvae, pupae, and adult. Each stage has a physiological time of development, the duration of the stage, symbolized as t_s, where s is the sth stage. The mean number of individuals in the sth stage is $\bar{N}_s$. The number of individuals v_s of one stage passing into the next during time period T is

$$v_s = \bar{N}_s \frac{T}{t_s}$$

Consequently, by measuring the weight gain of an individual during its sojourn in stage s as its weight at the end of the sth stage minus the weight at the end of the previous stage, and multiplying the weight gain by the number of individuals passing out of the sth stage during time period T, the production of the sth stage during period T is approximately

$$P_g(s) = v_s(W_s - W_{s-1})$$

The total production of the population due to growth is the sum of the productions of each stage.

16-2 PRODUCTION IN PLANTS

The fixation of light energy by plants forms the basis of the energy transfers in a community. This fixation is governed by the well-known chemical equation

$$6CO_2 + 6H_2O \xrightarrow{\text{light}} C_6H_{12}O_6 + 6O_2$$

Part of the carbohydrates produced by photosynthesis will be utilized by the plants in their metabolism, and the energy lost as heat. The general sequence of energy transfers in plant populations is

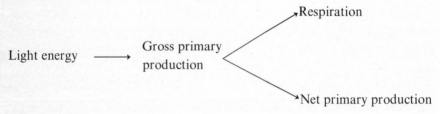

Usually only a small percentage of the incident sunlight is utilized and fixed as gross primary production, and in some plant species net primary production is only a small proportion of gross primary production. Two primary methods have been developed to estimate production in plant populations: (1) gas exchange, and (2) harvest methods. Gas exchange techniques can be used to estimate gross primary production, respiration, and net primary production. Harvest methods are limited to estimating net primary production.

In the gas exchange methods parts of the plant are enclosed in some sort of container, and air of known carbon dioxide content is pumped over them. Two containers are used, a dark container and a transparent container. The carbon dioxide content of the air is measured before entering each container and upon leaving the container. In the dark container carbon dioxide is added to the air because of respiration, while in the transparent container carbon dioxide is fixed by photosynthesis and liberated by respiration. The carbon dioxide losses or gains over a period of time can be converted into calories by utilizing the appropriate conversion tables in Brody (1945). The carbon dioxide gain per time period in the dark container represents the energy of gross primary production lost as respiration, and the carbon dioxide gain or loss in the transparent container is an estimate of net primary production. The sum of the two quantities is an estimate of gross primary production.

The gas exchange method is plagued by a variety of technical problems. Photosynthesis and respiration are dependent on the age of the plant, the part of the plant, temperature, humidity, and a variety of other factors. The techniques of gas exchange estimation are discussed in Newbould (1967). In general harvest methods have been used much more frequently in field studies of net primary production. They are simpler and perhaps more reliable. Sometimes harvest and gas exchange methods are combined, by estimating respiration with a dark container and net primary production by harvesting.

16-2.1 Estimation of Production in Herbaceous Plants by the Harvest Method

The most common procedure in estimating primary production in a grass or forb species is to divide the study area into quadrats and to take sample quadrats at random. Sets of samples are taken at different times, most commonly at the beginning and the end of the growing season. The total biomass for a given area at both samplings is estimated, above and below ground. The increase in biomass from one sample to the next is a rough indication of the net primary production of the plant population during the year. For annuals it may be a good estimate, but for perennials it probably will not be as good. In perennials it is usually necessary to separate new growth from material 1 year old and older. If loss of plant material to herbivores is significant, it should be taken into account. As in an animal population production will be equal to the change in biomass plus losses to herbivores and decay of older vegetation.

Kucera et al. (1967) estimated the annual production of big bluestem, *Andropogon gerardi*, and associated plants in a prairie preserve in Missouri. Above-ground biomass was estimated at the beginning of the growing season and at the end of the growing season by clipping all material above ground in a series of quadrats. All plant material in the first sample represented old growth, but in the second sample old and new plant materials were separated and the caloric content of each determined. The root system biomass was estimated at each sample period from core samples by using a method described in Dahlman and Kucera (1965). The results of their study are summarized in Table 16-3. The annual net primary production of the big bluestem population is 4.351×10^6 calories per square meter per year, or in weight equivalents 992 grams of dry biomass were produced in 1 year on 1 square meter of ground. The total biomass per square meter of the sample at the beginning of the growing season was 9.316×10^6 calories per square meter, compared with 11.324×10^6 calories per square meter at the end of the

Table 16-3 CALORIC EQUIVALENTS FOR BLUESTEM STANDING CROP, ROOT INCREMENT, OLD ROOTS, AND LITTER OF DIFFERENT AGES; GROSS BIOMASS ENERGY AND TOTAL NET ANNUAL ENERGY PRODUCTION. MEASURES ARE IN CALORIES PER SQUARE METER, FOR THE YEAR 1962

Plant material	Calories per gram	Gross energy First sample	Gross energy Second sample	Annual energy production (net)
Standing crop				
Foliage	4,071.7	...	1.962×10^6	1.962×10^6
Flower stalks	4,287.1			
Below ground				
Roots (new)	4,291.3	...	1.583×10^6	1.583×10^6
Rhizomes (new)	4,514.0	...	$.806 \times 10^6$	$.806 \times 10^6$
Below ground, old				
Roots	4,067.6	3.909×10^6	3.585×10^6	
Rhizomes	4,226.1	2.012×10^6	1.801×10^6	
Litter				
1 year old	4,007.6	1.632×10^6	$.840 \times 10^6$	
2–3 years old	3,871.2	1.419×10^6	$.824 \times 10^6$	
4 years old or more	3.032.4	$.344 \times 10^6$	$.150 \times 10^6$	
Total		9.316×10^6	11.324×10^6	4.351×10^6

SOURCE: From Kucera, C. L., Dahlman, R. C., and Koelling, M. R., Total Net Productivity and Turnover on an Energy Basis for a Tallgrass Prairie, *Ecology*, vol. 48, 1967.

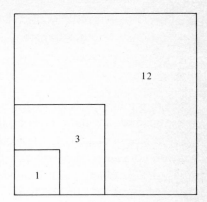

FIGURE 16-13
The arrangement of nested quadrats in Wiegert's method of determining optimum quadrat size for sampling vegetation to estimate biomass. The numbers refer to the relative areas of the three parts of a set. (*From Wiegert, 1962.*)

season. In taking into account net productivity, 2.008×10^6 calories per square meter were lost during the growing season, either through decay of litter or herbivore consumption of old vegetation. This is 54 percent of the net annual production.

In estimating productivity the size of the sample quadrats and the cost of taking each sample should be considered. This subject is discussed in Chap. 11. The mean square of a series of samples changes as the size of the sample quadrats changes, because of the patchy distribution of the plant species (see Chap. 5). The mean square is an estimate of the variance of the samples. The axioms of sampling suggest that the size and number of samples should be chosen to reduce the variance of the estimated biomass as much as possible, within the limitations of the money or time available.

The first step in the sampling program is to study the relationship between the size of the quadrat used and the variance of the estimate of the biomass obtained. Wiegert (1962) sampled in an old field in Michigan using nested quadrats of five different sizes (Fig. 16-13). Plant material was divided into four categories: grass, forbs, total green, and dead. Thirty sets of nested quadrats constituted the sample. The estimated mean biomass per .250 square meter and its variance V_m as estimated by each sample quadrat size are given in Table 16-4. A variance relative to the variance of the smallest quadrat, .016 square meter, is obtained by dividing the variance estimate for each size quadrat by the corresponding estimate for the size-1 quadrat. These relative variances will be indicated as V_r. The relative variance of the grass "species" in quadrat size 4 is $V_r = \frac{.318}{.968} = .329$. If cost is not a factor, the lowest variance of the estimated mean biomass per .250 square meter is obtained by using quadrat size 12, .187 square meter, with a relative variance of .143. If cost is a factor, one possible approach is to minimize the product $V_r C_r$, where C_r is the relative cost of taking a sample in a quadrat of size x relative to quadrat size 1. The relative cost is computed as

$$C_r = \frac{c_f + x c_v}{c_f + c_v}$$

where c_f is a fixed cost for each quadrat, consisting of time spent walking between quadrats, weighing material, and so forth, and c_r is the cost of sampling an area the size of quadrat 1. The quantity $x c_v$ represents the cost of sampling a quadrat of size x. The cost data for Wiegert's experiment are listed in Table 16-5. The product $V_r C_r$ is plotted against quadrat size (Fig. 16-14), and the lowest value of $V_r C_r$ on the graph is used as the "correct" sampling quadrat size to use in the estimation of biomass. For grass the quadrat size is .047 square meter, and for forbs .187 square meter.

Table 16-4 **ESTIMATED GRAMS OF BIOMASS OF FOUR CATEGORIES OF VEGETATION PER .250 SQUARE METER AND THE VARIANCE OF THE ESTIMATE AS DETERMINED WITH FIVE DIFFERENT-SIZED SAMPLE QUADRATS**

Quadrat size	Vegetation category	Variance of the estimate	Estimated grams per .250 m^2
1 (.016 m^2)	Grass	.968	4.68
	Forbs	2.079	5.99
	Total green	2.654	10.61
	Dead	21.535	45.79
3 (.047 m^2)	Grass	.243	3.82
	Forbs	1.895	4.42
	Total green	2.079	8.25
	Dead	20.707	41.09
4 (.063 m^2)	Grass	.318	3.92
	Forbs	1.743	4.82
	Total green	1.908	8.85
	Dead	16.738	36.04
12 (.187 m^2)	Grass	.139	3.60
	Forbs	.578	3.54
	Total green	.748	7.13
	Dead	11.473	37.70
16 (.250 m^2)	Grass	.150	3.72
	Forbs	.785	3.89
	Total green	.944	7.57
	Dead	11.654	38.32

SOURCE: From Wiegert, R. G., The Selection of an Optimum Quadrat Size for Sampling the Standing Crop of Grasses and Forbs, *Ecology*, vol. 43, 1962.

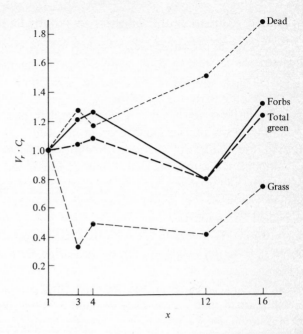

FIGURE 16-14

Graph of $V_r C_r$ against quadrat size to determine the lowest product and hence the quadrat size to use in sampling vegetation to estimate biomass. Areas of the quadrat sizes, in square meters, are: 1 (.016), 3 (0.47), 4 (.063), 12 (.188), and 16 (.250). (*From Wiegert, 1962.*)

Table 16-5   **COST DATA, IN MINUTES, FOR THE VARIOUS QUADRAT SIZES**

	Quadrat size, x				
	1	3	4	12	16
Fixed cost, c_f	10	10	10	10	10
Variable cost, $c_v x$	2	6	8	24	32
Total cost of quadrat	12	16	18	34	42
Relative cost, C_r	1.00	1.33	1.50	2.83	3.50

SOURCE: From Wiegert, R. G., The Selection of an Optimum Quadrat Size for Sampling the Standing Crop of Grasses and Forbs, *Ecolog*, vol. 43, 1962.

16-2.3   Estimation of Production in Woody Plants: Dimension Analysis

Woody plants present special difficulties in estimating productivity.   Although it may be feasible to chop down a random sample of trees or shrubs once or perhaps even twice in order to determine biomass, it is indeed an undertaking to do it repeatedly.

In dimension analysis a single sample of the woody plants is taken by cutting down the trees and digging up their root systems.   The total dry weight in grams of the tree or shrub is measured and separated into components due to roots, stem, branches, leaves, and so forth.   In addition, the diameter of the stem, annual width of wood increment, and circumference are also measured.   By taking a series of several samples, total dry weight can be related by regression equations to functions of the more easily measured statistics.   Therefore, in later samples these regression equations can be used to estimate the dry weight of a tree by measuring, for example, the diameter of the tree.

Dimension analysis has become a fairly complex subject, and is far more complicated than I have made it sound.   The reader, if interested, may refer to Whittaker and Woodwell (1968) for a discussion of the application of dimension analysis in estimating the productivity of an oak forest in New Jersey.   Whittaker and Woodwell used several types of measurements, and have compared the relative effectiveness of each.

16-2.4   Estimating Production in Phytoplankton

The reader is referred to Vollenweider (1965) for a discussion of the plethora of methods available for estimating respiration and net primary production in phytoplankton.

16-3   THE EXPLOITATION OF A POPULATION: PLAICE

One of the major problems facing ecologists is to develop methods of maximizing the yield of certain food sources.   The problem might also be looked at conversely as "minimizing the yield" of undesirable species, such as the spruce budworm or the European corn borer.   Resource management is the subject of a book by Watt (1968).   This section illustrates how one team of researchers modeled an exploited population, and how they applied to this model to plaice, *Pleuronectes platessa*, a flatfish, to determine the "best" fishing policies to use to increase the yield of plaice caught.   The purpose of the section is not so much to present general methods of modeling an exploited population, but rather to show how many of the ideas and methods discussed in earlier parts of this book can and have been applied to an economically important problem. Beverton and Holt's monograph (1957) on the modeling of North Atlantic exploited fish populations is one of the truly classic works in ecology. Their model is presented in its simplest form here, and the reader should refer to their monograph for more complete treatment and development. Although the models were developed primarily for demersal fish populations, they can probably be profitably applied to a terrestrial animal population with suitable modifications. For a further treatment of fish exploitation models the reader might begin by reading Paulik and Greenough (1966).

Men catching plaice are not particularly different from any other species of predator. To be successful a predator should exploit his prey in a way to maximize his yield, and at the same time avoid overexploitation, which would cause the prey species to decline or perhaps become extinct. Any action of the predator either directly or indirectly affects the density of the prey population, and hence the predator's yield. By modeling the prey population, the effect of changes in the predator's harvesting strategy may be determined by changing the relevant parameters of the model. The models presented below are the simplest situations simulated by Beverton and Holt, and apply to populations with constant birthrates and death rates. It is also assumed that the population is at equilibrium. More succinctly, the population density of the plaice is assumed to be fairly constant from year to year.

In a closed fish population the exploitable fish are subject to four influences: (1) recruitment of individuals to the exploited phase of the life cycle, (2) growth of individuals in the exploited phase of life, (3) capture of individuals in the exploited phase by fishing, and (4) natural death of individuals in the exploited phase. In terms of biomass the first two influences add to the exploitable biomass, and the last two decrease the available biomass.

The age of a fish at recruitment t_p is defined as the age at which an individual enters the fishing area. In general t_p is zero, but in some species, such as the plaice, the young occupy areas other than the adults and migrate to the adult grounds later, i.e., are recruits. In a deer population t_p is usually 0, and the number of recruits in a year is equal to the number of fawns born. The age at which individuals are first exploited is t_p'. These age relationships are illustrated in Fig. 16-15.

The recruits of a single year class of the plaice population become liable to encounters with fishing gear at age t_p. The gear is constructed so that fish can be captured only when they have reached the age t_p'. From age t_p to

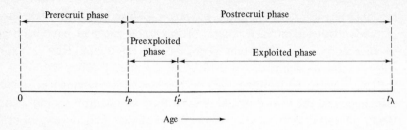

FIGURE 16-15
A figure to illustrate the age groupings used in the text discussion of the exploitation model of a plaice population.

t_p' the number of individuals in the year class is decreased by natural mortality only at a constant proportional rate M, or

$$\frac{dN}{dt} = -MN$$

where N is the density of the year class. The number of fish in the age class decreases exponentially. Some of the following steps in the derivation require calculus. The steps are presented to show how the model was derived. The methodology should be clear even if the mechanics are not. Solving the differential equation for N_t, the number of fish in the age class at time t

$$N_t = (\text{constant})e^{-Mt}$$

If $t = t_p$, N_t by definition is equal to the number of recruits R, so the constant, by the rules of solving differential equations, is equal to the number of recruits R.

$$N_t = Re^{-M(t-t_p)}$$

The number of fish remaining at age t_p' is

$$N_{t_p'} = R' = Re^{-M\rho}$$

where $\rho = t_p' - t_p$.

After age t_p' fish will be subject to both natural mortality and fishing mortality

$$\frac{dN}{dt} = -(F + M)N$$

where F is the mortality rate due to fishing in the age group $t_p' \leq t \leq t_\lambda$, where t_λ is the maximum age reached by the exploited fish. Solving this equation the number of fish in the age class remaining between ages t_p' and t is

$$N_t = R'e^{-(F+M)(t-t_p')}$$

The weight of an individual at any age between t_p' and t_λ is found from the Bertalanffy growth equation, and is

$$w_t = W_\infty \sum_{n=0}^{3} a_n e^{-nK(t-t_0)}$$

where $a_0 = +1$, $a_1 = -3$, $a_2 = +3$, and $a_3 = -1$. The parameters W_∞ and K were discussed in Sec. 16-1. The total weight of the year class at age t is

$$N_t w_t = R'W_\infty e^{-(F+M)(t-t_p')} \sum_{n=0}^{3} a_n e^{-nK(t-t_0)}$$

The rate at which fish of the age group are caught, the rate of yield, is the same as the rate of decrease due to fishing, except that the sign is positive. Denoting the yield in weight as Y_w, the rate of yield in weight from the year class is

$$\frac{dY_w}{dt} = FN_t w_t$$

or substituting for N_t and w_t

$$\frac{dY_w}{dt} = FR'W_\infty e^{-(F+M)(t-t_p')} \sum_{n=0}^{3} a_n e^{-nK(t-t_0)}$$

By another series of steps (see Beverton and Holt, 1957) the total yield obtained from a year class during its entire catchable life (i.e., from t_p' to t_λ) is found to be

$$Y_w = FRW_\infty e^{-M\rho} \sum_{n=0}^{3} \left[\frac{a_n e^{-nK(t_p'-t_0)}}{F+M+nK} \right] \left[1 - e^{-(F+M+nK)\lambda} \right]$$

where $\lambda = t_\lambda - t_p' =$ the catchable life-span. Because the population is in equilibrium, the total yield from the entire fish population is equal to the yield throughout the catchable life-span Y_w of one of the age classes. Other models of the mean length, weight, population biomass, and mean age were also derived.

The parameters of the model were determined by various methods discussed by Beverton and Holt. For the North Atlantic plaice population these parameters were found to be: $M = .10$, $t_p = 3.72$ years, $t_p' = 3.72$ years,

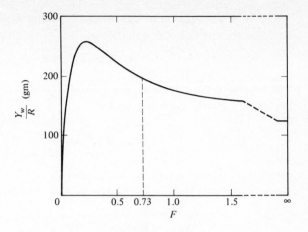

FIGURE 16-16
Yield per recruit in plaice as a function of the fishing rate. The vertical broken
line at $F = .73$ indicates the prewar fishing intensity. With t_p' equal to 3.72 the
peak of the graph indicates a maximum yield can be attained by decreasing the
fishing rate to about .20. (*From Beverton and Holt, 1957.*)

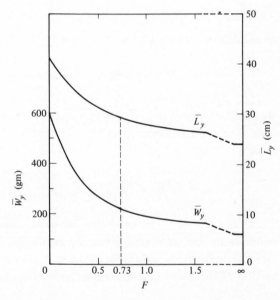

FIGURE 16-17
Mean weight $\overline{W}_y$ and mean length $\overline{L}_y$ as a function of the fishing rate F. The
exploitable age is $t_p = 3.72$ years. Both mean length and mean weight can be
increased by decreasing the fishing rate. (*From Beverton and Holt, 1957.*)

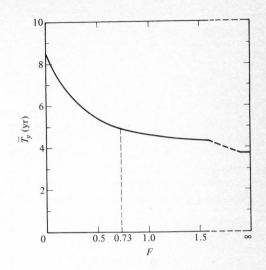

FIGURE 16-18
Mean age of plaice as a function of the fishing rate. The exploitable age is
$t_p' = 3.72$ years. The mean age increases as F decreases. (*From Beverton and Holt, 1957.*)

$t_\lambda = 15$ years, $t_0 = -.815$ years, $W_\infty = 2,867$ grams, $K = .095$, $t_p' - t_0 = 4.535$ years, $t_p' - t_p = 0$, and $t_\lambda - t_p' = 11.28$ years. The effect of changing the exploitation pattern, either by changing the rate of fishing F or by changing the age at which fish are first liable to be caught t_p', can be simulated by varying these two parameters in the models and noting the effects of the variation on yield and other catch characteristics.

The effect of changing the fishing rate F on the yield per recruit is shown in Fig. 16-16. The vertical broken line at $F = .73$ indicates the fishing intensity prior to World War II. Note that the yield per recruit would be increased if the fishing rate were decreased. Figure 16-17 illustrates the effect of fishing mortality on the mean length and weight of the fish caught, and Fig. 16-18 on the mean age of the fish caught. Figure 16-19 indicates the effect of changing the age at which fish are first exploited on the yield per recruit. For plaice this is done by changing the mesh size of the nets. The yield per recruit can be increased by exploiting fish older than those exploited during the prewar years. The fishing rate and age of exploitation can both be varied, as in Fig. 16-20. Points of equal yield are connected by lines, as in a topographic map. These diagrams are called *isopleth diagrams*. The

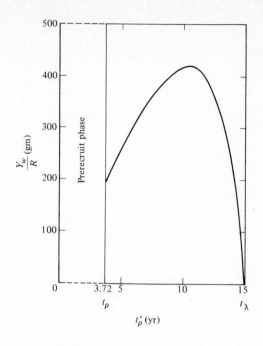

FIGURE 16-19
Yield per recruit of plaice as a function of the age at which the fish are first exploited. The fishing rate is $F = .73$. The yield per recruit can be substantially increased by increasing the age of first exploitation considerably beyond the prewar value of 3.72 years. (*From Beverton and Holt, 1957.*)

point P indicates the prewar values of F and t_p'. Clearly the yield per recruit can be greatly increased by increasing F and t_p', so that P will fall in the upper right-hand corner of the diagram.

Beverton and Holt (1957) went on to modify their models to take into consideration the effects of density, competition, spatial dispersion, and other factors. Although every exploited population is unique, it is commonly observed that the maximum yield from an exploited population is achieved by harvesting the older individuals, as was true of the plaice population.

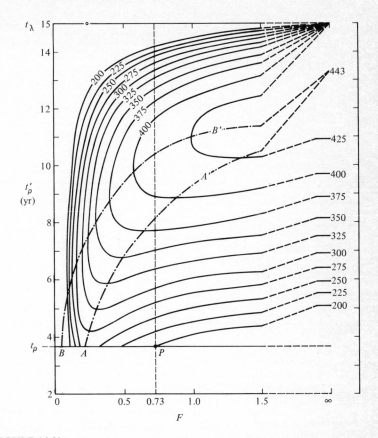

FIGURE 16-20

Yield per recruit of plaice as a function of both fishing intensity and age of first exploitation. Points of equal yield are connected by lines or isopleths. The point P indicates the prewar values of F and t_p'. The line AA' joins the maxima of yield-mortality curves and the line BB' joins the maxima of yield-mesh curves. The maximum yield per recruit can be achieved by exploiting old fish at a very high rate, as in the upper right-hand corner of the diagram. (*From Beverton and Holt, 1957.*)

17

ENERGY FLOW AND NUTRIENT CYCLING THROUGH AN ECOSYSTEM

An ecosystem consists of a network of functional relationships between a community of plants and animals and the physical environment, and within the community itself. The ecosystem also includes the cycles of nutrients and directional flow of energy through the community and physical environment. Part Four of this book dealt with the numerical relationships between the species of the ecosystem. The primary purpose of this chapter is to describe the flow of energy and matter through the various links of the ecosystem.

17-1 ENERGY FLOW THROUGH AN ECOSYSTEM

The flow of energy through an ecosystem can be represented in its simplest form by the diagram in Fig. 17-1. Energy enters the ecosystem as sunlight. The sunlight may strike vegetation, or be lost directly as heat. From the sunlight striking the plants, i.e., the energy entering the primary producers box in the diagram, part will be used in photosynthesis and part will be lost as heat. Of the total energy entering the ecosystem a vast proportion, usually over 90 percent, is lost directly or indirectly as heat. Part of the energy incorporated in

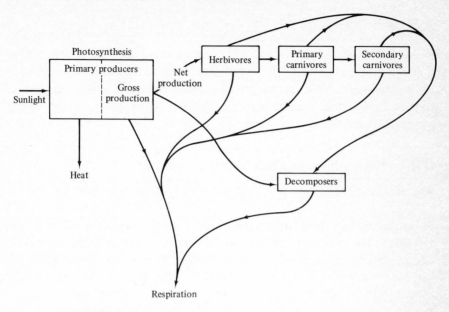

FIGURE 17-1
A diagrammatic representation of the flow of energy through an ecosystem divided
into a series of components.

organic compounds is used in the metabolism of plants, and is shunted off into
the respiration funnel of the diagram.   Some of the net primary production is
eaten by the herbivores and some enters the decomposer box.   A proportion
of the energy entering the herbivore component is lost to respiration, part to the
decomposers, and some passes to the primary carnivores.   The same is true of
the primary carnivores.   Because the secondary carnivores are the last
component in Fig. 17-1 to which enough energy flows to maintain the
populations, energy from this level flows either to the decomposers or is lost
as respiration.   Each of the boxes in the diagram represents a trophic level,
and a transfer of energy from one trophic level to another is associated with an
efficiency of transfer.

Figure 17-1 is, of course, unrealistic, because the energy transfers in
a community take the form of a web among species rather than a nice, neat
series of transfers between distinct trophic levels.   This complexity can be
dealt with either by assigning indeterminate species on a proportional basis
to each of the trophic levels, or by breaking up the trophic levels into single
species populations or guilds of species functionally related in their energy in-
takes.   Most often the latter course is preferable, because it avoids the necessity

of assigning species to subjectively created trophic levels. In practice it is impossible to measure energy flow through all the species of a community. Most studies, by necessity, are confined to those few plant and animal populations responsible for the largest proportion of the energy transferred in the community. Although most communities consist of a large number of species, a very few species may account for the majority of the energy moving through the system.

17-1.1 Energy Flow in a Salt Marsh

The first step in creating a graphic model of energy flow through an ecosystem is to determine the primary members of the food web and the food web interactions. Teal (1962) studied the flow of energy in a salt marsh in Georgia. The plant and animal life of such a marsh is relatively simple. The predominant species of plant is cord grass, *Spartina* sp. In deeper water algae are also important primary producers. The food web for the marsh is shown in Fig. 17-2. Cord grass is fed upon by several herbivorous insects, but the most important species are the plant hopper *Prokelisia marginata* and the grasshopper *Orchelimum fidicinium*. Much of the cord grass dies and is decomposed by bacteria. The bacteria digest the cellulose, and leave behind detritus rich in protein. This detritus, along with algal detritus, is utilized by a group of bottom feeders including several species of crabs, annelids, and nematodes. Both the herbivorous insects and the detritus feeders suffer from some predators.

The energy entering the marsh ecosystem was estimated by Teal to be 600,000 kilocalories per square meter per year. The cord grass utilized 34,580 kilocalories per square meter per year of this energy as gross primary production (Table 17-1), and the algae contributed another 1,800 kilocalories per square meter per year. A total of 36,380 kilocalories per square meter per year was utilized by the primary producers in gross primary production, about 6.1 percent of the calories entering the ecosystem as light. Of the 34,580 kilocalories per square meter per year gross primary production of the cord grass 6,585 kilocalories per square meter per year became net primary production, about 23 percent of gross primary production. The remaining 28,000 kilocalories per square meter per year were lost to respiration. Net primary production of the algae was 1,620 kilocalories per square meter per year. Of the net primary production of the cord grass, 305 kilocalories per square meter per year were eaten by the herbivores. The grasshoppers respired 18.6 kilocalories per square meter per year and produced 10.8 kilocalories per square meter per year, an assimilation rate of 29.4 kilocalories per square

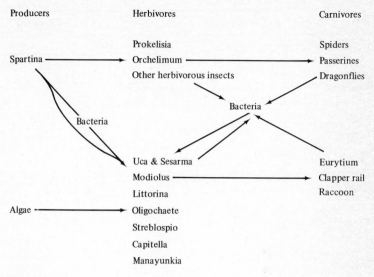

FIGURE 17-2
Food web of a Georgia salt marsh, with groups listed in their approximate order of importance. (*From Teal, 1962.*)

Table 17-1 ENERGETICS DATA FOR THE CORD GRASS OF A GEORGIA SALT MARSH

Season		Short spartina, 42% total area	Levee spartina, 58% total area
Winter, 2 months at 10°C	Standing crop	300 fresh g/m²	750 g/m²
	Respiration	235 kcal/m²	580 kcal/m²
Spring, 3 months at 17.5°C	Standing crop	600 g/m²	1,350 g/m²
	Respiration	1,250 kcal/m²	2,800 kcal/m²
Summer, 4 months at 26°C	Standing crop	705 g/m²	3,225 g/m²
	Respiration	6,450 kcal/m²	29,600 kcal/m²
Autumn, 3 months at 20°C	Standing crop	900 g/m²	1,800 g/m²
	Respiration	3,240 kcal/m²	6,480 kcal/m²
Production		2,570 kcal/m²/yr	8,970 kcal/m²/yr

Marsh average: net production = 6,580 kcal/m²/yr; respiration = 28,000 kcal/m²/yr; gross production = 34,580 kcal/m²/yr.

SOURCE: From Teal, J. M.: Energy Flow in the Salt Marsh Ecosystem of Georgia, *Ecology*, vol. 43, 1962.

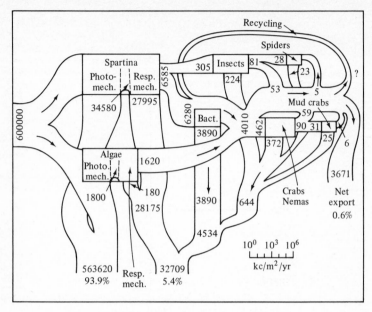

FIGURE 17-3
Energy flow diagram for a Georgia salt marsh. (*From Teal, 1962.*)

meter per year. Of the ingested food 70 kilocalories per square meter per year were egested to enter the decomposer cycle. The plant hoppers respired 205 kilocalories per square meter per year and produced 70 kilocalories per square meter per year. The net production efficiencies of the grasshoppers and plant hoppers were 37 and 25 percent, respectively. The two species of herbivores together produced 81 kilocalories per square meter per year, a gross production efficiency of $\frac{81}{305} = 27.6$ percent. Fifty-three percent of these 81 kilocalories per square meter per year entered the decomposer cycle, and 28 kilocalories per square meter per year were eaten by predators, mostly spiders.

The remaining cord grass net primary production died and settled to the bottom, where it was degraded by bacteria. Of the 6,585 kilocalories per square meter per year net primary production of the cord grass, 3,890 kilocalories per square meter per year were respired by bacteria, the remainder entering the herbivore and detritus trophic levels. Approximately 1,620 kilocalories per square meter per year were contributed to the detritus feeders by the algae. Of the total of 5,510 kilocalories per square meter per year

Table 17-2 SUMMARY OF SALT-MARSH ENERGETICS

Input as light	600,000 kcal/m^2/year
Loss in photosynthesis	563,620 kcal/m^2/year; 93.9%
Gross production	36,380; 6.1% incident light
Producer respiration	28,175; 77% gross production
Net production	8,205 kcal/m^2/year
Bacterial respiration	3,890 kcal/m^2/year; 47% net production
Other consumers respiration	644 kcal/m^2/year
Total energy dissipation by consumers	4,534 kcal/m^2/year; 55% net production
Export	3,671 kcal/m^2/year; 45% net production

SOURCE: From Teal, J. M.: Energy Flow in the Salt Marsh Ecosystem of Georgia, *Ecology*, vol. 43, 1962.

of detritus, 3,671 kilocalories per square meter per year were lost to the ecosystem by the washing of the detritus out of the marsh by wave action.

The total skein of energy transfers is illustrated in Fig. 17-3, and summarized in Table 17-2. Only 1.37 percent of the total incident sunlight becomes net primary production, and 45 percent of the net production is lost to the surrounding water by tide action. The bacteria consume six times more of the net primary production than all the remaining animals.

17-1.2 The Decomposers

In the salt-marsh ecosystem 96 percent of the net primary production of the cord grass enters the decomposer cycle or is lost by the movement of dead plant material out of the ecosystem by wave action. Commonly, if not usually, a vast proportion of net primary production does not pass to the herbivores but to the decomposers. The discrepancy between the herbivore and carnivore levels and the decomposers may not be as great in other ecosystems, but herbivores rarely consume all or even the majority of the net primary production produced by the plants. The substantial fall of leaves in a forest every year is strong evidence of the vast amounts ,of energy entering decomposer cycles. In addition, the feces and dead bodies of the herbivores and carnivores eventually wind up in the decomposer cycle.

In a forest litter falls to the forest floor and accumulates. , The litter is broken down into minerals and simple organic compounds, and the energy is lost in respiration. Rarely does litter continue to accumulate, as it must have once in the Carboniferous coal forests. If litter does not accumulate, an equilibrium of some sort has been reached where the amount of energy entering the decomposer cycle is matched by the energy lost in respiration.

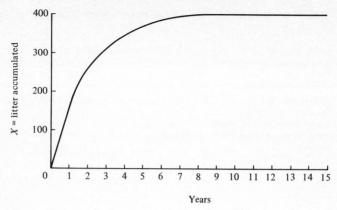

FIGURE 17-4

Litter accumulation as a function of time beginning with no litter at all. The parameters of the model used to estimate the litter accumulation were $L = 200$ grams per square meter per year and $k = .5$.

The change in litter ΔX during some interval of time Δt can be expressed by the simple difference

$$\frac{\Delta X}{\Delta t} = \text{income for the interval} - \text{loss for the interval}$$

(Olson, 1963). As the interval of time becomes infinitely small, the difference equation becomes the differential equation

$$\frac{dX}{dt} = L - kX \tag{17-1}$$

where L is the steady income of litter, X is the amount of litter present, and k is a constant instantaneous rate of litter decay. If litter is continuously added at a constant rate throughout the year and decays at a constant rate, Eq. (17-1) has the solution

$$X_t = \left(\frac{L}{k}\right)(1 - e^{-kt})$$

If we start with no litter at all, the function $1 - e^{-kt}$ approaches 1 as time increases, and an equilibrium amount of litter equal to L/k is reached. Suppose that $L = 200$ grams of litter per square meter per year, and $k = .5$. After 2 years the accumulated litter will be $X_2 = \frac{200}{.5}(1 - e^{-1.0}) = 252.85$ grams of litter per square meter. Figure 17-4 illustrates the theoretical curve of litter accumulation if $k = .5$. As the value of k decreases (litter decomposes more slowly), litter accumulation reaches equilibrium more slowly and the equilibrium amount of litter increases. If $k = .5$, X equilibrium $= \frac{200}{.5} = 400$ grams per square meter. However, if $k = .1$, X equilibrium $= \frac{200}{.1} = 2000$ grams per square meter. Figure 17-5 illustrates the litter accumulation equilibrium and decomposition rates in a number of terrestrial ecosystems.

A simple experiment is used to determine k by artificially eliminating L. A given amount of litter is placed in a sack or some other type of enclosure, and the litter left after regular intervals of time is measured. Eliminating L from Eq. (17-1), $dX/dt = -kX$, a negative version of the familiar exponential curve of Chap. 1. Solving the equation for X_t

$$X_t = X_0 e^{-kt}$$

where X_0 is the initial amount of litter. The value of k can be estimated from a set of observed data points by transforming Eq. (17-2) into a linear regression equation; $\ln(X/X_0) = -kt$. The negative exponential curves of litter remaining at time t in the absence of any incoming litter are illustrated in Fig. 17-6 for $k = .2$ and $k = .5$.

In most temperate forests litter does not accumulate at a constant, continuous rate, but during a single burst in the fall. By assuming that L occurs as a single, instantaneous addition once a year, the above equations can be modified to take into account the periodicity of litter addition. If we designate J_n as the annual peak in litter after the addition of litter each year, the peak in the nth year is

$$J_n = \left(\frac{L}{k'}\right)(1 - e^{-kn})$$

where $k' = 1 - e^{-kt}$. Litter will decay at a negative exponential rate after this peak, declining to minimum value F_n just before the addition of litter during the following year.

$$F_n = (1 - k')J_n$$

If the curve of litter accumulation is calculated and plotted (Fig. 17-7), it will appear to be sawtoothed. With time both J_n and F_n approach equilibrium values; J_n equilibrium $= L/k'$, and F_n equilibrium $= (L/k') - L$.

The time required for litter accumulation to reach 50 percent of equilibrium is equal to approximately $.693/k$. Ninety-five percent of equilibrium is reached at approximately $3/k$. On some soils the rate of decomposition may be so slow that literally centuries pass before litter accumulation reaches equilibrium. Olson (1958) found that the decomposition

492

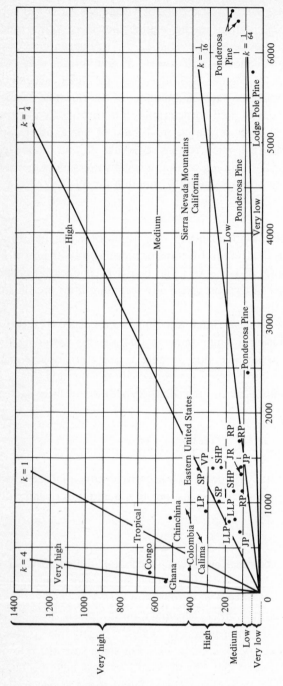

FIGURE 17-5

Estimation of the decomposition rate k for carbon in evergreen forests, from the ratio of annual litter production L to the steady-state accumulation of the forest floor. The data were compiled from various sources. *(From Olson, 1963.)*

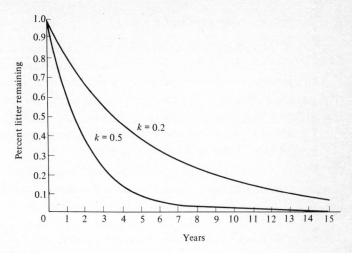

FIGURE 17-6 ·
Percent of litter remaining as a function of time for two different rates of decay:
$k = .2$ and $k = .5$. No new litter is being added.

rate of litter in sand dune soils near Lake Michigan was about .003.
Therefore, the time required to reach 95 percent of equilibrium litter
accumulation is on the order of 1,000 years. This slow decay rate appears
to be fairly characteristic of many ecosystems in temperate climates, and
explains, in part, why the development of a humus layer and a climax forest
is so slow during succession.

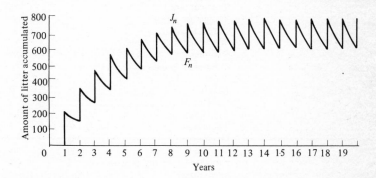

FIGURE 17-7
Litter accumulation as a function of time if litter addition occurs once a year and is
assumed to be instantaneous. J_n is the maximum litter during 1 year, and F_n is
the minimum amount of litter during 1 year.

17-1.3   The Primary Producers

The rate of net primary production differs from community type to community type.   Table 17-3, compiled by Whittaker and Likens from a variety of sources, is a compendium of average net primary productivities in various terrestrial and aquatic ecosystems.   Productivities range from low values of less than 200 grams dry weight per square meter per year in tundra and desert to 2,000 grams dry weight per square meter per year in marshes, swamps, and tropical forests.   Some intensively raised crops under ideal conditions may produce over 1,000 grams dry weight per square meter per year, and a few tropical crops have been known to produce more than 3,000 grams dry weight per square meter per year.   However, there is great variability in rates of net primary production in any given community, depending upon environmental factors such as temperature, rainfall, and various nutrients.

Table 17-3   NET PRIMARY PRODUCTIVITY FOR A VARIETY OF MAJOR ECOSYSTEM TYPES. DATA ASSEMBLED FROM A VARIETY OF SOURCES

Ecosystem	Area, 10^6 km^2	Net productivity per unit area as dry g/m^2/year	
		Range	Mean
Lake and stream	2	100–1500	500
Swamp and marsh	2	800–4000	2000
Tropical forest	20	1000–5000	2000
Temperate forest	18	600–2500	1300
Boreal forest	12	400–2000	800
Woodland and shrubland	7	200–1200	600
Savanna	15	200–2000	700
Temperate grassland	9	150–1500	500
Tundra and alpine	8	10–400	140
Desert scrub	18	10–250	70
Extreme desert, rock, and ice	24	0–10	3
Agricultural land	14	100–4000	650
Total land	149	...	730
Open ocean	332	2–400	125
Continental shelf	27	200–600	350
Attached algae and estuaries	2	500–4000	2000
Total ocean	361	...	155
Total earth	510	...	320

SOURCE: From Whittaker, R. H., "Communities and Ecosystems," Macmillan Company, New York, 1970.

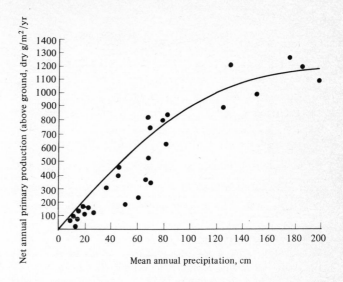

FIGURE 17-8
Relation of net annual primary production to rainfall.   Data from various sources;
peak productions of unstable communities are excluded.   (*From Whittaker, 1970.*)

Moisture appears to be the most important environmental variable in
terrestrial ecosystems.   The influence of mean annual rainfall on net primary
production is illustrated in Fig. 17-8.   Generally as moisture increases so does
net primary production.   With increasing moisture the rate of net primary
production plateaus, and beyond 140 centimeters of precipitation per year
net primary production depends on other factors.

Rates of net primary production generally decrease with decreasing
temperature, although the rates of decrease are different for coniferous and
deciduous trees.   Whittaker (1970) drew the following general conclusions
about the effects of temperature on net primary production from his studies
on productivity in the Great Smoky Mountains:

1   Net primary productions of stable forests of conifers and deciduous
trees, curves *A* and *B* of Fig. 17-9, under relatively favorable moisture
conditions are convergent over a wide range of warm-temperature to
cold-temperature climates.
2   There is a point beyond cool climates at which productivity decreases
more rapidly toward the low values of alpine and arctic climates.

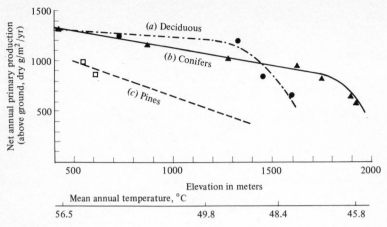

FIGURE 17-9
Relation of forest net production to the elevation gradient in the Great Smoky Mountains, Tennessee, for (a) broadleaf deciduous forests of moist environments, (b) evergreen coniferous forests of hemlock, spruce, and fir of moist environments, (c) pine forests of dry environments. (*From Whittaker, 1970.*)

3 Although net primary production of climax deciduous and evergreen forest is similar in warm-temperature climates, in the transition to the arctic and alpine forests the conifers have adaptive advantage in the form of higher productivity.

4 In pine forests of dry environments, curve *C* of Fig. 17-9, the inter-arctic and alpine forests, the conifers have adaptive advantage in the in net primary production with elevation.

There is a general tendency for net primary production to increase with succession. There are many exceptions to this rule, however. In abandoned fields net primary productivity peaks during the first year or so, and falls off afterward to a lower level. In the development of a forest the productivity gradually rises to a high point near or at the climax. It has also sometimes been observed that net primary production decreases from a maximum just before the climax is reached.

17-2 NUTRIENT CYCLING

The cycling of nutrients through an ecosystem is conceptually similar to the flow of energy through the same ecosystem, with one fundamental difference. Energy enters the ecosystem as light, and is rapidly lost as heat. In contrast,

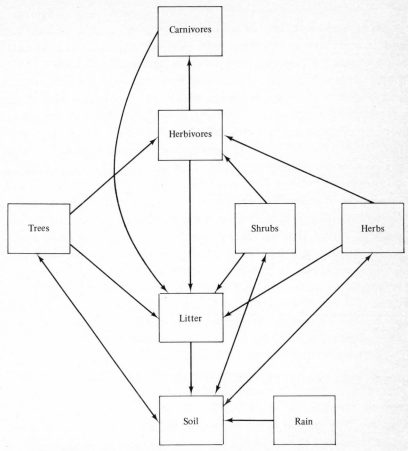

FIGURE 17-10
Diagrammatic illustration of the flow of a nutrient through the components of a simple ecosystem. The flow from one component to another may be one-way or two-way. Each arrow represents a rate.

nutrients are not lost or gained except to and from adjacent ecosystems. These nutrients are constantly cycled through the ecosystem, barring losses to water runoff, harvesting by man, and migration of animals or plant seeds and pollen. In the jargon of the systems analyst energy flow is an open system and nutrient cycling is a closed system.

As in energy flow studies an ecosystem can be broken down into a series of components or compartments. For example, a forest can be divided into the compartments in Fig. 17-10. Each arrow indicates the flow of nutrients from one compartment or component to another. The division of the ecosystem into components is purely subjective and a figment of the scientist's good judgment.   For example, herbivores receive nutrients from tree, shrub, and herb components, and lose nutrients to carnivore and litter components.   A flow may be two-way, as indicated by double-headed arrows in the figure.   Nutrients are picked up from the soil by the trees, but nutrients are lost from the tree component to the soil by the leaching of nutrients from the trees by rainwater.   Nutrients in rainfall represent an input of nutrients to the ecosystem, and water runoff represents a major loss of nutrients.

Each arrow in Fig. 17-10 represents a rate, such as grams of phosphorus per year, or micrograms of zinc per week, or some other convenient measure of the rate of nutrient flow from one compartment to another.   The pool or amount of nutrient in each of the compartments is estimated, as is the amount of nutrient moving from one compartment to another during some specified period of time, often a year.   The cycling of calcium in a British pine forest during a year based on data in Ovington (1962) and Whittaker (1970) is illustrated in Fig. 17-11.   Each rate in the diagram is expressed as kilograms per hectare per year.   The rates were estimated by direct measurement.   An excellent example of the study of the distribution and cycling of minerals in a temperate deciduous forest may be found in Duvigneaud and Denaeyer-De Smet (1970).

The amounts and rates of movements of different nutrients differ. Sometimes nutrients are differentiated as having a "tight" circulation or a "loose" circulation.   Circulation of a limited nutrient is almost always tight because the majority of the nutrient is tied up in the biomass of the ecosystem.   The fraction of the nutrient in the soil is subject to rapid turnover, and the amount present in the soil is greatly affected by the turnover rates in the plants.   In most cases calcium (as in Fig. 17-11) is in loose circulation, because it is abundant and not a limiting nutrient.   The limiting nutrient or nutrients may differ from area to area.   The addition of a limiting nutrient to the ecosystem often results in a great increase in productivity.   The addition of phosphorus to aquatic habitats where phosphorus is often limiting sometimes results in huge algal blooms.   It has been postulated that tropical forests have relatively tight nutrient cycling for many elements compared to temperate forests.   In a tropical forest the largest proportion of several critical elements may be contained in trees and herbs.   If the trees are cut and removed, the land rapidly loses its fertility.   Decomposition rates in the tropics appear to be much higher than in temperate forests, indicating a rapid turnover of nutrients in tropical forests.   For a more comprehensive review of the distribution of nutrients in several types of terrestrial ecosystems the reader may consult several of the articles in Reichle (1970).

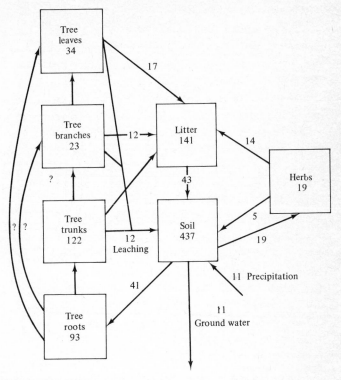

FIGURE 17-11
Cycling of calcium through a British pine forest.   Each arrow indicates a rate in kilograms per hectare per year.   The pool of calcium in each component per hectare is included in each box.   Unknown rates are indicated by a question mark.   (*Data from Ovington, 1962, and Whittaker, 1970.*)

17-2.1   An Experimental Study of Nutrient Cycling

Patten and Witkamp (1967) have studied the cycling of cesium-134 in an experimental ecosystem.   Their ecosystem was roughly equivalent to a decomposer cycle, and consisted of five compartments: litter, microflora, leachate, soil, and millipedes (Fig. 17-12).   The arrows in the figure illustrate all the possible pathways of cesium-134 transfer from one compartment to another. Each path is determined by a rate λ_{ij} expressed in the experiments as the percent of the radioactivity in the ith compartment transferred to the jth compartment per day.   These rates were assumed to be constant.   Several experiments of increasing complexity were conducted.   The following systems were established: (*1*) sterile white oak leaves, (*2*) leaves and microflora,

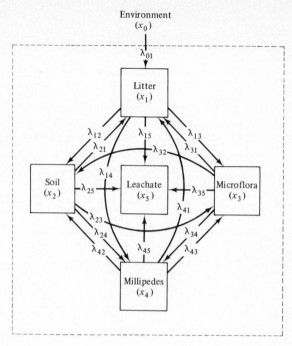

FIGURE 17-12
Diagram of the five compartments of Patten and Witkamp's cesium-134 nutrient cycling experiment. The arrows indicate all possible routes of cesium transfer and the associated rate constants λ_{ij}. (*From Patten and Witkamp, 1967.*)

(3) leaves, microflora, and millipedes, and (4 to 6) same as 1 to 3 but with soil. In all the experiments sterile water was dripped on the leaves as a leachate.

The amount of radioactivity introduced in the leaves and distributed among the five compartments will not change significantly with time because the half-life of cesium-134 is 2.07 years. The total activity in the ecosystem at any time t is a constant X. The amount of radioactivity in any of the five compartments at time t can be represented as $X_j(t)$. Because no radioactivity is lost during the experiment, $\sum X_j(t) = X$. If the mass of the jth compartment is $m_j(t)$, a measure of the concentration of the radioactivity at time t is

$$x_j(t) = \frac{X_j(t)}{m_j(t)}$$

The rate of change of radioisotope concentration in each of the compartments is a balance of incomes and losses, or

$$\frac{dx_j}{dt} = \sum_i \lambda_{ij} x_i - \sum_i \lambda_{ji} x_j \qquad i, j = 1, 2, 3, 4, 5 \qquad i \neq j$$

For a simple, two-compartment model $dx_1/dt = \lambda_{21}x_2 - \lambda_{12}x_1$ and $dx_2/dt = \lambda_{12}x_1 - \lambda_{21}x_2$. With five compartments there are five differential equations. In these experiments data were gathered in the form of the fraction of radioactivity in each compartment relative to the total system activity; i.e., $y_j = x_j/X$. In percentage terms

$$\frac{dy_j}{dt} = \sum_i \lambda_{ij} y_i - \sum_i \lambda_{ji} y_i$$

The data gathered are the values of $y_j(t)$. These experimental values of $y_j(t)$ for each of the experiments are represented by data points in Fig. 17-13. The parameters λ_{ij} are then estimated. Sometimes these rate parameters can be determined by direct measurement. Sometimes, as in this case, some of the rates are not easily estimated. Patten and Witkamp (1967) estimated these parameters by the simultaneous solution of a series of differential equations for each system on an analog computer. The reader should refer to their article for a discussion of the methods involved. The technique consists of fitting curves to the data points and choosing the set of solution of the $\hat{\lambda}_{ij}$ giving the most satisfactory fit of the curves to the data points. This is the source of the curves fitted to the data points in Fig. 17-13. The rate parameters for each of the six systems are summarized in Table 17-4.

The simplest system investigated was a two-compartment system of litter and leachate. The radioactive cesium was leached from the leaves by the water at a rate of 3.7 percent per day; i.e., $\lambda_{15} = .037$ per day. The experimental values of $y_j(t)$ and the fitted curves are shown in Fig. 17-13. In the second experiment a three-compartment system was used: litter, soil, and leachate. In this experiment the total rate of loss from the litter was the same as in the first experiment, but 3.6 percent went to the soil and .1 percent to the leachate. Cesium-134 once in the soil was not transferred back to the litter or the leachate. Cesium-134 is trapped by the soil and, therefore, the soil is referred to as a *sink*. In the third experiment the system consisted of three components: litter, microflora, and leachate. Again the rate of loss from the litter was 3.6 percent per day. However, when soil was added as a fourth compartment, the soil and microflora had a synergistic effect, increasing the total loss of radioactivity from the litter to 6 percent per day.

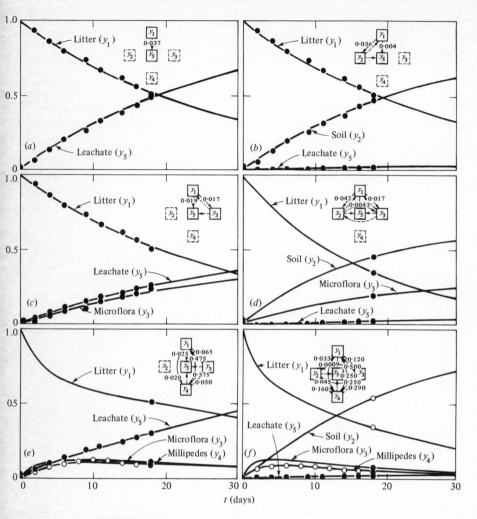

FIGURE 17-13
Estimated curves of changes in relative distribution y_j of radiocesium in the six experimental combinations of the five compartments. Means of three replicates represent the data points. The open circles represent estimates of relative cesium concentrations in millipedes based on radioassays of animals which died. Insets show compartment combinations and rate constants for each indicated transfer route. Pathways over which no cesium was transferred are represented by broken arrows.

In the total five-compartment system the loss of radioactivity from litter to other compartments was increased to 19.9 percent per day. Patten and Witkamp utilized the model of the five-compartment system to investigate by simulation the effect of adding radioactivity to the system at a constant rate. When radioactivity was added to the litter compartment at a constant rate, the amount of radioactivity in the litter, microflora, and millipedes compartments came to equilibrium in time. Because the soil and leachate act as sinks, the radioactivity of these two compartments increased with time, tending toward infinity as a limit. Patten and Witkamp were also able to estimate equilibrium concentrations, turnover, concentrations, and system stability, i.e., the ability of the ecosystem to react to a change in radio-activity input per day.

In this particular model the rates of transfer of nutrient from one compartment to another were assumed to be constant proportions of the

Table 17-4 **TRANSFER RATES FOR THE CESIUM TRANSFER EXPERIMENTS OF PATTEN AND WITKAMP**

Parameters	Combinations					
	I	II	III	IV	V	VI
λ_{12}		.036		.042		.033
λ_{13}			.017	.017	.065	.120
λ_{14}					.020	.045
λ_{15}	.037	.001	.019	.0013	.025	.0009
Sum	.037	.037	.036	.060	.110	.199
λ_{21}		0		0		0
λ_{23}				0		0
λ_{24}						0
λ_{25}		0		0		0
Sum		0		0		0
λ_{31}			0	0	0	.500
λ_{32}				0		0
λ_{34}					.375	.250
λ_{35}			0	0	0	0
Sum			0	0	.375	.750
λ_{41}					.475	.250
λ_{42}						.160
λ_{43}					.050	.250
λ_{45}					0	0
Sum					.525	.660

SOURCE: From Patten, B. C., and Witkamp, M., Systems Analysis of [134]Cesium Kinetics in Terrestrial Microcosms, *Ecology*, vol. 48, 1967.

radioactivity in each compartment.   However, in an aquatic system studied by Whittaker (1961) the rates were not constant, because of changes in the species compositions and changes in the physiology of the organisms.

17-3   ENERGY FLOW AND NUTRIENT CYCLING IN A GRASSLAND ECOSYSTEM

It seems only fitting to end this book by presenting an example of the measurement of energy and nutrient flow through a terrestrial ecosystem in order to tie together some of the loose ends of Chaps. 16 and 17.   Van Hook (1971) studied a grassland ecosystem in Tennessee.   The grassland was dominated by two species of grass, *Festuca arundinacea* and *Andropogon virginicus*, and a few species of forbs.   The herbivores of the field were represented by *Melanoplus sanguinipes* (a grasshopper), *Conocephalus fasciates* (a grasshopper), and a catchall class Hemiptera-Homoptera.   *Pteronemobius fasciates*, a cricket, represented an omnivore component, and a wolf spider, *Lycosa punctulata*, was the primary carnivore.

Plant biomass, both live and dead, was estimated by sampling the vegetation at 6-week intervals for 1 year on .25-meter-square quadrats at six randomly chosen locations.   The insects and spiders were sampled by quadrat cages at weekly intervals from April to December.

17-3.1   Energy Flow

This grassland ecosystem was divided into a series of 12 components. Figure 17-14 illustrates these 12 components and all the possible routes of energy transfer.   Each arrow represents a rate.   The problem is to determine the energy in each component and the rates of transfer from one component to another.

The total biomass of the grass material in the grassland over a period of a year, and the estimated biomass of the five arthropod components for a year, were determined.   For each arthropod species an energy budget was constructed.   Ingestion was estimated by the radioactive tracer method discussed in Chap. 16 by allowing the herbivores to feed on radioactively tagged plants.   The energy flux from a plant component to an herbivore component is complicated, because of the food preferences of the herbivores. Van Hook calculated the energy flux per day from component i to component j, EF_{ij}, as

$$C_j \lambda_{ji} = EF_{ij} = \frac{(x_i)(w_{ij})(p_j)(x_j)(CE_i)}{(\sum x_i)(w_{ij})} \qquad (17\text{-}3)$$

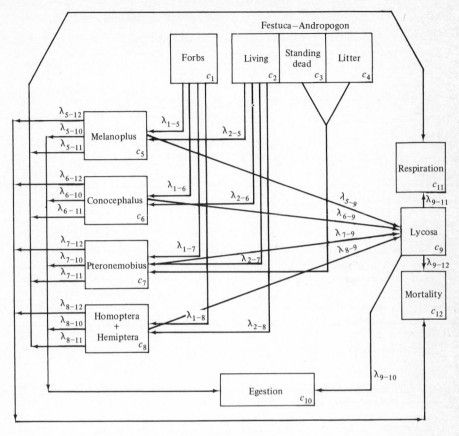

FIGURE 17-14
A model of the grassland arthropod community showing the major compartments (C_i) and the pathways of energy flux (λ_{ij}).   The arrows indicate the direction of flow. (*From Van Hook, 1971.*)

where:

$x_i =$ the proportion of the total available food composed of component i

$w_{ij} =$ the feeding preference of the jth component for the ith component

$p_j =$ the quantity of food ingested by a consumer or predator expressed as percent of dry body weight per day

$x_j =$ the biomass dry weight of the jth compartment

$CE_i =$ the caloric equivalent of the food component consumed.

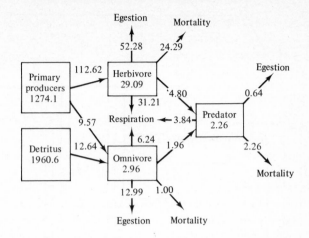

FIGURE 17-15
Annual energy budget of energy fluxes and net production in kilocalories per square meter by the arthropod and vegetation components of an eastern Tennessee grassland ecosystem. The values inside compartments represent net production, except for the detritus component. The detritus value equals standing crop of standing dead and litter. The values on the arrows represent annual energy fluxes. Arrows indicate the direction of flow. (*From Van Hook, 1971.*)

The change in energy in any of the compartments in Fig. 17-15 is equal to the total gains from other components minus the losses, or

$$\Delta C_i = \sum \lambda_{ij} C_i - \sum \lambda_{ji} C_j \qquad (17\text{-}4)$$

The transfer rates between producers and consumers and the predator were estimated with Eq. (17-3), and expressed in calories per meter square per day. The energy transferred to egestion, compartment 10, was estimated by multiplying the proportion of dry body weight egested per day by the biomass of the compartment in grams per meter square times the caloric equivalents in calories per gram. The rates of energy loss to respiration of the consumers and the predator were estimated from the equation

$$O_2 = aW^B$$

(see Sec. 16-1). Because temperature affects the rate of metabolism, the rate of respiration was corrected for temperature and expressed as

$$O_2 = aW'^B e^{.0693(T-20)}$$

where T is in degrees centigrade with 20°C as a reference point.

The mortality compartment includes energy losses to predatory mortality other than *Lycosa*, exuviae, and reproductive products. The size of the component was estimated as

$$C_{12} = \sum_{i=5}^{9} \lambda_{i,\,12}\, C_i$$

where $\lambda_{i,\,12}\, C_i$ equals the net flux to compartment 12 from compartment i. This flux was estimated to be

$$\lambda_{i,\,12}\, C_i = \sum_{i=1}^{3} \lambda_{ij}\, C_i - \left(\lambda_{i,\,9}\, C_9 + \sum_{j=10}^{11} \lambda_{ij}\, C_j \right)$$

where $\lambda_{ij}\, C_i$ equals the total input from compartments 1 through 3, $\lambda_{i,\,9}\, C_9$ equals the total loss to the predator, and $\lambda_{ij}\, C_j$ equals losses due to excretion and respiration. In other words, each $\lambda_{i,\,12}\, C_i$ represents the energy of the arthropod population not accounted for by predators, respiration, or excretion.

The model, Eq. (17-4), was evaluated weekly for 35 weeks. If conditions remain relatively constant during a week, the weekly transfer rates will be seven times the daily rates. The total 35-week growing season transfer rates are the sum of the 35 weekly rates. A necessary amount of fudging and extrapolation of data comes in at this point. The annual energy fluxes from compartment i to j in kilocalories per square meter are listed in Table 17-5.

Table 17-5 ANNUAL ENERGY FLUXES IN KILOCALORIES PER SQUARE METER THROUGH THE ARTHROPOD COMPARTMENTS OF A GRASSLAND ECOSYSTEM

Compartment i	Compartment j							
	C_5	C_6	C_7	C_8	C_9	C_{10}	C_{11}	C_{12}
C_1	7.83	1.56	1.76	.67	—	—	—	—
C_2	82.60	14.74	7.81	5.22	—	—	—	—
C_3	—	—	—	—	—	—	—	—
C_4	—	—	12.64	—	—	—	—	—
C_5	—	—	—	—	3.04	38.82	24.71	23.14
C_6	—	—	—	—	1.12	10.27	5.12	1.22
C_7	—	—	—	—	1.96	12.99	6.24	1.00
C_8	—	—	—	—	.64	3.19	1.38	−.07
C_9	—	—	—	—	—	.64	3.84	2.26
Total	90.43	16.30	22.21	5.89	6.76	65.91	41.29	27.69

— means transfer not existing or not pertinent.

SOURCE: From Van Hook, R. I., Energy and Nutrient Dynamics of Spider and Orthopteran Populations in a Grassland Ecosystem, *Ecol. Monogr.*, vol. 41, 1971.

Table 17-6 TROPHIC LEVEL COMPARISONS OF ENERGY FLOW AND NET PRODUCTION FOR THE ARTHROPOD AND VEGETATIONAL COMPONENTS OF A GRASSLAND ECOSYSTEM. THE SUBSCRIPTS n AND $n-1$ DENOTE THE RESPECTIVE TROPHIC LEVELS

Trophic level	A_n/I_n	R_n/I_n	P_n/I_n	NP_n/I_n	I_n/I_{n-1}	A_n/A_{n-1}	R_n/R_{n-1}	I_n/NP_{n-1}	NP_n/R_n
Herbivore	.54	.28	.04	.26	—	—	—	.09	.93
Omnivore	.41	.28	.09	.13	—	—	—	.01	.47
Predator	.90	.57	—	.33	.05	.09	.10	.21	.59

— means ratios not applicable.

Assimilation efficiency (A_n/I_n) = ratio of assimilated energy to ingested energy. Respiration efficiency (R_n/I_n) = ratio of energy expended in metabolism to ingested energy. Ecological efficiency (P_n/I_n) = ratio of energy passed on to the next higher trophic level of the ingested energy. Ecological growth efficiency (NP_n/I_n) = ratio of the total energy available to be passed on to the next higher trophic level to the ingested energy. Transfer efficiency (I_n/I_{n-1}) = ratio of energy ingested in trophic level n to that ingested in trophic level $n-1$. Progressive efficiency (A_n/A_{n-1}) = ratio of energy assimilated in trophic level n to that assimilated in trophic level $n-1$. Respiratory ratio (R_n/R_{n-1}) = ratio of energy used in metabolism in trophic level n to that used in trophic level $n-1$. Consumption efficiency (I_n/NP_{n-1}) = ratio of energy ingested by consumers to the net production by the preceding trophic level. Secondary production/respiration (NP_n/R_n) = ratio of energy accumulated to the energy expended in metabolism.

SOURCE: From Van Hook. R. I., Energy and Nutrient Dynamics of Spider and Orthopteran Populations in a Grassland Ecosystem, *Ecol. Monogr.*, vol. 41, 1971.

Utilizing the harvest method Van Hook estimated the net primary productivity of the grassland to be 1,274.1 kilocalories per square meter during 1968. The net secondary productivity of the herbivores was calculated for 1 week as

$$NP_{s,t} = \sum_{i=5}^{8} (\lambda_{i,9} C_9 + \lambda_{i,12} C_{12})$$

where the transfer rates are for 1 week, not 1 day as before. The total annual net secondary production is

$$NP_s = \sum_{t=1}^{35} NP_{s,t}$$

The annual energy budget for the grassland is shown in Fig. 17-15. The figures in the boxes are the annual net productions, and the numbers on the arrows are the annual transfer rates. A number of transfer efficiencies are listed in Table 17-6.

17-3.2 Nutrient Cycling

Van Hook (1971) studied the cycling of potassium, calcium, and sodium in the grassland ecosystem. The same fundamental compartments model was used, but the respiration compartment was removed. Rates of nutrient transfer for *Pteronemobius*, *Melanoplus*, *Conocephalus*, and *Lycosa* were determined in the laboratory at a range of temperatures with radioactive tracers (see Sec. 16-1). The rates of transfer per day from compartment i to compartment j, NE_{ij}, are

$$NE_{ij} = \frac{(x_i)(w_{ij})(p_j)(x_j)(Q_i)}{(\sum x_i)(w_{ij})}$$

where Q_i equals the concentration of the nutrient in the ith component. The rate of nutrient loss through egestion in milligrams of nutrient per square meter per day NE_{jk} was estimated as

$$NE_{jk} = (k'Q_j + p_1 Q_i I_j)x_j$$

where:

k' = the biological elimination rate of the element.

Q_j = the concentration of the element in the consumer.

I_j = the ingestion rate for the consumer in milligrams per milligram dry weight of animal per day.

Table 17-7 ANNUAL FLUXES OF POTASSIUM IN MILLIGRAMS PER SQUARE METER THROUGH THE ARTHROPOD COMPARTMENTS OF A GRASSLAND ECOSYSTEM

Compartment j

Compartment i	C_5	C_6	C_7	C_8	C_9	C_{10}	C_{11}	C_{12}
C_1	4.55	1.12	1.36	.85	—	—	—	—
C_2	47.76	13.41	7.11	6.42	—	—	—	—
C_3	—	—	—	—	—	—	—	—
C_4	—	—	10.79	—	—	—	—	—
C_5	—	—	—	—	.45	2.69	—	49.17
C_6	—	—	—	—	.08	.44	—	14.01
C_7	—	—	—	—	.47	5.36	—	13.43
C_8	—	—	—	—	.42	?	—	6.85
C_9	—	—	—	—	—	.55	—	.87
Total	52.31	14.53	19.26	7.27	1.42	9.04	—	84.33

— means transfers not existing or not pertinent.

SOURCE: From Van Hook, R. I., Energy and Nutrient Dynamics of Spider and Orthopteran Populations in a Grassland Ecosystem, *Ecol. Monogr.*, vol. 41, 1971.

Table 17-8 TROPHIC LEVEL COMPARISONS OF NUTRIENT FLUXES THROUGH ARTHROPOD AND VEGETATION COMPONENTS OF A GRASSLAND ECOSYSTEM. THE SUBSCRIPTS n AND $n-1$ DENOTE THE RESPECTIVE TROPHIC LEVELS

Trophic level	A_n/I_n	P_n/I_n	I_n/I_{n-1}	A_n/A_{n-1}	I_n/NP_{n-1}
		Sodium			
Herbivore	.91	.03	—	—	.08
Omnivore	.75	.14	—	—	.01
Predator	.65	—	.04	.03	.05
		Calcium			
Herbivore	.16	.01	—	—	.11
Omnivore	.46	.02	—	—	.01
Predator	.35	—	.01	.02	.04
		Potassium			
Herbivore	.95	.01	—	—	.02
Omnivore	.72	.02	—	—	.01
Predator	.61	—	.02	.01	.02

— means ratios not applicable.

SOURCE: From Van Hook, R. I., Energy and Nutrient Dynamics of Spider and Orthopteran Populations in a Grassland Ecosystem, *Ecol. Monogr.*, vol. 41, 1971.

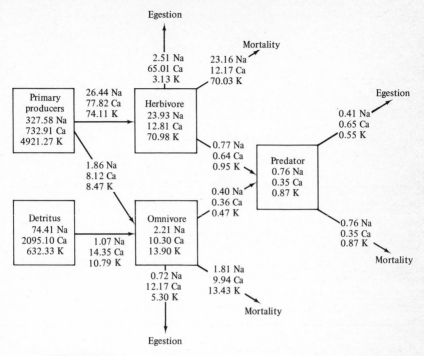

FIGURE 17-16
Annual nutrient budget in milligrams per square meter for sodium, calcium, and potassium showing the maximum standing crops of nutrients inside compartments and nutrient fluxes on the arrows through the arthropod and vegetation components of an eastern Tennessee grassland ecosystem. (*From Van Hook, 1971.*)

The pool of each nutrient in each compartment was estimated monthly. The monthly rates were extrapolated from the daily rates, and added to obtain an annual nutrient transfer. The annual rates for potassium are listed in Table 17-7. The annual nutrient budget is illustrated in Fig. 17-16. The values in the boxes are the maximum nutrient concentrations of the standing crop. Efficiencies are presented in Table 17-8. Assimilation rates of sodium and potassium are both high compared with the assimilation rates of calcium. This observation suggests that sodium and potassium are limiting nutrients, particularly to the herbivores, but that calcium is in abundant supply.

APPENDIX: MATHEMATICAL SYMBOLS

NOTATION

$a = b$	a is equal to b
$a \approx b$	a is approximately equal to b
$a \sim b$	a is asympotically equal to b
$a \equiv b$	a is defined to be b
e	2.7182818
$\ln$ and $\log_e$	logarithm to the base e; natural log
$\log_{10}$ or $\log$	logarithm to the base 10
e^x	an exponential; e raised to the xth power
$\exp[f(x)]$	$e^{f(x)}$
a^{-1}	the reciprocal of a, equal to $1/a$
a^{-2}	$1/a^2$
$a^{1/2}$	the square root of a
$a^{-1/2}$	$1/a^{1/2}$
$f(x)$	function of the variable x
$f(x, y)$	function of the variables x and y
ΔB	a discrete change in the variable B
$\Delta B/\Delta t$	a discrete change in the variable B given a change Δt in t

dx/dt	the instantaneous change in x given an instantaneous change in t
$\int_a^b f(x)$	integral of the function $f(x)$; equal to the area beneath the curve given by $f(x)$ between the points a and b
∞	infinity
π	3.14159265
$n!$	n factorial; $n(n-1)(n-2)(n-3) \ldots (2)(1)$
$a > b$	a is greater than b
$a < b$	a is less than b
$a \geq b$	a is greater than or equal to b
$|a|$	absolute value of a; e.g., $|-3| = |3| = 3$
$\sum a_i$	sum of the numbers a_i
$\sum \sum a_{ij}$	sum of a_{ij} over both columns and rows
a_i	the value of a in the ith row
a_j	the value of a in the jth column
a_{ij}	the value of a in the ith row and jth column

MATRIX NOTATION

$\mathbf{A}$	the matrix $\mathbf{A}$
$\mathbf{AB}$	postmultiplication of the matrix $\mathbf{A}$ by the matrix $\mathbf{B}$
$\mathbf{x}$	column vector $\mathbf{x}$
$\mathbf{x}'$	row vector $\mathbf{x}'$
$\mathbf{A}^{-1}$	the inverse of $\mathbf{A}$
$|\mathbf{A}|$	the determinant of $\mathbf{A}$
$\mathbf{I}$	the identity matrix

PROBABILITY AND STATISTICS

$\binom{n}{x}$	$\dfrac{n!}{x!(n-x)!}$
$E(X)$	the expected value of the variable X
$\text{var}(X)$	the variance of the variable X
$\Pr(x)$	the probability of the event x
$\Pr(x|y)$	conditional probability; probability of event x given event y has happened
$P_x(t)$	probability of event x at time t
$\Pr(1.7842 \leq \mu \leq 2.4833) = .95$	the probability of the true population parameter μ lying in the interval 1.7842 to 2.4833 is 95 percent; a confidence interval
μ	population mean

σ^2	population variance
$\bar{x}$	sample mean
s^2	sample variance
$SE(k)$	standard error of the estimate of the parameter k
$\hat{k}$	estimate of the parameter k
z	a standardized normal variate
t	a statistic to be compared with the t distribution
F	a statistic to be compared with the F distribution
χ^2	a statistic to be compared with the chi-square distribution
H_0:	null hypothesis
H_1:	alternative hypothesis
α	the probability of rejecting the null hypothesis when it is true

SELECTED GREEK LETTERS

α	alpha	ζ	zeta	ρ	rho
β	beta	θ	theta	σ	sigma
γ	gamma	λ	lambda	Σ	Capital sigma
Γ	Capital gamma	μ	mu	τ	tau
δ	delta	ξ	xi	ϕ	phi
Δ	Capital delta	π	pi	χ	chi
ε	epsilon	Π	Capital pi	ω	omega

LITERATURE CITED

ABRAMOWITZ, M., and SEGUN, I. A. (eds.): "Handbook of Mathematical Functions with Formulas, Graphs, and Mathematical Tables," Dover Publications, Inc., New York, 1965.

ANDERSON, T. W.: "Introduction to Multivariate Statistical Analysis," John Wiley and Sons, New York, 1958.

ANDREWARTHA, H. G., and BIRCH, L. C.: "The Distribution and Abundance of Animals," University of Chicago Press, Chicago, 1954.

ANSCOMBE, F. J.: Sampling Theory of the Negative Binomial and the Logarithmic Series Distributions, Biometrika, 37:358–382 (1950).

BAILEY, N. J. T.: On Estimating the Size of Mobile Populations from Recapture Data, Biometrika, 38:293–306 (1951).

BAILEY, N. J. T.: "The Mathematical Theory of Epidemics," Hafner Publishing Company, New York, 1957.

BAILEY, N. J. T.: "The Elements of Stochastic Processes with Applications to the Natural Sciences," John Wiley and Sons, New York, 1964.

BAILEY, N. J. T.: Stochastic Birth, Death, and Migration Processes for Spatially Distributed Populations, *Biometrika*, **55**:189–198 (1968).

BARNETT, V. D.: The Monte Carlo Solution of a Competing Species Problem, *Biometrics*, **18**:76–103 (1962).

BARTLETT, M. S.: Deterministic and Stochastic Models for Recurrent Epidemics, *Proc. Third Berkeley Symp. Math. Stat. Probab.*, **4**:81–109 (1956).

BARTLETT, M. S.: Measles Periodicity and Community Size, *J. R. Stat. Soc.*, A, **120**:48–70 (1957).

BARTLETT, M. S.: "Stochastic Population Models in Ecology and Epidemiology," Methuen and Company Ltd., London, 1960.

BERNADELLI, H.: Population Waves, *J. Burma Res. Soc.*, **31**:1–18 (1941).

BERRY, G.: A Mathematical Model Relating Plant Yield with Arrangement for Regularly Spaced Crops, *Biometrics*, **23**:505–515 (1967).

BERTALANFFY, L. VON: Untersuchungen über dies Gesetzlichkeit des Washstums. 1. Teil. Algemeine Grundlagen der Theorie; mathematicsche und physiologische Gesetzlichkeiten des Wachstums bein Wassertieren, *Arch. Entwicklungsmech.*, **131**:613–652 (1934).

BEVERTON, R. J. H., and HOLT, S. J.: "On the Dynamics of Exploited Fish Populations," Fishery Investigations, Series II, Vol. XIX, H.M.S.O., London, 1957.

BEYER, W. H. (ed.): "Handbook of Tables for Probability and Statistics," 2d ed., Chemical Rubber Company, Cleveland, 1968.

BIRCH, L. C.: The Intrinsic Rate of Natural Increase of an Insect Population, *J. Ecol.*, **17**:15–26 (1948).

BIRCH, L. C.: Experimental Background to the Study of the Distribution and Abundance of Insects. II. The Relation between Innate Capacity for Increase in Numbers and the Abundance of Three Grain Beetles in Experimental Populations, *Ecology*, **34**:712–726 (1953).

BIRD, F. T., and BURK, J. M.: Artificially Disseminated Virus as a Factor Controlling the European Spruce Sawfly, *Diprion hercyniae* (Htg.), in the Absence of Introduced Parasites, *Can. Entomol.*, **93**:228–238 (1961).

BLISS, C. I.: The Analysis of Insect Counts as Negative Binomial Distributions, *Proc. X Int. Congr. Entomol.*, **2**:1015–1032 (1958).

BLISS, C. I., and FISHER, R. A.: Fitting the Negative Binomial Distribution to Biological Data and a Note on the Efficient Fitting of the Negative Binomial, *Biometrics*, **9**:176–200 (1953).

BOSSERT, W. H.: "Project TACT," Memorandum 85, Applied Mathematics, Harvard University, Cambridge, 1968.

BRADFORD, E., and PHILIP, J. R.: Stability of Steady Distributions of Asocial Populations Dispersing in One Dimension, *J. Theor. Biol.*, **29**:13–26 (1970).

BRAY, J. R., and CURTIS, J. T.: Ordination of the Upland Forest Communities of Southern Wisconsin, *Ecol. Monogr.*, **27**:325–349 (1957).

BRODY, S.: "Bioenergetics and Growth," Reinhold, New York, 1945.

BROWN, R. T., and CURTIS, T. T.: Upland Conifer-Hardwood Forest of Northern Wisconsin, *Ecol. Monogr.*, **22**:217–234 (1952).

BURNETT, T.: Effects of Temperature and Host Density on the Rate of Increase of an Insect Parasite, *Am. Nat.*, **85**:337–352 (1951).

CHAPMAN, D. G.: Population Based on Change of Composition Caused by a Selective Removal, *Biometrika*, **42**:279–290 (1955).

CHAPMAN, D. G., and MURPHY, G. I.: Estimates of Mortality and Population from Survey-Removal Records, *Biometrics*, **21**:921–935 (1965).

CLARK, P. J., and EVANS, F. C.: Distance to Nearest Neighbor as a Measure of Spatial Relationships in Populations, *Ecology*, **35**:445–453 (1954).

COCHRAN, W. G.: "Sampling Techniques," 2d ed., John Wiley and Sons, New York, 1963.

COHEN, J. E.: "A Model of Simple Competition," Harvard University Press, Cambridge, 1968.

COHEN, J. E.: Alternate Derivations of a Species-Abundance Relation, *Am. Nat.*, **102**:165–172 (1968).

COLE, L. C.: The Measurement of Interspecific Association, *Ecology*, **30**:411–424 (1949).

COLE, L. C.: The Population Consequences of Life History Phenomena, *Q. Rev. Biol.*, **29**:103–137 (1954).

CONNELL, J. H.: Effects of Competition, Predation by *Thais lapillus*, and Other Factors on Natural Populations of the Barnacle *Balanus balanoides*, *Ecol. Monogr.*, **31**:61–106 (1961).

CORBERT, A. S.: The Distribution of Butterflies in the Malay Peninsula, *Proc. R. Entomol. Soc. Lond. A*, **16**:101–116 (1942).

CORMACK, R. M.: A Test for Equal Catchability, *Biometrics*, **22**:330–342 (1966).

CORMACK, R. M.: The Statistics of Capture-Recapture Methods, *Ann. Rev. Oceanogr. Mar. Biol.*, **6**:455–506 (1969).

CROCKER, R. L., and MAJOR, J.: Soil Development in Relation to Vegetation and Surface Age at Glacier Bay, Alaska, *J. Ecol.*, **43**:427–448 (1955).

CROMBIE, A. C.: The Effect of Crowding upon the Oviposition of Grain-Infesting Insects, *J. Exp. Biol.*, **19**:311–340 (1942).

CROMBIE, A. C.: On Intraspecific and Interspecific Competition in Larvae of Graminivorous Insects, *J. Exp. Biol.*, **20**:135–151 (1944).

DAHLMAN, R. C., and KUCERA, C. L.: Root Productivity and Turnover in Native Prairie, *Ecology*, **46**:84–89 (1965).

DARWIN, J. H., and WILLIAMS, R. M.: The Effect of Time of Hunting on the Size of a Rabbit Population, *N. Z. J. Sci.*, **7**:341–352 (1964).

DEBACH, P., and SMITH, H. S.: The Effect of Host Density on the Rate of Reproduction of Entomophagous Parasites, *J. Econ. Entomol.*, **34**:741–745 (1941).

DEBACH, P., and SUNDBY, R. A.: Competitive Displacement between Ecological Homologues, *Hilgardia*, **34**:105–166 (1963).

DEMPSTER, J. P.: The Population Dynamics of the Moroccan Locust (*Dociostaurus marocannus* Thunberg) in Cyprus, *Anti-Locust Bull.*, no. 27 (1957).

DOBZHANSKY, T., and WRIGHT, S.: Genetics of Natural Populations X. Dispersion Rates of *Drosophila pseudoobscura*, *Genetics*, **28**:304–340 (1943).

DUBLIN, L. I., and LOTKA, A. J.: On the True Rate of Natural Increase, *J. Am. Stat. Assoc.*, **20**:305–339 (1925).

DUVIGNEAUD, P., and DENAEYER-De SMET, S.: Biological Cycling of Minerals in Temperate Deciduous Forests, in Reichle, D. E. (ed.), "Ecological Studies: Analysis and Synthesis. 1. Analysis of Temperate Forest Ecosystems," Springer-Verlag, New York, pp. 199–225, 1970.

ELTON, C., and NICHOLSON, M.: The Ten Year Cycle in Numbers of Lynx in Canada, *J. Anim. Ecol.*, **11**:215–244 (1942).

EVANS, D. A.: Experimental Evidence Concerning Contagious Distributions in Ecology, *Biometrika*, **40**:186–211 (1953).

EVANS, F. G. C.: An Analysis of the Behavior of *Lepidochitona cinereus* in Response to Certain Physical Features of the Environment, *J. Anim. Ecol.*, **20**:1–10 (1951).

FISHER, R. A., CORBETT, A. S., and WILLIAMS, C. B.: The Relation between the Number of Species and the Number of Individuals in a Random Sample of an Animal Population, *J. Anim. Ecol.*, **12**:42–58 (1943).

GATES, C. E.: Simulation Study of Estimators for the Linear Transect Sampling Method, *Biometrics*, **25**:317–328 (1969).

GAUSE, G. F.: "The Struggle for Existence," Williams and Wilkins, Baltimore, 1934.

GLEASON, H. A.: The Individualistic Concept of Plant Association, *Torrey Bot. Club Bull.*, **53**:7–26 (1926).

GOOD, I. J.: "Probability and the Weighing of Evidence," Griffin, London, 1950.

GOOD, I. J.: The Population Frequencies of Species and the Estimation of Population Parameters, *Biometrika*, **40**:237–264 (1953).

GOWER, J. C.: Some Distance Properties of Latent Root and Vector Methods Used in Multivariate Analysis, *Biometrika*, **53**:325–328 (1966).

GREENWOOD, M., BRADFORD HILL, A., TOPLEY, W. W. C., and WILSON, J.: "Experimental Epidemiology," Medical Research Council Special Report No. 209, H.M.S.O., London, 1936.

GREIG-SMITH, P.: The Use of Random and Contiguous Quadrats in the Study of the Structure of Plant Communities, *Ann. Bot. Lond. N.S.*, **16**:293–316 (1952).

GRIFFITHS, K. J., and HOLLING, C. S.: A Competition Submodel for Parasites and Predators, *Can. Entomol.*, **101**:785–818 (1969).

GUILD, W. J. M.: Variation in Earthworm Numbers within Field Populations, *J. Ecol.*, **21**:169–181 (1952).

HAIRSTON, N. G.: Species Abundance and Community Organization, *Ecology*, **40**:404–416 (1959).

HAIRSTON, N. G.: On the Relative Abundance of Species, *Ecology*, **50**:1091–1094 (1969).

HAIRSTON, N. G., ALLAN, J. D., COLWELL, R. K., FUTUYMA, D. J., HOWELL, J., MATHIAS, J. D., and VANDERMEER, J. H.: The Relationship between Species Diversity and Stability. An Experimental Approach with Protozoa and Bacteria, *Ecology*, **49**:1091–1101 (1969).

HOLLAND, P. W.: "Some Properties of a Biogeography Model," Memorandum NH-107, Dept. of Statistics, Harvard University, Cambridge, 1968.

HOLLING, C. S.: Some Characteristics of Simple Types of Predation and Parasitism, *Can. Entomol.*, **91**:385–398 (1959).

HOLLING, C. S.: The Functional Response of Predators to Prey Density and Its Role in Mimicry and Population Regulation, *Mem. Entomol. Soc. Can.*, no. 45 (1965).

HOLLING, C. S.: The Functional Response of Invertebrate Predators to Prey Density, *Mem. Entomol. Soc. Can.*, no. 48 (1966).

HUFFAKER, C. B.: Experimental Studies on Predation: Dispersion Factors and Predator-Prey Oscillations, *Hilgardia*, **27**:343–383 (1958).

HUTCHESON, K.: A Test for Comparing Diversities Based on the Shannon Formula, *J. Theor. Biol.*, **29**:151–154 (1970).

IWAO, S.: On a Method of Estimating the Rate of Population Interchange between Two Areas, *Res. Pop. Ecol.*, **5**:44–50.

JOLLY, G. M.: Explicit Estimates from Capture-Recapture Data with Both Death and Dilution—Stochastic Model, *Biometrika*, **52**:225–247 (1965).

KENDALL, D. G.: Deterministic and Stochastic Epidemics in Closed Populations, *Proc. Third Berkeley Symp. Math. Stat. Probab.*, **4**:149–165 (1956).

KERSHAW, K. A.: Association and Covariance Analysis of Plant Communities, *J. Ecol.*, **49**:643–654 (1961).

KERSHAW, K. A.: "Quantitative and Dynamic Ecology," American Elsevier Publishing Company, New York, 1964.

KEYFITZ, N.: "An Introduction to the Mathematics of Populations, Addison-Wesley, Reading, Mass., 1968.

KILBURN, P. D.: Analysis of the Species-Area Relation, *Ecology*, **47**:831–843 (1966).

KLEKOWSKI, R. Z., PRUS, T., and ZYROMSKA-RUDZKA, H.: Elements of Energy Budget of *Tribolium castaneum* (Hbst.) in Its Developmental Cycle, in Petrusewicz, K. (ed.), "Secondary Productivity of Terrestrial Ecosystems (Principles and Methods)," Warszawa, Krakow, 1967.

KUCERA, C. L., DAHLMAN, R. C., and KOELLING, M. R.: Total Net Productivity and Turnover on an Energy Basis for a Tallgrass Prairie, *Ecology*, **48**:536–541 (1967).

KULLBACK, S.: "Information Theory and Statistics," John Wiley and Sons, New York, 1959.

LESLIE, P. H.: On the Use of Matrices in Certain Population Mathematics, *Biometrika*, **33**:183–212 (1945).

LESLIE, P. H.: Some Further Notes on the Use of Matrices in Population Mathematics, *Biometrika*, **35**:213–245 (1948).

LESLIE, P. H.: A Stochastic Model for Studying the Properties of Certain Biological Systems by Numerical Methods, *Biometrika*, **45**:16–31 (1958).

LESLIE, P. H.: The Properties of a Certain Lag Type of Population Growth and the Influence of an External Random Factor on a Number of Such Populations, *Physiol. Zool.*, **32**:151–159 (1959).

LESLIE, P. H.: A Stochastic Model for Two Competing Species of *Tribolium* and Its Application to Some Experimental Data, *Biometrika*, **49**:1–25 (1962).

LESLIE, P. H., and GOWER, J. C.: The Properties of a Stochastic Model for Two Competing Species, *Biometrika*, **45**:316–330 (1958).

LESLIE, P. H., and GOWER, J. C.: The Properties of a Stochastic Model for the Predator-Prey Type of Interaction between Two Species, *Biometrika*, **47**:219–234 (1960).

LESLIE, P. H., PARK, T., and MERTZ, D. B.: The Effect of Varying the Initial Numbers on the Outcome of Competition between Two *Tribolium* Species, *J. Anim. Ecol.*, **37**:9–23 (1968).

LEVINS, R.: "Evolution in Changing Environments: Some Theoretical Explorations," Monographs in Population Biology, Princeton University Press, Princeton, 1968.

LEWIS, E. G.: On the Generation and Growth of a Population, *Sankya*, **6**:93–96 (1943).

LEWONTIN, R. C.: Selection for Colonizing Ability, pp. 77–94 in Baker, H. G., and Stebbins, G. L. (eds.), "Genetics of Colonizing Species," Academic Press, New York, 1965.

LLOYD, M., ZAR, J. H., and KARR, J. R.: On the Calculation of Information-Theoretical Measures of Diversity, *Am. Midl. Nat.*, **79**:257–272 (1968).

LOTKA, A. J.: "Elements of Physical Biology," Williams and Wilkins, Baltimore, 1925.

MacARTHUR, R. H.: Fluctuations of Animal Populations, and a Measure of Community Stability, *Ecology*, **36**:533–536 (1955).

MacARTHUR, R. H.: On the Relative Abundance of Bird Species, *Proc. Nat. Acad. Sci. U.S.*, **43**:293–295 (1957).

MacARTHUR, R. H.: Population Ecology of Some Warblers of Northeastern Coniferous Forests, *Ecology*, **39**:599–619 (1958).

MacARTHUR, R. H.: Theory of the Niche, pp. 159–176 in Lewontin, R. C. (ed.), "Population Biology and Evolution," Syracuse University Press, Syracuse, 1968.

MacARTHUR, R. H., and LEVINS, R.: The Limiting Similarity, Convergence, and Divergence of Coexisting Species, *Am. Nat.*, **101**:377–385 (1967).

MacARTHUR, R. H., and MacARTHUR, J. W.: On Bird Species Diversity, *Ecology*, **42**:594–598 (1961).

MacARTHUR, R. W., RECHER, H., and CODY, M.: On the Relation between Habitat Selection and Species Diversity, *Am. Nat.*, **100**:319–325 (1966).

MacARTHUR, R. H., and WILSON, E. O.: An Equilibrium Theory of Insular Zoogeography, *Evolution*, **17**:373–387 (1963).

MacARTHUR, R. H., and WILSON, E. O.: "The Theory of Island Biogeography," Princeton University Press, Princeton, 1967.

MCGILCHRIST, C. A.: Analysis of Competition Experiments, *Biometrics*, **21**:975–985 (1965).

MCGUIRE, J. U., BRINDLEY, T. A., and BRANCROFT, T. A.: The Distribution of European Corn Borer Larvae *Pyrausta nubilalis* (Hbn) in Field Corn, *Biometrics*, **13**:65–78 (1957).

MCINTOSH, R. P.: The Continuum Concept of Vegetation, *Bot. Rev.*, **33**:130–187 (1967).

MACKENZIE, J. M. D.: Fluctuations in the Numbers of British Tetragonids, *J. Anim. Ecol.*, **21**:128–153 (1952).

MARSHALL, D. R., and JAIN, S. K.: Interference in Pure and Mixed Populations of *Avena fatua* and *A. barbata*, *J. Ecol.*, **57**:251–270 (1969).

MEAD, R.: A Mathematical Model for the Estimates of Interplant Competitions, *Biometrics*, **23**:189–205 (1967).

MILLER, R. S.: Pattern and Process in Competition, *Advan. Ecol. Res.*, **4**:1–74 (1967).

MILLER, R. S., and STEPHEN, W. J. D.: Spatial Relationships in Flocks of Sandhill Cranes (*Grus canadensis*), *Ecology*, **47**:323–327 (1966).

MORAN, P. A. P.: The Statistical Analysis of Sunspot and Lynx Cycles, *J. Anim. Ecol.*, **18**:115–116 (1949).

MORAN, P. A. P.: The Statistical Analysis of Game-Bird Records, *J. Anim. Ecol.*, **21**:154–158 (1952).

MORAN, P. A. P.: The Statistical Analysis of the Canadian Lynx Cycle, *Aust. J. Zool.*, **1**:163–173 (1953).

MORISITA, M.: Measuring the Dispersion of Individuals and Analysis of the Distributional Patterns, *Mem. Fac. Sci. Kyushu Univ. Ser. E (Biol.)*, **2**:215–235 (1959).

MORRIS, R. F.: A Sequential Sampling Technique for Spruce Budworm Egg Surveys, *Can. J. Zool.*, **32**:302–313 (1954).

MORRIS, R. F.: The Development of Sampling Techniques for Forest Insect Defoliators, with Particular Reference to the Spruce Budworm, *Can. J. Zool.*, **33**:225–294 (1955).

MORRIS, R. F.: (ed.): The Dynamics of Epidemic Spruce Budworm Populations, *Mem. Entomol. Soc. Can.*, no. 31 (1963).

MURIE, A.: "The Wolves of Mount McKinley," Fauna of the National Parks of the U.S., Fauna Series, no. 5, U.S. Govt. Printing Off., Washington, D.C., 1944.

NEWBOULD, P. J.: "Methods of Estimating the Primary Production of Forests," IBP Handbook, no. 2, Blackwell, Oxford, 1967.

NICHOLSON, A. J.: The Self-Adjustment of Populations to Change, *Cold Spring Harbor Symp. Quant. Biol.*, **22**:153–173 (1957).

NICHOLSON, A. J., and BAILEY, V. A.: The Balance of Animal Populations, *Proc. Zool. Soc. Lond.*, **1935**:551–598.

NIELSON, M. M., and MORRIS, R. F.: The Regulation of the European Spruce Sawfly Numbers in the Maritime Provinces of Canada from 1937 to 1963, *Can. Entomol.*, **96**:773–785 (1964).

OLSON, J. S.: Rates of Succession and Soil Changes on Southern Lake Michigan Sand Dunes, *Bot. Gaz.*, **119**:125–170 (1958).

OLSON, J. S.: Energy Storage and the Balance of Producers and Decomposers in Ecological Systems, *Ecology*, **44**:322–331 (1963).

ORIANS, G. H.: A Capture-Recapture Analysis of a Shearwater Population, *J. Anim. Ecol.*, **27**:71–86 (1958).

OVINGTON, J. D.: Quantitative Ecology and the Woodland Ecosystem Concept, *Advan. Ecol. Res.*, **1**:103–192 (1962).

PAINE, R. T.: Natural History, Limiting Factors and Energetics of the Opisthobranch *Navanax inermis*, *Ecology*, **46**:603–619 (1965).

PAINE, R. T.: Food Web Complexity and Species Diversity, *Àm. Nat.*, **100**:65–76 (1966).

PARIS, O. H.: Vagility of P^{32}-Labeled Isopods in Grassland, *Ecology*, **46**:635–648 (1965).

PARK, T.: Experimental Studies of Interspecies Competition. I. Competition between Populations of the Flour Beetles *Tribolium confusum* Duval and *Tribolium castaneum* Herbst, *Ecol. Monogr.*, **18**:265–308 (1948).

PARK, T.: Experimental Studies on Interspecies Competition. II. Temperature, humidity, and Competition in Two Species of *Tribolium*, *Physiol. Zool.*, **27**:177–238 (1954).

PATTEN, B. C., and WITKAMP, M.: Systems Analysis of ^{134}Cesium Kinetics in Terrestrial Microcosms, *Ecology*, **48**:813–824 (1967).

PAULIK, G. J., and GREENOUGH, J. W., JR.: Management Analysis for a Salmon Resource System, pp. 215–252 in Watt, K. E. F. (ed.), "Systems Analysis in Ecology," Academic Press, New York, 1966.

PEARSON, P. G.: Small Mammals and Old Field Succession on the Piedmont of New Jersey, *Ecology*, **40**:249–255 (1959).

PETRUSEWICZ, K., and MACFAYDEN, A.: Productivity of Terrestrial Animals: Principles and Methods, IBP Handbook, no. 13, Blackwell Scientific Publications, Oxford, 1970.

PHILLIPSON, J.: Secondary Productivity in Invertebrates Reproducing Once in a Lifetime, pp. 459–475 in Petrusewicz, K. (ed.), "Secondary Productivity of Terrestrial Ecosystems (Principles and Methods), Warszawa, Krakow, 1967.

PIELOU, E. C.: A Single Mechanism to Account for Regular, Random, and Aggregated Populations, *J. Ecol.*, **48**:575–584 (1960).

PIELOU, E. C.: Segregation and Symmetry in Two-Species Populations as Studied by Nearest Neighbor Relations, *J. Ecol.*, **49**:255–269 (1961).

PIELOU, E. C.: Runs of One Species with Respect to Another in Transects through Plant Populations, *Biometrics*, **18**:579–593 (1962a).

PIELOU, E. C.: The Use of Plant-to-Neighbor Distances for the Detection of Competition, *J. Ecol.*, **50**:357–367 (1962b).

PIELOU, E. C.: The Spatial Pattern of Two-Phase Patchworks of Vegetation, *Biometrics*, **20**:156–167 (1964).

PIELOU, E. C.: The Spread of Disease in Patchily-Infected Forest Stands, *Forest Sci.*, **11**:18–26 (1965).

PIELOU, E. C.: Species-Diversity and Pattern-Diversity in the Study of Ecological Succession, *J. Theor. Biol.*, **10**:370–383 (1966a).

PIELOU, E. C.: The Measurement of Diversity in Different Types of Biological Collections, *J. Theor. Biol.*, **13**:131–144 (1966b).

PIELOU, E. C.: "An Introduction to Mathematical Ecology," John Wiley and Sons, New York, 1969.

PIMENTEL, D.: The Influence of Plant Spatial Patterns on Insect Populations, *Ann. Entomol. Soc. Am.*, **54**:61–69 (1961).

POOLE, R. W.: Temporal Variation in the Species Diversity of a Woodland Caddisfly Fauna from Central Illinois, *Trans. Ill. Acad. Sci.*, **63**:383–385 (1971).

PRATT, D. M.: Analysis of Population Development in *Daphnia* at Different Temperatures, *Biol. Bull.*, **85**:116–140 (1943).

PRESTON, F. W.: The Commonness and Rarity of Species, *Ecology*, **29**:254–283 (1948).

PRESTON, F. W.: The Canonical Distribution of Commonness and Rarity. Part 1, *Ecology*, **39**:185–215 (1962).

PRICE, P. W.: Characteristics Permitting Coexistence among Parasitoids of a Sawfly in Quebec, *Ecology*, **51**:445–454 (1970).

QUARTERMAN, E.: Early Plant Succession on Abandoned Cropland in the Central Basin of Tennessee, *Ecology*, **38**:300–309 (1957).

QUENOUILLE, M. H.: Approximate Tests of Correlation in Time Series, *J. R. Stat. Soc.*, **B11**:68–84 (1949).

RABINOVICH, J. E.: Vital Statistics of *Synthesiomyia nudiseta* (Diptera: Muscidae), *Ann. Entomol. Soc. Am.*, **73**:749–752 (1970).

REICHLE, D. E.: Radioisotope Turnover and Energy Flow in Terrestrial Isopod Populations, *Ecology*, **48**:351–366 (1967).

REICHLE, D. E. (ed.): "Ecological Studies: Analysis and Synthesis. 1.   Analysis of Temperate Forest Ecosystems," Springer-Verlag, New York, 1970.

RICHARDS, O. W., and WALOFF, N.: Studies on the Biology and Population Dynamics of British Grasshoppers, *Anti-locust Bull.*, no. 17 (1954).

ROBSON, D. S., and REGIER, H. A.: Sample Size in Peterson Mark-Recapture Experiments, *Trans. Am. Fish. Soc.*, **93**:215–226 (1964).

ROHLF, F. J.: The Effect of Clumped Distributions in Sparse Populations, *Ecology*, **50**:716–721 (1969).

ROOT, R. B.: The Niche Exploitation Pattern of the Blue-Grey Gnat Catcher, *Ecol. Monogr.*, **37**:317–350 (1967).

SAKAI, K., TAKASHI, N., HIRAIZUMI, Y., and IYANA, S.: Studies in Competition in Plants and Animals. IX.   Experimental Studies on Migration in *Drosophila melanogaster*, *Evolution*, **12**:93–101 (1958).

SEARLE, S. R.: "Matrix Algebra for the Biological Sciences (Including Applications in Statistics)," John Wiley and Sons, New York, 1966.

SIMBERLOFF, D. S.: Experimental Zoogeography of Islands: A Model for Insular Colonization, *Ecology*, **50**:296–314 (1969).

SIMBERLOFF, D. S., and WILSON, E. O.: Experimental Zoogeography of Islands.   A Two-Year Record of Colonization, *Ecology*, **51**:934–937 (1970).

SIMPSON, E. H.: Measurement of Diversity, *Nature*, **163**:688 (1949).

SINIFF, D. B., and SKOOG, R. O.: Aerial Sampling of Caribou Using Stratified Random Sampling, *J. Wildl. Manage.*, **28**:391–401 (1964).

SKELLAM, J. G.: Random Dispersal in Theoretical Populations, *Biometrika*, **38**:196–218 (1951).

SNEDECOR, G. W., and COCHRAN, W. G.: "Statistical Methods," 6th ed., Iowa State University Press, Ames, 1967.

SOKAL, R. R.: Pupation Rate Differences in *Drosophila melanogaster*, *Univ. Kans. Sci. Bull.*, **46**:697–715 (1966).

SOKAL, R. R., and ROHLF, F. J.: "Biometry: The Principles and Practice of Statistics in Biological Research," W. H. Freeman and Company, San Francisco, 1969.

TAYLOR, N. W.: A Mathematical Model for *Tribolium confusum* Populations, *Ecology*, **48**:290–294 (1967).

TAYLOR, N. W.: A Mathematical Model for Two *Tribolium* Populations in Competition, *Ecology*, **49**:843–848 (1968).

TEAL, J. M.: Energy Flow in the Salt Marsh Ecosystem of Georgia, *Ecology*, **43**:614–624 (1962).

TINBERGEN, L.: The Natural Control of Insects in Pinewoods. I.   Factors Influencing the Intensity of Predation by Songbirds, *Arch. Neerl. Zool.*, **13**:265–343 (1960).

TUXEN, R. (ed.): "Handbook of Vegetation Science," manuscript.

ULLYETT, G. C.: Oviposition by *Ephestia kühniella* Zeller, *J. Entomol. Soc. S. Afr.*, **8:**53–59 (1945).

UTIDA, S.: Studies on Experimental Population of the Azuki Bean Weevil, *Callosobruchus chinensis* (L.). I. The Effect of Population Density on the Progeny Populations, *Mem. Coll. Agric. Kyoto*, **48:**1–30 (1941).

UTIDA, S.: Interspecific Competition between Two Species of Bean Weevils, *Ecology*, **34:**301–307 (1953).

UTIDA, S.: Population Fluctuation, An Experimental and Theoretical Approach, *Cold Spring Harbor Symp. Quant. Biol.*, **22:**139–151 (1957).

VANDERMEER, J. H.: The Competitive Structure of Communities: An Experimental Approach with Protozoa, *Ecology*, **50:**362–371 (1969).

VAN HOOK, R. I.: Energy and Nutrient Dynamics of Spider and Orthopteran Populations in a Grassland Ecosystem, *Ecol. Monogr.*, **41:**1–26 (1971).

VOLLENWEIDER, R. A.: Calculation Models of Photosynthesis-Depth Curves and Some Implications Regarding Day Rate Estimates in Primary Production Measurements, pp. 427–457 in Goldman, C. R. (ed.), "Primary Productivity in Aquatic Environments," University of California Press, Berkeley, 1965.

WANGERSKY, P. J., and CUNNINGHAM, W. J.: Time Lag in Population Models, *Cold Spring Harbor Symp. Quant. Biol.*, **22:**329–338 (1957).

WATERS, W. E.: Sequential Sampling in Forest Insect Surveys, *Forest Sci.*, **1:**68–79 (1955).

WATT, A. S.: Pattern and Process in the Plant Community, *J. Ecol.*, **35:**1–22 (1947).

WATT, A. S.: The Effect of Excluding Rabbits from Grassland. A. (*Xerobrometum*) in Breckland, 1937–60, *J. Ecol.*, **50:**181–198 (1962).

WATT, K. E. F.: A Mathematical Model for the Effect of Densities of Attacked and Attacking Species on the Number Attacked, *Can. Entomol.*, **91:**129–144 (1959).

WATT, K. E. F.: The Effect of Population Density on Fecundity in Insects, *Can. Entomol.*, **92:**674–695 (1960).

WATT, K. E. F.: Community Stability and the Strategy of Biological Control, *Can. Entomol.*, **97:**887–895 (1965).

WATT, K. E. F.: "Ecology and Resource Management," McGraw-Hill Book Company, New York, 1968.

WHITE, E. G., and HUFFAKER, C. B.: Regulatory Processes and Population Cyclicity in Laboratory Populations of *Anagasta kühniella* (Zeller) (Lepidoptera: Phycitidae). I. Competition for Food and Predation, *Res. Popul. Ecol.*, **11:**57–83 (1969).

WHITTAKER, R. H.: A Criticism of the Plant Association and Climatic Climax Concepts, *Northwest Sci.*, **25:**17–31 (1951).

WHITTAKER, R. H.: A Consideration of Climax Theory: The Climax as a Population and Pattern, *Ecol. Monogr.*, **23:**41–78 (1953).

WHITTAKER, R. H.: Vegetation of the Great Smoky Mountains, *Ecol. Monogr.*, **26:**1–80 (1956).

WHITTAKER, R. H.: Experiments with Radiophosphorus Tracer in Aquarium Microcosms, *Ecol. Monogr.*, **31:**157–188 (1961).

WHITTAKER, R. H.: Classification of Natural Communities, *Bot. Rev.*, **28:**1–239 (1962).

WHITTAKER, R. H.: Gradient Analysis of Vegetation, *Biol. Rev.*, **49:**207–264 (1967).

WHITTAKER, R. H.: "Communities and Ecosystems," Macmillan Company, New York, 1970.

WHITTAKER, R. H., and WOODWELL, G. M.: Dimension and Production Relations of Trees and Shrubs in the Brookhaven Forest, New York, *J. Ecol.*, **56**:1–25 (1968).

WIEGERT, R. G.: The Selection of an Optimum Quadrat Size for Sampling the Standing Crop of Grasses and Forbs, *Ecology*, **43**:125–129 (1962).

WILLIAMS, C. B.: Comparing the Efficiency of Insect Traps, *Bull. Entomol. Res.*, **42**:513–517 (1951).

WILLIAMS, E. J.: The Analysis of Competition Experiments, *Aust. J. Biol. Sci.*, **15**:509–525 (1962).

WILLIAMS, W. T., and LAMBERT, J. M.: Multivariate Methods in Plant Ecology. I. Association-Analysis in Plant Communities, *J. Ecol.*, **47**:83–101 (1959).

WILLIAMS, W. T., and LAMBERT, J. M.: Multivariate Methods in Plant Ecology. II. The Use of an Electronic Digital Computer for Association Analysis, *J. Ecol.*, **48**:689–710 (1960).

WILLIAMS, W. T., and LAMBERT, J. M.: Multivariate Methods in Plant Ecology. III. Inverse Association-Analysis, *J. Ecol.*, **49**:717–729 (1961).

WILLIAMS, W. T., LAMBERT, J. M., and LANCE, G. N.: Multivariate Methods in Plant Ecology. V. Similarity Analyses and Information-Analysis, *J. Ecol.*, **54**:427–445 (1966).

WILLIAMS, W. T., LANCE, G. N., WEBB, L. J., TRACEY, J. G., and DALE, M. B.: Studies in the Numerical Analysis of Complex Rain-Forest Communities. III. The Analysis of Successional Data, *J. Ecol.*, **57**:515–535 (1969).

WILSON, E. B., BENNETT, C., ALLEN, M., and WORCHESTER, J.: Measles and Scarlet Fever in Providence, R.I., 1929–34 with Respect to Age and Size of Family, *Proc. Am. Phil. Soc.*, **80**:357–476 (1939).

WILSON, E. O., and SIMBERLOFF, D. L.: Experimental Zoogeography of Islands: Defaunation and Monitoring Techniques, *Ecology*, **50**:267–278 (1969).

ZWÖLFER, H.: The Structure of the Parasite Complexes of Some Lepidoptera, *Z. Angew. Entomol.*, **51**:346–357 (1963).

INDEX